SOME USEFUL APPROXIMATE CONVERSION FACTORS

1 mi = 1,600 m
1 mi = 5,280 ft
1 ft = 0.30 m
1 lb = 4.5 newtons
1 kg = 14.6 slugs
1 calorie = 4.2 joules
1 amu (atomic mass unit) = 1.66×10^{-27} kg
1 kilowatthour = 3.6×10^{6} joules
1 AU (astronomical unit) = 9.3×10^{7} mi
= 1.5×10^{11} m
1 angstrom (Å) = 10^{-10} m

SOME ABBREVIATIONS USED IN THIS TEXT

m = meter
km = kilometer
cm = centimeter
mm = millimeter
ft = foot
yd = yard
in. = inch
mi = mile
s = second
lb = pound
N = newton
kg = kilogram
amu = atomic mass unit
J = joule
cal = calorie
C = coulomb
A = ampere
V = volt
W = watt
kWh = kilowatthour
hz = hertz
eV = electron-volt

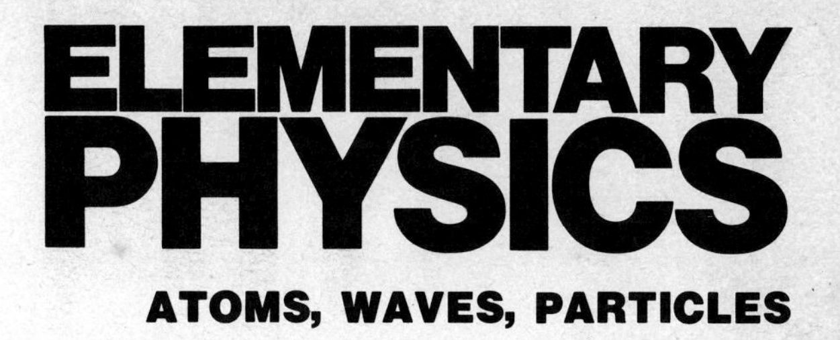
ELEMENTARY
PHYSICS
ATOMS, WAVES, PARTICLES

SECOND EDITION

ELEMENTARY PHYSICS

ATOMS, WAVES, PARTICLES

GEORGE A. WILLIAMS

Professor of Physics
University of Utah

McGRAW-HILL BOOK COMPANY
New York St. Louis
San Francisco Auckland
Düsseldorf Johannesburg
Kuala Lumpur London
Mexico Montreal
New Delhi Panama
Paris São Paulo
Singapore Sydney
Tokyo Toronto

Library of Congress Cataloging in Publication Data

Williams, George Abiah, date
Elementary physics.

Includes bibliographies.
1. Physics. I. Title.
QC21.2.W54 1975 530 75-20375
ISBN 0-07-070402-3

ELEMENTARY
PHYSICS: Atoms, Waves, Particles

1234567890VHVH798765

This book was set in
Helvetica Light
by Black Dot, Inc.
The editors were
Robert A. Fry
and Michael LaBarbera;
the designer was
Nicholas Krenitsky;
the production supervisor was
Leroy A. Young.
New drawings were done by
Vantage Art, Inc.
Von Hoffmann Press, Inc.,
was printer and binder.

TO MY CHILDREN

CONTENTS

PART THREE: ELECTRICITY: THE FUNDAMENTAL EXPERIMENTAL LAWS

PART FOUR: ATOMS, MATTER, AND ELECTRONS

PREFACE

At almost every college and university there is a belief that an educated student should have some understanding of science. Since awareness of the "energy crisis" has spread, this belief has intensified. One of the more difficult jobs in college teaching is to translate that belief into a workable course or courses that can be handled by students of limited backgrounds, and still provide a meaningful experience. In recent years there have been many books written with this goal in mind. This book, a revision of the author's previous attempt, takes into account his experience in using the first edition and feedback from various instructors and students who used the first edition.

The goal of this book, similar to that of its predecessor, is to arrive at an understanding of contemporary physics using as little time on background material as possible. This is, therefore, not principally a book about "everyday physics," but a serious attempt to cover topics of importance to physicists today. This is important, since today the interactions of physics with society are not usually at the level of everyday science, but involve very sophisticated technical and economic questions. Some exposure to the ideas of science behind these questions is needed for the intelligent judgment that will be required of all citizens, not just senators and congressmen.

The mathematical level of the book has been kept as low as possible, although, in the author's opinion, a course in physics with no mathematics gives a student a false idea of what science is really all about. In order to meet the needs of the student who is not mathematically inclined, the math level has been "toned-down" from the first edition. Some of the more mathematical sections have been eliminated. In many places equations in words, rather than symbols, are used to facilitate the understanding by the verbally, rather than mathematically, oriented student.

A number of major changes have been made from the first edition. A new chapter (17) on special relativity has been added. A discussion of the second law of thermodynamics has been added to Chapter 7, and a new chapter (8) on the energy problem is included, as well as a new chapter (13) on states of matter. Chapters 6-10 in the first edition on the atom, atomic masses, and the periodic chart have been condensed into one, Chapter 12. The material on circular motion and gravity has been moved forward and written as a single chapter (4). In general the material added, condensed, or omitted responds to the comments of instructors and students who have used the first edition.

The chapter end material has been heavily revised; many of the problems are new and all have been class tested. The end material includes questions, problems graduated in difficulty, and a self-test on the terms, ideas, and names encountered in the chapter. One item of selected reading, mostly of a biographical or historical nature, is included in most chapters. Additional readings will be found in the instructor's guide.

The author wishes to thank the many instructors who have commented on various aspects of the first edition. In particular he would like to acknowledge Stephen Kral for contributions to the questions and problems. George Cassiday and Robert Goble have read Chapters 22 and 23 and suggested ways in which these chapters can be brought up to date. All of the typing was done by Jan Moffat.

George A. Williams

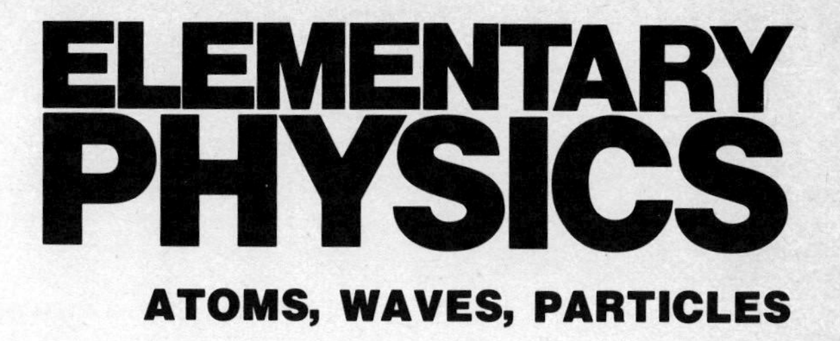

ATOMS, WAVES, PARTICLES

WHY PHYSICS?

Many students who take a course for which this book is designed will do so because it is required. Why are there required science courses? What role does physics play in the scheme of things? These are valid questions to be asked. In this chapter we will attempt to answer some of them.

I once gave the first lecture of this course to a class, and described the course content as "the physics every senator should know." For the rest of the quarter, whenever I covered a sticky point, there came from the first row the remark, "What's a senator got to know *that* for?" The best answer is that senators make many decisions involving science that will affect everyone. If the decisions are wrong, we will all suffer. And all of us will, or should, vote for these senators. So we must have understanding as well as they.

1.1 IS SCIENCE THE ANSWER?

Until recently, most people in our society thought that any problem of a technological nature could be solved if only enough time, effort, and money were put into solving it. Some examples of spectacular achievements brought about by science and technology working together are: sending men to the moon, conquering the crippling disease polio, and constructing the atomic bomb. You can probably think of others.

Until now, technology has grown as far and as fast as man has been willing to go. But now we may be forced to make choices, because our world can no longer survive the effects of unlimited growth of technology. The recent decision by the United States not to build an SST is one of the few cases in which we have not done something that was technologically possible. It is becoming clear that science and technology *cannot* solve all the problems, at least not in the old way.

Consider the much praised and much abused automobile. We can probably reduce the pollution from automobiles by a noticeable amount if we work hard on the problem. Suppose air pollution from each car is reduced by 50 percent. But if the number of cars doubles, we will be right back where we started. Or we might decide to use electric cars—only to find that there is a problem of pollution from the electric power plants which supply the electricity to charge the batteries of these cars.

Let us try another approach—limit the number of cars. But this would mean either limiting the number of people or limiting their freedom to own a car. The problem then becomes a political issue, not a scientific or a technological matter. We find that we must make a political decision (are we to legislate who can have what kind of car?) to solve a technological problem (air pollution). If our political decisions about these things are to be the right ones, the entire electorate, as well as the decision makers, will have to understand some of the technical facts. If we do not understand the technical problems involved, we may demand solutions that seem to be all right in the short run, but in fact are very damaging in the long run.

An example of such a situation is the overuse of chemical fertilizers that increase the size of crops for a short while, but in the long run damage the soil and lead to the pollution of water by nitrates and phosphates. When these chemicals are dumped into a river or lake, they cause an increase in the growth of algae (tiny one-celled plants). In turn, these algae use up all the oxygen in the water, causing further trouble since fish need oxygen to survive. Another example is the overuse of insecticides such as DDT that are not broken down biologically and so persist in the environment. Some kinds of birds are already in danger of extinction because of the effects of DDT on their bodies. They lay fewer eggs, and those they do lay have such thin shells that they break before the young birds hatch. There have also been episodes in which the use of pesticides has been more effective against predators than against the pest it was designed to eradicate. The result is an explosion of the population of the pest species.

A present-day example is the use of nuclear reactors to produce electricity. We should be very careful that we understand clearly all the consequences of the proliferation of reactors and how to deal with them. The amount of damage that could be done to the human race is too great to allow for any mistakes or oversights.

Much damage has already been done to the world we live in as a result of too little examination of all the consequences—long-term as well as short-term—of the widespread use of particular inventions such as the automobile or DDT. Even more damage can be done now by attempting to solve these problems with hasty, emotional decisions based on too little information.

It has been suggested that the solution to all these problems is to stop technology in its tracks and forget about science. But this cannot be the way out. The problems will not go away if we turn our backs on science. They can be solved only if more rather than fewer people are scientifically literate.

Scientific literacy includes an understanding of what is meant by a *scientific* (or *natural*) law. We have mentioned the myth that science can solve any problem as long as enough money and equipment are provided. But no amount of money or effort will allow us to change a law of nature. Nature's laws, in general, tell us what we can and cannot do. An example of such a law is one called the *law of conservation of energy*. It says that energy can never be created or destroyed, but just changed from one form to another.

Our society is experiencing what has been called an "energy crisis." We use electric energy when we turn on a light or a motor. We use chemical energy when we burn gasoline in our cars to create mechanical energy. We use energy obtained from chemical or electric energy to heat our homes. Most of the sources from which we obtain energy, such as natural gas or petroleum, are beginning to be in short supply in the world. So there is a growing shortage of acceptable sources of energy to meet an ever-increasing demand for it. The shortage, then, is one side of the "energy crisis."

Another side of the "crisis" has to do with the effect that tremendous worldwide energy use might have on the environment. Suppose we do find a way to make enough usable energy. Suppose this can be done without intolerable amounts of air pollution or other contamination. It is a fact that almost all the energy we use turns up as heat in the environment. The law of energy conservation tells us we cannot change that. Since energy can neither be created nor destroyed, where will we put it after we have used it to do work?

If the rest of the world, in the process of technological growth, approaches the rate of energy use in this country, there will be a large increase in the total amount of energy used in the world. This world total could easily become a significant part of the energy we receive from the sun. If we dump into our environment an amount of heat greater than about 1 percent of the heat energy received from the sun, a number of serious consequences may follow. All this heat cannot simply be swept under a rug. Even the oceans are not large enough to take up the heat without eventually warming up our climate. If the climate were to become too warm and the ice and snow of the North and South Poles melted, the levels of the oceans would rise significantly and many of the major cities of the earth would be drowned. At present rates, this could occur within 150 years—perhaps sooner.

Science can tell us what we might or can do about the "energy crisis." It will not tell us what we *ought* to do. Unless the citizens of this country and the rest of the world are informed, the decisions about what will be done will probably be the wrong ones. Citizens who do not know the facts are likely to take the point of view that scientists are just alarmists and can safely be ignored. Senators may adopt a similar point of view.

In a sense, our world has operated up to now with what can be called "nomad economics." When land has been used up, farmers have moved on. We have behaved as though the world's resources were limitless. In recent years, more and more people have come to realize the limitations of our earth. It is becoming clear that we must adopt "spaceman economics," because earth is really a spaceship, in the sense that it is totally isolated, and we are very unlikely to obtain any resources whatsoever from any other source. When the earth's resources are gone, they will be gone. Period. We must therefore preserve, conserve, recycle, and in every way be careful with what we have. Science can help show us *how* to do this.

But science cannot do the work. This work must be done by each one of us, or it will not be done at all.

1.2 PHYSICS: WHAT IS IT?

Physics is the most basic of all the sciences. Some aspect of physics is involved in, and fundamental to, every other branch of science.

Physicists are found in many environments—universities, government laboratories, and industrial laboratories. In general, what they are doing is studying some aspect of how the universe is put together. Many are interested in finding, identifying, and studying the fundamental building blocks of all matter—if such exist! Others attempt to understand the forces that hold matter together. Still others are interested in the electrical, magnetic, thermal, mechanical, and other properties of matter. Another group is interested in the properties of matter at very low temperatures, and also in devising clever means to reach still lower temperatures. There are those interested in the structure of stars, galaxies, and the universe. Others are interested in the physics of the earth, geophysics. There are physicists working in the border regions between physics and chemistry, physics and biology, and physics and the engineering sciences. In fact, anywhere the fundamentals of nature are being studied, there are physicists.

One of the major areas of physics, and one which will occupy a significant fraction of our time, is the study of the structure of matter and its constituent parts. Nineteenth-century physics and chemistry proved that matter consists of atoms as basic particles. Twentieth-century physics has determined the structure of the atom in terms of smaller particles. Also, in the twentieth century, much about the structure of the atomic nucleus has been learned, to the point where atomic reactors and thermonuclear bombs have been constructed, taking advantage of this knowledge. To arrive at the modern view of the nature of matter, it has been necessary to achieve significant developments in other areas of physics, among them the nature of light. We shall find that progress in understanding the nature of light in the early part of the twentieth century has played a crucial role in our understanding of the structure of the atom.

Each of the areas we shall discuss has developed through a long process of trial and error, and experiment and theory, before the present state was achieved. This is a dynamic process, and physics today is a dynamic science. It continues to evolve and to advance in new areas. By the study of the history of some of the completed developments we may understand something about how the unfinished business of today's physics is being handled.

THE LAWS OF PHYSICS

Much of the most important work in physics involves the discovery of the laws of physics. These laws are not passed by a legislature. They are statements, usually mathematical, which summarize our knowledge about a particular area of physics or class of experiment. These laws are acquired by the study of how the universe behaves, until some generalization is possible. Then the generalization or law is tested to see that everything it predicts actually happens as predicted. We shall study several examples of such laws. The primary purpose is, as observed by Richard Feynman, "not to show how clever we are to have discovered them, but rather to marvel at a nature which can obey such elegant and simple laws."

The method by which these laws are set up and tested is called the scientific method. It is useful to outline this method, even though in actual practice the process is not always followed in a coherent manner. The creative process is not always capable of being pigeonholed. The essential idea is to propose a model and see if it works. The model is of the "suppose this is the way things are put together" nature. The next step is to calculate all possible consequences of the model proposed. These consequences are tested by experiment or observation. The model is considered valid if all possible consequences turn out to agree with the observation. Even one prediction of a model which does not agree with experiment invalidates the model, at least in part. We will encounter a number of model-building steps in this book, although we will not always discuss the process in a formal way.

1.3 WHY STUDY PHYSICS?

Man is a curious animal. Since the beginning of our time on earth we have sought answers to the "why" of the natural phenomena which surround us on all sides. Many of the answers proposed credited natural phenomena to the machinations of assorted supernatural deities. As modern science has developed, more and more natural phenomena have been understood in terms of sensible models concerning how the world is put together. An understanding of physics is essential to the understanding of the world around us and how it behaves. Anyone who shows a curiosity about the environment can justify the study of physics as helping partially to satisfy this curiosity.

The development of modern physical science, to which physics is basic, is one of the major intellectual achievements of the human race. Surely, no person who lays claim to being educated can completely ignore a major fraction of the intellectual achievement of Western civilization. Furthermore, the development of understanding in science has had a major influence in other areas of knowledge. Philosophy today wrestles with some of the problems posed by science.

Another reason for studying physics or any of the sciences is the ever-growing interaction between science and society. Before World War II, the achievements of physics were of primary interest to those workers in the laboratory who were directly involved. The fact that discoveries in physics paved the way for the atomic bomb propelled physics into an ever-increasing interaction with the society around it.

TABLE 1.1 Some areas of physics, and an idea of what they do. Many physicists do work that overlaps two or three of these areas

Discipline	Area of interest
Solid-state physics	Properties of solid matter. Transistors evolved from this area of work
Low-temperature physics (cryogenics)	Properties of matter at very low temperatures and the study of techniques of reaching low temperatures. Superconductors are an important area of interest
High-energy–cosmic ray physics	The study of interactions of nuclear particles at very high energies, directed toward the understanding of the fundamental nature of the elementary building blocks of the universe
Astrophysics	The application of physical principles to the processes that go on in stars and galaxies. Things like quasars and pulsars have been of recent interest here
Cosmology	The study of the origins of the universe, using Einstein's theory of general relativity as a basic tool. "Black holes" were predicted using these ideas
Nuclear and reactor physics	The study of the properties of nuclei. Directed mostly these days toward those properties of interest to the design of nuclear reactors
Plasma physics	The study of gases at very high temperatures. Essential to the eventual practical use of fusion (hydrogen bomb) energy in a controlled manner

The decisions to make the bomb, and later use it, were political decisions. Yet these were political decisions about scientific matters made by people who were not trained in science. More and more decisions of this kind are being made. If they are to be made with intelligence, it is important that the people making them have some understanding and appreciation of science.

The newspapers carry almost daily stories about science or the

TABLE 1.1 *(Continued)*

Discipline	Area of interest
Geophysics	The application of physical principles to the study of the structure of the earth (and moon). Understanding the causes of and possible prediction of earthquakes is an important area
Biophysics	The application of physical principles to biological systems; for example, the conduction of impulses in nerves
Applied or engineering physics	As the name indicates, the application of the developments in practical engineering situations
Chemical physics	Physical techniques applied to the solution of chemical problems; at the borderline of the two fields
Optics	Primarily these days the study of lasers. One interest is in more powerful lasers, for work in thermonuclear fusion
High-polymer physics	A study of mechanical and other properties of high polymers (plastics)
Medical physics	The primary interest here is the interaction of nuclear radiation with the human body

interaction of science with society. To read these with intelligence requires some understanding of science, its language, and its goals. The United States spent $20 billion in a race to put a man on the moon. Much has been made of the benefits which will accrue to science as a result of this project. In some quarters, however, the project has been described as a "moondoggle," and the suggestion made that the money should have been spent elsewhere.

It should be clear that science in general, and physics in particular, are an important part of the intellectual and political life of our time. In every area of our society decisions are being made about science which affect not only science and scientists, but everyone. And yet most of the people making these decisions are not scientifically trained.

A few years ago, roughly one-third or more of Congress had legal training. Only one person, in both houses of Congress, had an advanced degree in any scientific discipline. It will probably always be this way, and therefore the nonscientists who must make decisions involving science and scientists should have some understanding of what science is all about. This book seeks to show a little of what physics is all about.

But perhaps the most important reason for studying physics is that physics is an exciting subject in its own right. Any challenge of the unknown makes an exciting story. Very few of us ever expect to set foot on Mt. Everest, but we can all appreciate from their accounts the struggle of Hillary and Tenzing to reach the top. Similarly, the struggle to wrest from nature some understanding of its basic organization can be exciting even for those who never expect to participate in the struggle. Toward the end of this book we shall discuss some areas where the struggle is still going on and the outcome is still in doubt.

SUMMARY

Science can tell us what the laws of nature permit and do not permit. The choice as to what we do, within the limits of the possible, is a political, not a scientific, choice. No effort of technology will allow us to circumvent a scientific law. Some literacy in the laws of science is necessary to make meaningful political decisions in technological matters.

Physics is the study of energy, light, electricity, and matter.

The laws of physics are concise summaries of the results of many experiments. They tell us how some part of our world behaves.

One of the primary reasons for the study of physics involves not only its impact on technology, but also the ever-increasing number of governmental decisions that concern some aspect of science.

SELECTED READING

Meadows, D. H., D. L. Meadows, J. Randus, and W. W. Behrens, III: "The Limits to Growth," Signet Books, The New American Library, New York, 1972. This is one of the truly important books of recent times. No educated person should be unaware of its message.

QUESTIONS

1 Ask yourself, do you believe that science can do anything it sets out to do?

2 How is it possible that there is an "energy crisis" if the law of energy conservation is valid?

3 If you were told that you must cut your personal use of energy in half, what would you give up?

4 If a scientist comes upon a discovery that might be used in ways the United States would disapprove of, should he suppress the discovery?

5 Do you think that the United States SST should have been built? Why, or why not?

6 List as many scientific achievements as possible that have had an impact on your life. (An example would be the eyeglasses some of us wear that are an outgrowth of scientific studies of the properties of light.)

7 Has physics had an impact in your area of academic interest? How?

8 As you go through this course, watch the newspapers or news magazines for articles about the impact of science on your daily life.

SELF-TEST

____________ technology
____________ science
____________ scientific law
____________ conservation of energy
____________ scientific model

1 A scientific law which states that energy is never created or destroyed

2 The application of the facts of science to produce useful things

3 A summary of the results of many experiments

4 A proposal about how some aspect of nature works. Subject to experimental test of its consequences

5 The study of how all aspects of the universe work and are put together

MECHANICS

The study of motion has been of interest to physical scientists for thousands of years. The Greek philosopher Aristotle described the motion of a falling body as the tendency of that object to seek its "natural place." The philosopher Zeno proved logically that motion cannot exist! With the rebirth of physical science in the Renaissance, the first area where significant results were achieved was in the study of motion. The major heroes of this search were Galileo Galilei and Isaac Newton, although there were many other contributors. One major point of this study was the attempt to understand the motions of the planets and their moons.

Newton's laws of motion, or mechanics, have been found to hold in a wide variety of physical situations. Again and again we shall apply these results to varying situations. Only when scientists have probed deeply into realms as small as the atom and smaller have modifications to Newton's mechanics been found necessary. Toward the end of this book we shall discuss some of these modifications.

Part 1, then, sets the stage. It introduces some basic concepts and vocabulary and shows how these are to be used in a number of simple situations. The student should be aware that the terms and ideas presented here will be used many times and should treat them much like vocabulary in the study of language, that is, as necessary building blocks which must be mastered in order to go on with the subject.

Mechanics is the scientific description of motion. Much of physics involves the study of forces, and in most cases the visible evidence for the existence of a force is that something moves. Therefore the tool for the study of the physics involved in a force is the observation of the motion caused by that force.

Chapter 2 is a discussion of the description of motion and a definition of the basic quantities. Chapter 3 relates observed motions to forces, and discusses the relationship between weight and mass. Chapter 4 is a discussion of circular motion, with application to the motion of planets and artificial satellites.

2 THE DESCRIPTION OF MOTION

In this chapter we shall develop the ideas and terms needed to describe objects in motion. One cannot read a foreign language without knowing at least some vocabulary. Similarly, one cannot study physics without learning its "vocabulary" and "grammar." The terms and ideas introduced here are part of the "vocabulary" of physics. Solving problems is similar to studying vocabulary in a language. Until one can use a word in a sentence, it has not been mastered; until basic physical concepts can be applied in the solution of simple problems, they have not been mastered.

The analogy of the study of physics to the study of a foreign language has one more parallel. Any student who has ever studied a language knows that it is impossible to learn by cramming just before exams. The only type of study which works is regular study, so that knowledge of the language is slowly built up. Similarly, physics is a "study-as-you-go" subject. Since there is a sequential build-up of ideas, each of which forms the basis of the next, new ideas must be studied as they come up.

2.1 THE NATURE OF MOTION

Mechanics is the study of motion and its causes. Motion implies that something has moved or changed its position. Also involved is the fact that time elapses while the motion is accomplished. Therefore at the root of the description of motion are the ideas of distance and time. All the other concepts we shall use—velocity, speed, and acceleration—will be developed from the fundamental ideas of length and time. These terms are probably familiar to the reader in an everyday context; in this chapter a more precise meaning will be given to them.

We describe objects in motion every day. For example, how fast did your car travel on the way to school: 30 miles/hour? 40 miles/hour? When you read a weather report, how is the motion of the wind described? When you arrange to meet someone, you gauge the distance involved and the time it will take. Each of these ideas involves statements about distance and time. Therefore a discussion of motion must begin with a discussion of distance, or length, and time.

THE MEASUREMENT OF LENGTH

Everyone understands from his own experience what is meant by length. We know that a football field is 100 yards (yd) long, that it is somewhat more than 3,000 miles from New York to San Francisco, and that most people are between 5 and 6 feet (ft) tall. Each of these statements is a statement about the length of something.

Take the football field, for example. What is the precise meaning of the statement that it is 100 yards long? It means that in some way, direct or indirect, the football field has been compared with a stick or other object whose length has been defined as 1 yard and that, when this comparison was made, the field was found to be 100 times the distance defined as 1 yard. Note that there are two separate ideas involved in this operation. The first is the agreement that a stick of a certain length shall be called 1 yard, and that all other sticks which are identical with it shall be called 1 yard also. This is the choice of a standard, or of a unit, of measurement. The second is the actual comparison between such a stick and the object whose length we wish to describe.

Historically, many standards of length have been used. The ancient Egyptians used the distance between the elbow and the tip of the middle finger and called this length the cubit. Unhappily for the usefulness of this system, one man's cubit was not the same as another's. The English foot, with which we are more familiar, originally was related to the length of the king's foot but, as no two kings' feet are identical, this unit presented the same difficulties. To be widely useful, a standard of length must be fixed, unvarying, and easily reproducible in any laboratory.

The primary system of units used in this book is the metric system, also called the mks or SI system.† The unit of length in the metric system is the meter. The metric system is replacing the English system in all industrialized countries, and the United States will be forced to follow suit soon, in order that our products will be able to compete in the export market.

The metric system was created during the French Revolution, in order to create a completely logical system which would avoid many of the awkward features of the other measuring systems. All units of length in the metric system have a very simple relationship to the basic unit, the meter (m). The centimeter (cm) is $\frac{1}{100}$ of a meter, the millimeter (mm) is $\frac{1}{1{,}000}$ of a meter, and the kilometer (km) is 1,000 meters. Each differs from the basic unit by a power of 10. The unit names are based on Latin prefixes which indicate the nature of the unit. For instance, *kilo-* means one thousand, and *centi-* means one-hundredth.

†The mks system of units is based on the meter, kilogram, and second as the basic quantities. SI stands for Systèm International d'Unites. The SI system is the currently accepted version of the mks system.

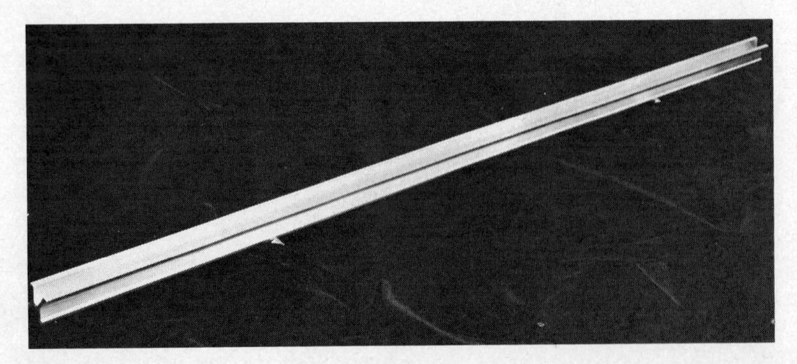

FIGURE 2.1 One of the international standard meters at the National Bureau of Standards, Washington, D. C. *(Courtesy National Bureau of Standards.)*

To continue the logical basis of the system, the meter was defined in terms of the dimensions of the earth. It was intended to be one ten-millionth of the distance from pole to equator along the meridian passing through Paris. From the results of the original measurement, a bar of platinum-iridium alloy with two scratches on it was constructed. The distance between these two marks was defined as the standard meter. The original bar is kept at the International Bureau of Weights and Measures, near Paris. Careful copies have been made and distributed to other standards laboratories, including the National Bureau of Standards in Washington, D.C. More precise measurements revised the

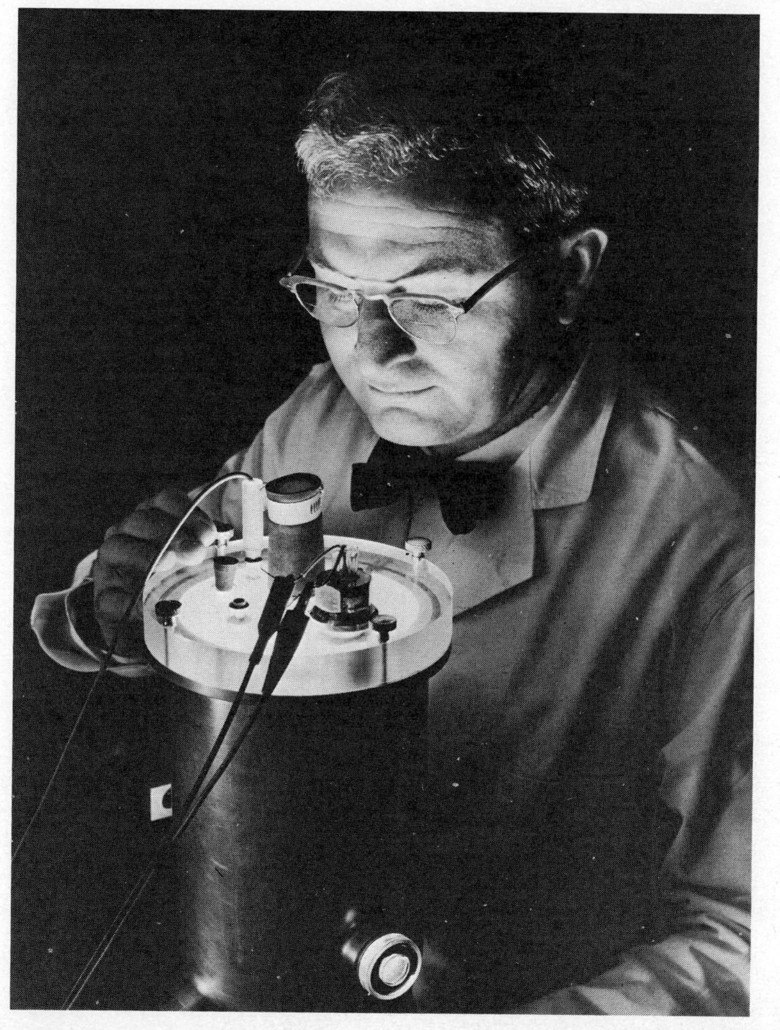

FIGURE 2.2 The standard of length. The wavelength of the orange-red light emitted by the krypton gas in this apparatus is the international standard of length. *(Courtesy National Bureau of Standards.)*

TABLE 2.1 Some metric units of length. This table shows some of the common prefixes used in the metric system and their meaning

$$1 \text{ kilometer (km)} = 1{,}000 \text{ meters (m)}$$

$$1 \text{ decimeter (dm)} = \frac{1}{10} \text{ meter}$$

$$1 \text{ centimeter (cm)} = \frac{1}{100} \text{ meter}$$

$$1 \text{ millimeter (mm)} = \frac{1}{1{,}000} \text{ meter}$$

$$1 \text{ micrometer } (\mu\text{m})^{\dagger} = \frac{1}{1{,}000{,}000} \text{ meter}$$

$$1 \text{ nanometer (nm)} = \frac{1}{1{,}000{,}000{,}000} \text{ meter}$$

†The micrometer was formerly referred to as a micron (μ) and is still frequently encountered in this form.

length of the Paris meridian, but the length of the standard meter has not been changed.

Recently, a new standard of length was chosen, capable of a far greater degree of accuracy and less susceptible to destruction or damage than the standard meter. This standard is based on precise measurements of the wavelength of light. For ordinary purposes the length of the meter has not been changed. The new standard allows more precise and accurate measurements to be made.

Although the foot was originally defined in the same way as the meter, that is, as a carefully measured distance between marks on a bar, its present legal definition in the United States is in terms of the standard meter: 1 yard, which is 3 feet, is defined as exactly 0.9144 meter. Therefore the standard meter is at present the legal basis of all units of length.

THE MEASUREMENT OF TIME

Length, in brief, is a static thing. It tells us nothing about motion. To discuss motion, we need to include a statement about time as well as distance.

In general, any physical phenomenon which repeats itself in a regular way can be used to define a unit of time. At present, the unit of time for the entire world is the second (s), which is defined as $\frac{1}{86{,}400}$ of the average length of 1 day. The repetitive phenomenon here is the daily motion of the sun and stars. This definition of the second assumes that the earth's rate of rotation, which determines the length of the day, is a constant. In fact, this is not quite true; the earth is slowing down because of a frictional effect of the tides. The period of rotation, and thus the day, is becoming longer by about 0.001 second each century.

Although this may seem an immeasurably small amount, the cumulative effect of this small change over many centuries is quite significant.

This change is verified by the time and place of total eclipses of the sun that are found recorded in ancient writings. A total eclipse is a rare event; it was of great import to people who looked to the heavens for signs, and occurrences of eclipses have been carefully documented for centuries. But today we can calculate precisely the time and place of ancient eclipses based on the present motion of the earth. When this is done, there is found to be a discrepancy of several hours, and therefore a discrepancy of thousands of miles in the location of the eclipse. This occurs because the unit of time used in the calculation is based on the present rate of rotation of the earth.

Therefore the rotation of the earth is not a good time standard, since it does not remain constant. For this reason other time standards have been designated. The most accurate of these are called *atomic clocks*, because they depend on the timing of fundamental vibrations of atoms or molecules. The frequency of these vibrations is believed not to change in the course of time, as the rotation of the earth does, and so it is more accurate to use them to keep time. We shall not be concerned in detail with these refinements; for our purposes, the second is exactly what we are accustomed to call it: $\frac{1}{86,400}$ of a day, or $\frac{1}{60}$ of a minute (min).

2.2 SPEED AND VELOCITY

Before we can accurately describe objects in motion, we must introduce three concepts: *speed* and *velocity*, which are closely related, and *acceleration*. Acceleration will be left to Section 2.4. Before you begin to read this section, ask yourself what the terms speed and velocity mean to you. When you finish this section, see how much your ideas have changed.

Our most common encounter with the term speed occurs when we talk of the speed of a car. We say that a car has a speed of 50 miles/hour. We have a dial, a speedometer, which tells us the speed of the car at

TABLE 2.2 A comparison of some metric and English units of length

1 meter	= 39.37 inches (in.)
1 kilometer	= 0.62 mile (mi)
1 centimeter	= 0.39 inch
1 millimeter	= 0.039 inch = approximately $\frac{1}{25}$ inch
1,609 meters	= 1 mile
1.609 kilometers	= 1 mile
30.5 centimeters	= 1 foot (ft)
2.54 centimeters	= 1 inch

FIGURE 2.3 Photographic zenith tube. This telescope is used to time precisely the passage of stars across the zenith at the United States Naval Observatory. In this fashion the length of the day, as measured by the rotation of the earth, is determined. *(Official United States Naval Observatory photograph.)*

any given moment. What, precisely, do we mean by this term? Speed is defined as the distance traveled during some convenient interval of time. This translates into an equation:

$$\text{Speed} = \frac{\text{distance traveled}}{\text{time for that distance to be traveled}} \tag{2.1}$$

Clearly we can use any convenient unit for distance (length) and any convenient unit of time. In this book we will use primarily the metric system, in which speed is measured in meters per second. We will also have occasion to use miles per hour and feet per second, because you are at present more familiar with these units. We want to achieve familiarity with the relationships between the two sets of units.

To obtain an idea of the magnitude of the numbers use these comparisons: 60 miles/hour = 88 feet/second, and 1 meter/second = about 3 feet/second = about 2 miles/hour.

Is it true that to measure the speed of a car you must drive for 1 hour and then see how far you have gone? Obviously not. The speedometer on your car is designed to tell you how far the car would travel in 1 hour if you continued at a steady pace. The speed at a particular moment, as measured by the speedometer, is called the *instantaneous speed*. The speed you would find by taking the total distance for a long trip, and dividing by the time the trip took, is called the *average speed*. Instantaneous speed (or velocity) is best defined as average speed measured over a very short interval of time.

What about velocity? Speed and velocity are frequently used interchangeably in everyday speech. In scientific work a distinction is made. Speed is a measure of how fast an object is traveling (without reference to direction). Velocity includes the idea of direction as well. Suppose, for example, you hop into your car and travel at 50 miles/hour for 2 hours. Where will you be? Obviously you must first know what direction you have been traveling. A statement about velocity should include a statement that the velocity is up, down, east, west, north, south, etc. A space traveler would indicate velocity as being "away from the sun and toward the star Vega." Only when direction is not of interest is speed the appropriate term.

Frequently algebraic signs are used to indicate direction. We may agree that velocities upward will be designated positive. Then velocities downward will be expressed as negative numbers. In this context the sign does not indicate less than zero, but direction. Similarly if we agree that north is the positive direction, velocities directed toward the south are negative.

Suppose we think of a car that is moving at a steady velocity of 40 miles/hour. At the end of 1 hour it will be 40 miles from its starting point, at the end of 2 hours 80 miles, and at the end of 3 hours 120 miles.

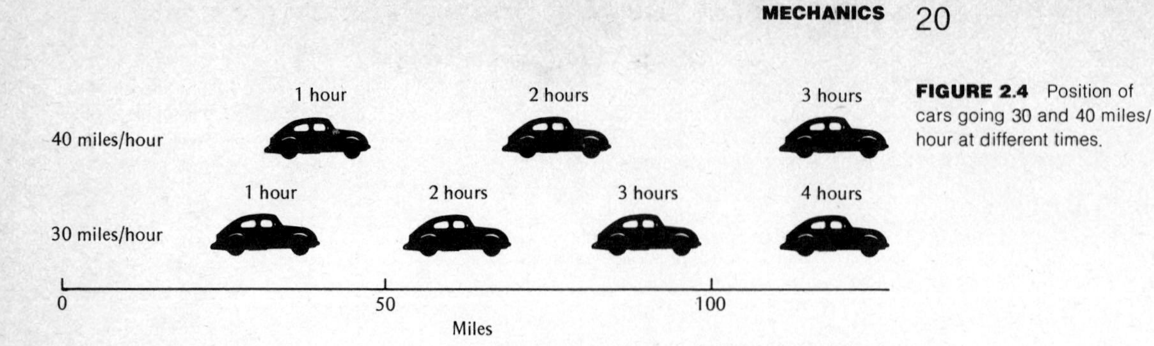

FIGURE 2.4 Position of cars going 30 and 40 miles/hour at different times.

When we look at these numbers we see that doubling the time doubles the distance. Tripling the time triples the distance. For this case of steady velocity (or speed) we say that the distance is proportional to the time. Sometimes this is written as

Distance $\propto$ time

Let us look at this idea by considering two cars, one traveling at 40 miles/hour and the other at 30 miles/hour. Figure 2.4 is one way of showing how far these cars go in different periods of time. But there is another way of making such comparisons, which is frequently more convenient. This is called graphing, and is shown in Fig. 2.5. The data (in this case, 30 miles in 1 hour, 60 miles in 2 hours, etc.) are put on the graph in a position determined by the time (horizontally) and the distance (vertically). Each dot on this graph represents the same information shown by one of the little cars in Fig. 2.4. The information for the 30-miles/hour car is connected by a solid line, and that for the 40-miles/hour car by a dotted line. We see two things: the points for each

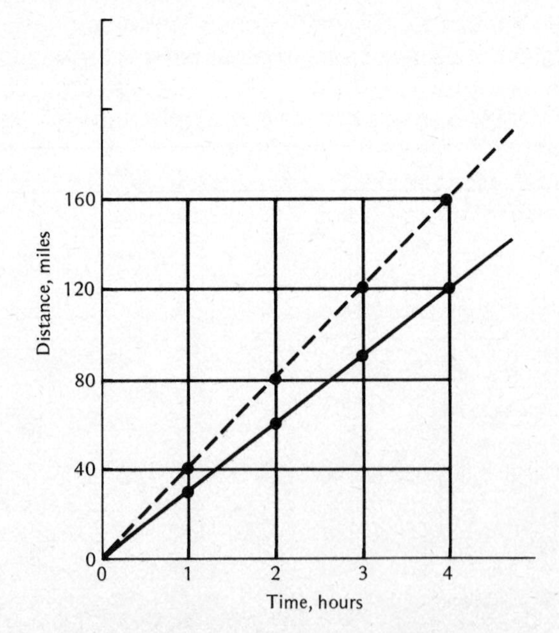

FIGURE 2.5 A graph of position as a function of time for a car going 40 miles/hour (dotted line), and one going 30 miles/hour (solid line).

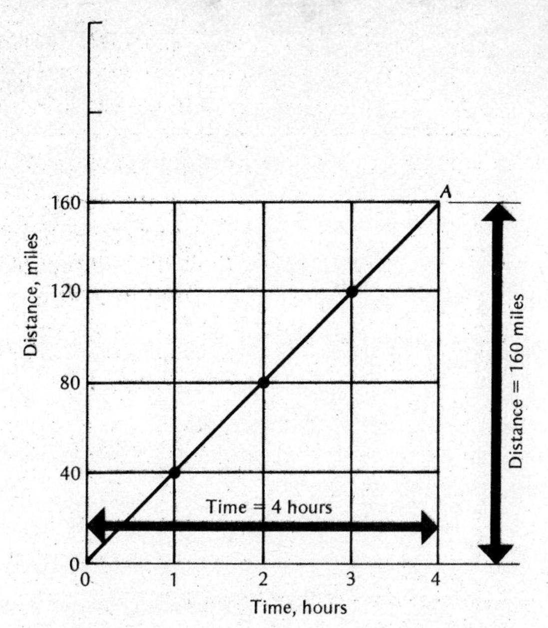

FIGURE 2.6 As the distance traveled in a given time interval increases, the slope of the line on the graph becomes steeper.

car fall on a straight line, and the slope of the 40-miles/hour line is steeper. Both of these facts show the utility of graphs. Whenever we find that two quantities are proportional to each other, we find their graph to be a straight line. Conversely, whenever we find the graph of two quantities to be a straight line, we can conclude that these two quantities are proportional.

Figure 2.6 shows why the slope is greater for the faster car. To calculate the velocity, we take the distance traveled at point *A* and divide by the elapsed time. The larger the velocity, the longer the distance for this time. If the arrow representing distance is longer, then the line of the graph will be steeper.

We have discussed the idea that velocity is defined by

$$\text{Velocity} = \frac{\text{distance}}{\text{time}} \tag{2.2}$$

Now suppose you were told that a car traveled 40 miles/hour for 5 hours, and asked how far it had traveled. You could probably answer, without difficulty, 200 miles. That is, for steady velocities,

$$\text{Distance} = \text{velocity} \times \text{time} \tag{2.3}$$

Similarly you would calculate the time elapsed as

$$\text{Time} = \frac{\text{distance}}{\text{velocity}} \tag{2.4}$$

Such equations are frequently written in an algebraic shorthand in which letters are used to represent the various quantities. This gives

$$v = \frac{d}{t} \tag{2.2}$$

$$d = vt \tag{2.3}$$

and

$$t = \frac{d}{v} \tag{2.4}$$

for the three equations written in words above. It is well to keep clearly in mind the words which go with such letters and the restrictions on each equation. [Those above are restricted to velocities (or speeds) that are constant, or to cases in which the velocity is an average velocity.] It is also important to note that only one of these need be remembered. The others are a simple rearrangement.

It is only possible to obtain a numerical solution for the equations above if two of the three quantities are known. Let us show this for two simple cases.

Example 2.1

A rifle bullet leaves the barrel of a gun with a velocity of 300 meters/second. How far does it go in $\frac{3}{4}$ second?

Using Equation 2.3 we have

$$\begin{aligned} \text{Distance} &= \text{velocity} \times \text{time} \\ &= 300\ \frac{\text{meters}}{\text{second}} \times 0.75\ \text{second} \\ &= 225\ \text{meters} \end{aligned}$$

Example 2.2

How long does it take the same bullet to travel 400 meters?

From Equation 2.4 we have

$$\begin{aligned} \text{Time} &= \frac{\text{distance}}{\text{velocity}} \\ &= \frac{400\ \text{meters}}{300\ \text{meters/second}} \\ &= 1.33\ \text{seconds} \end{aligned}$$

2.3 UNITS CONVERSION

One problem which frequently recurs is the conversion of a quantity expressed in one set of units to some other set. To illustrate a procedure to do this we will convert a velocity of 60 miles/hour to kilometers per hour and to meters per second.

We need to look in the tables in the front of this book and find that (rounded off)

1 mile = 1.6 kilometers
1 mile = 1,600 meters
1 hour = 3,600 seconds

These quantities are *identities*. That is, they express the same physical quantity in different units.

To convert 60 miles/hour to kilometers per hour we set up the relationship

$$60\ \frac{\text{miles}}{\text{hour}}\left(\qquad\right) = \frac{\text{kilometers}}{\text{hour}}$$

In the empty parentheses we place an identity—the one relating miles to kilometers. We put miles on the bottom, so miles will cancel on the left. We put kilometers on top, because we want to end up with kilometers on top.

$$60\ \frac{\cancel{\text{miles}}}{\text{hour}}\left(\frac{1.6\ \text{kilometers}}{1\ \cancel{\text{mile}}}\right) = 96\ \frac{\text{kilometers}}{\text{hour}}$$

The conversion of 60 miles/hour to meters per second requires two identities.

$$60\ \frac{\cancel{\text{miles}}}{\cancel{\text{hour}}}\underbrace{\left(\frac{1{,}600\ \text{meters}}{1\ \cancel{\text{mile}}}\right)\left(\frac{1\ \cancel{\text{hour}}}{3{,}600\ \text{seconds}}\right)}_{\text{identities}} = 26.7\ \frac{\text{meters}}{\text{second}}$$

2.4 ACCELERATION

Situations in which speed or velocity are not changing are not too common. Most situations of physical interest occur where velocity is not constant. If we want to discuss changing velocities, we must introduce a new concept called *acceleration*.

Acceleration is defined as the rate at which velocity (or speed) is changing. For instance, a drag racer starting at rest may reach a speed of about 200 miles/hour in less than 8 seconds. Clearly the name of the game in drag racing is acceleration—the ability to increase speed as quickly as possible. In the example just discussed the speed of the racer went from 0 to 200 miles/hour in 8 seconds. Therefore the speed must have increased by 25 miles/hour during each second. The acceleration is 25 miles/hour each second.

Two hundred miles/hour is roughly one hundred meters/second. So we could calculate the acceleration as

$$\text{Acceleration} = \frac{100 \text{ meters/second}}{8 \text{ seconds}}$$
$$= 12.5 \text{ meters per second each second}^{\dagger}$$

The meaning of this is that the speed of the drag racer increased by 12.5 meters/second during each of the 8 seconds. In fact, for a real drag racer the acceleration would not be a constant, but we will ignore that additional complexity.

So we can define acceleration as

$$\text{Acceleration} = \frac{\text{change in velocity (or speed)}}{\text{time during which the change took place}} \quad (2.5)$$

We can make a graph of velocity versus time for our drag racer. This is shown in Fig. 2.8. We see that a straight line results. Therefore velocity is directly proportional to time for a situation with constant acceleration. If we make a plot of distance and time for the same case, we will not find a straight line. This shows that the relationship between distance and time is not a simple proportion when there is constant acceleration. We will see later what this relationship is.

†This is sometimes written meters per second per second, and abbreviated m/second² (or m/s²).

FIGURE 2.7 Every aspect of the design of a drag racer is calculated to allow maximum acceleration. *(Courtesy Ernest Baxter, Black Star.)*

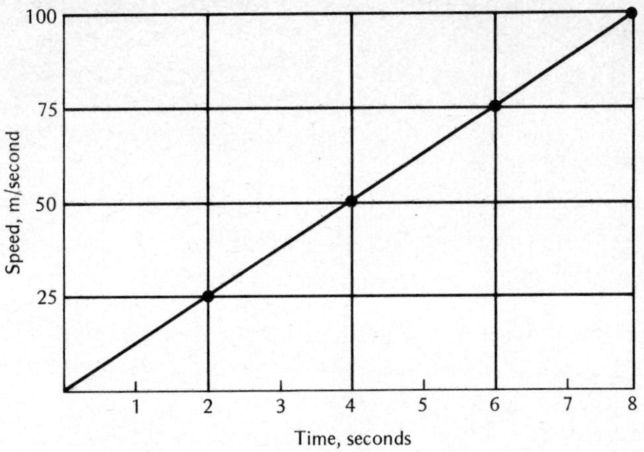

FIGURE 2.8 A graph of speed as a function of time, for a drag racer.

Example 2.3

A passenger car can accelerate at about 3 meters per second each second. How long does it take from rest to achieve a velocity of 20 meters/second (about 40 miles/hour)?

$$\text{Acceleration} = \frac{\text{change in velocity}}{\text{elapsed time}}$$

$$3\ \text{meters/second}^2 = \frac{20\ \text{meters/second}}{t}$$

$$t = \frac{20\ \text{meters/second}}{3\ \text{meters/second}^2}$$

$$t = 6\tfrac{2}{3}\ \text{seconds}$$

2.5 AN EXAMPLE: MOTION OF A FALLING OBJECT

Where in nature do we encounter examples of motion with constant velocity or constant acceleration? One very simple example occurs: the motion of a falling object. The understanding of this problem was central to the beginnings of modern physics. In the early 1600s Galileo Galilei (1564–1642) and Isaac Newton (1642–1727) considered the problem of falling bodies in "scientific" or definitive terms.

FIGURE 2.9 Galileo Galilei (1564–1642). *(Courtesy University of Pennsylvania Library.)*

Before Galileo's time physical science was pretty well dominated by the ideas of the Greek philosopher Aristotle (384–322 B.C.). There were others before Galileo who challenged Aristotle's ideas, but they had small impact. Aristotle believed the free fall of an object to be an example of what he called natural motion. He believed that each of the four elements—earth, air, fire, and water—had its natural place:

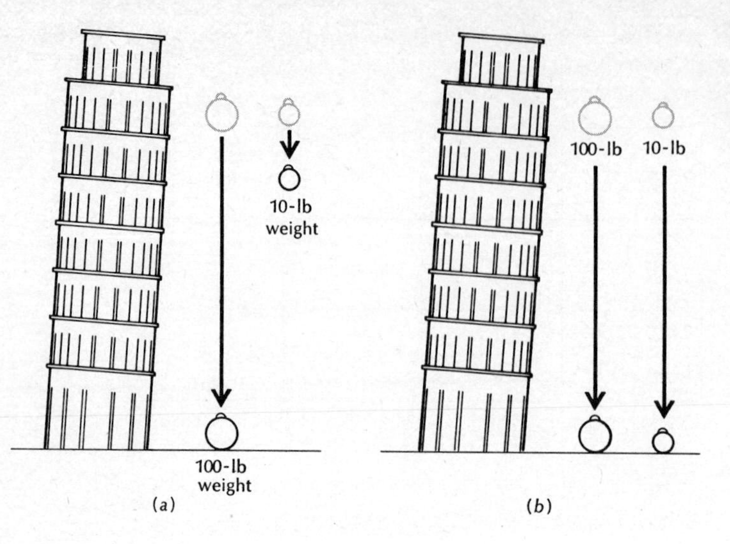

FIGURE 2.10 *(a)* Aristotle's version of two falling bodies. The heavier object falls faster than the light one. *(b)* The result of Galileo's hypothetical experiment. The heavy and the light objects arrive at the ground at almost precisely the same time.

fire was above air, air above water, and water above earth. The way any object behaved was thought to depend on the proportion of the various elements in it. The natural motion of an object was toward its natural place. Natural motions were therefore either upward (a balloon filled with hot air) or downward (a stone, which was mostly earth). Aristotle contrasted this with what he called forced, or violent, motion which was the motion of objects in a horizontal direction due to pushing and pulling or to forces.

One of Aristotle's conclusions from his ideas was that, in its natural motion downward, the *velocity* of a body was *constant* and proportional to its weight. That is, if two objects, one twice as heavy as the other, were dropped, the heavier one would fall at twice the velocity of the lighter one. This is a prediction which might have easily been tested experimentally. One of the distinctions between science before and after the seventeenth century lay in whether such predictions were followed up with experimental tests. Aristotle did no experimental test. Because of his scientific reputation, and later because of the backing of the Christian church, Aristotle's hypothesis was not successfully challenged for many centuries.

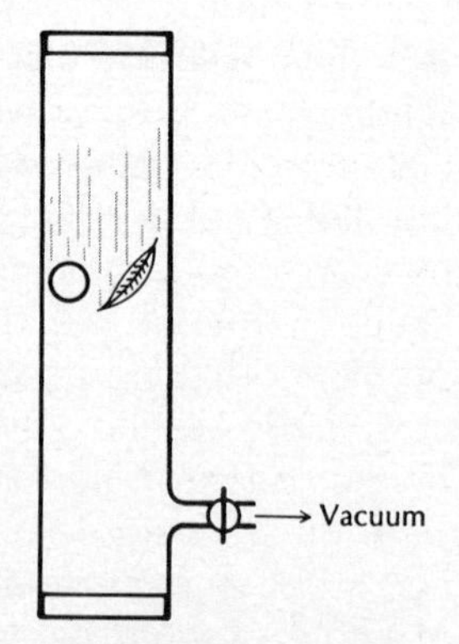

FIGURE 2.11 In a vacuum a feather and a lead weight fall in precisely the same manner.

Galileo is said to have performed an experiment to test Aristotle's ideas. The story may be legend, but the experiment it describes could have been performed, and Galileo was well aware of the results. Aristotle predicted that a 100-pound (lb) weight would reach the ground from a high place in a time 10 times shorter than a 10-pound weight. Galileo is said to have dropped weights from the Leaning Tower of Pisa to test this hypothesis. When this experiment is performed, the heavy weight actually does strike the ground slightly sooner, but by no means in a time 10 times shorter than the light weight. The small difference ob-

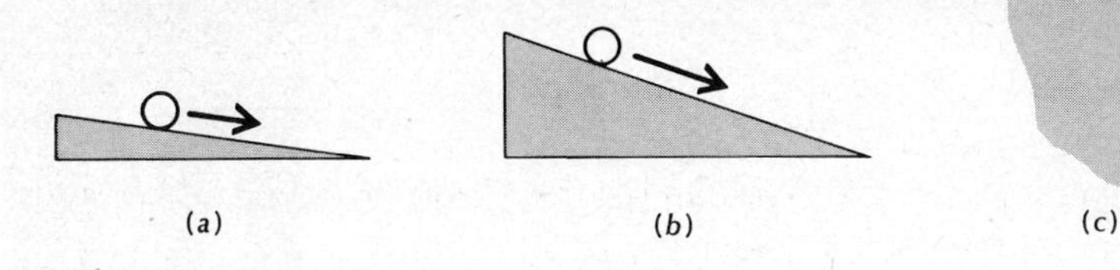

FIGURE 2.12 A ball rolling down three planes of increasing steepness.

served to occur can be attributed to the resistance of the air to the falling object. If the same experiment is performed in a vacuum, even a feather will fall as fast as a lead weight.

However, disproving Aristotle's proposition was not sufficient to describe the motion of a freely falling body. Galileo's further achievement lay in formulating a description and proving by experiment that it was the correct one.

The accurate measurement of the velocity of a falling body was not possible in Galileo's time. It was possible of course to measure distance, but the techniques available for the measurement of small intervals of time were too crude. Galileo's invention of the pendulum clock was possibly an outgrowth of his desire to find more accurate ways to measure time. In some of his experiments he used a water clock, weighing the amount of water flowing from a vessel in a given time and taking the weight of the water as proportional to the time elapsed. Because of the difficulty in making a direct measurement of time, it was necessary for him to find some other way to test a description of freely falling bodies. Galileo considered the motion of objects rolling down an inclined plane, and showed that as he made the plane steeper and steeper he would approach free-fall motion.

It is possible to make two simple assumptions as to the nature of the motion of a freely falling body. Aristotle chose one: the velocity is constant. (This is not equivalent to assuming that the velocity is proportional to the weight.) If the velocity of a falling body is not constant, the next most simple assumption is that the acceleration is a constant. Although the simplest assumptions do not always agree with the facts, they often are the most convenient starting points for the development of a physical theory and, in a surprising number of cases, turn out to be the correct description. If, under experimental test, both of these simple

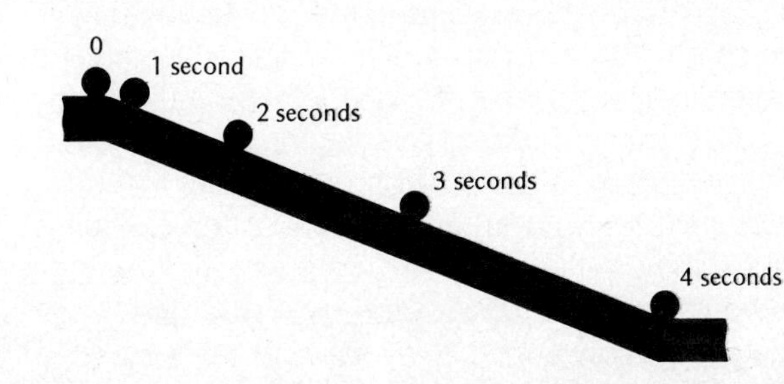

FIGURE 2.13 Position of a ball rolling down an incline at 1, 2, 3, and 4 seconds after being released.

assumptions had turned out to be false, we would have had to investigate more complex assumptions.

Figure 2.13 shows a ball at random points as it rolls down a plane, and Fig. 2.14 is a photograph of a falling ball. The ball is given a multiple exposure by flashing a light every $\frac{1}{30}$ second. We can examine the two proposals about falling bodies by examining either the ball rolling on the incline or the falling ball. We can immediately rule out constant velocity, because that requires equal distances to be traveled for equal time intervals. Both of the objects shown are traveling greater and greater distances in each time interval. Therefore their velocity is not constant—it is increasing as the objects are accelerated.

We can show, by direct measurement on Fig. 2.14, that the acceleration is constant. First we calculate the velocity for each numbered time interval in the photograph. We do this by remembering that velocity is distance divided by time. We then measure the distance with a millimeter ruler, and divide by the time interval $\frac{1}{30}$ second. This will give a table of velocities, in millimeters per second (Table 2.3). Then we look at the change in velocity from one time interval to the next. Acceleration is change in velocity divided by time. If the change in velocity between each pair of time intervals is a constant, the falling ball undergoes a constant acceleration. This is precisely what we find, and what is found in all other experiments on falling objects.

We can summarize all this as follows. All falling objects (if we ignore the complication of air resistance) fall with a constant acceleration. The numerical value of this acceleration is found to be the same, no matter what the weight of an object and no matter what material it is made of. The second half of this statement has been shown experimentally to be true to an accuracy of better than one part in a billion, in the elegant and delicate experiments of Robert Dicke at Princeton University. These experiments are not direct measurements of falling objects, so we cannot easily describe them here.

The constant acceleration of a falling body is called the *acceleration due to gravity*. It is given the symbol g. As we will see, it varies by about $\frac{1}{2}$ percent from place to place on the earth's surface. Its numerical value is approximately 9.8 meters per second each second, or 32 feet per second each second. This means that during each second the velocity of a falling body increases by 9.8 meters/second, or 32 feet/second. The acceleration due to gravity is much different if we travel to the moon. There it is approximately 1.5 meters per second each second, or 5 feet per second each second. This means that on the moon falling objects acquire velocity more slowly. They appear to float, as in slow motion, as you may have noticed on telecasts from the moon.

Two questions can be asked about falling objects. After a certain period of time, how fast are they falling and how far have they fallen? The velocity is easily calculated using the definition of acceleration, where now we know the acceleration but not the velocity.

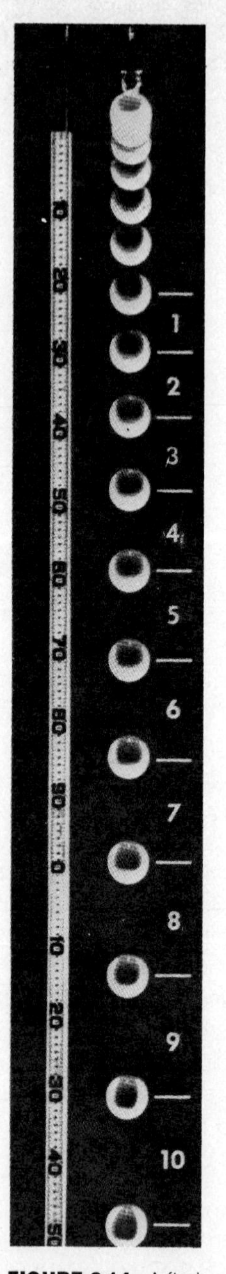

FIGURE 2.14 A flash photograph of a falling object. The scale on the left is in centimeters, and the time between successive positions of the ball is $\frac{1}{30}$ second. *(Courtesy PSSC, "Physics," D. C. Heath and Company, 1965.)*

Time interval	Distance, millimeters	Velocity, millimeters/second	Acceleration, millimeters per second each second
1	7.5	225	
2	8.5	255	900
3	9.5	285	900
4	10.6	318	990
5	11.9	357	1,170
6	13.0	390	990
7	13.9	417	810
8	14.9	447	900
9	16.0	480	990
10	17.1	513	990

TABLE 2.3 Measurements made on Fig. 2.14 and calculations of velocity and acceleration. The values for acceleration are not all exactly equal, because of errors occurring in the measurement of the distances. You should make enough of these measurements and calculations so that you understand the construction of this table.

$$\text{Acceleration} = \frac{\text{change in velocity}}{\text{elapsed time}} \tag{2.5}$$

Change in velocity = acceleration × time

or, starting at rest ($v = 0$),

$$v = at \quad \text{or} \quad v = gt \tag{2.6}$$

How far an object starting at rest falls in a given period of time is a somewhat more difficult question. We have shown before that

Distance = average velocity × time

But what is the average velocity for something that is accelerated? This turns out to be a simple average of the velocity at the beginning and the velocity at the end.

$$\text{Average velocity} = \frac{\text{beginning velocity} + \text{final velocity}}{2}$$

We have shown in Equation 2.6 that the final velocity is given by $v = gt$ if the beginning velocity is zero. Therefore, for a falling body starting at rest,

$$\text{Distance} = \left(\tfrac{1}{2}gt\right)t$$

$$d = \tfrac{1}{2}gt^2 \tag{2.7}$$

Clearly this works for any acceleration, not just falling bodies. Here we have used g to stand for the specific acceleration of falling bodies.

SYMMETRY

Figure 2.15 shows a boy throwing a stone in the air. The stone reaches its highest point and then falls back toward the earth. At the top of its motion the velocity must be zero, because it is changing from velocity

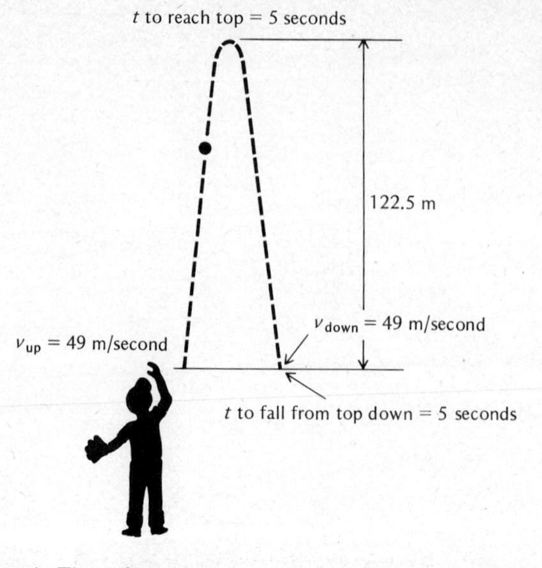

FIGURE 2.15 This drawing shows the symmetry between the motion of a ball on its way up and on its way down.

upward to velocity downward. Therefore we can treat the stone at the top as a wholly new problem, as though it had just been dropped at that point. Equation 2.7 gives the relationship between distance and time for this falling stone. It also gives the relationship between distance and time for the stone on its way up. The two halves of the problem are symmetric. This is true for all aspects of this problem.

Example 2.4

How far does an object fall on the moon in 4 seconds?

$$d = \tfrac{1}{2}gt^2$$
$$= \tfrac{1}{2}(1.5 \text{ meters/second}^2)(4 \text{ seconds})^2$$
$$d = 12 \text{ meters} \qquad \text{since } 4^2 = 16$$

On earth the same object would fall:

$$d = \tfrac{1}{2}(9.8 \text{ meters/second}^2)(4 \text{ seconds})^2$$
$$d = 78.4 \text{ meters}$$

2.6 ANOTHER EXAMPLE

In Fig. 2.16 we see police officers measuring the skid marks from a car. Their purpose is to determine the speed of the car before its brakes were applied. How can this be done? In Table 2.4 we give a list of stopping distances for various speeds on wet or dry roads. Examine the stopping

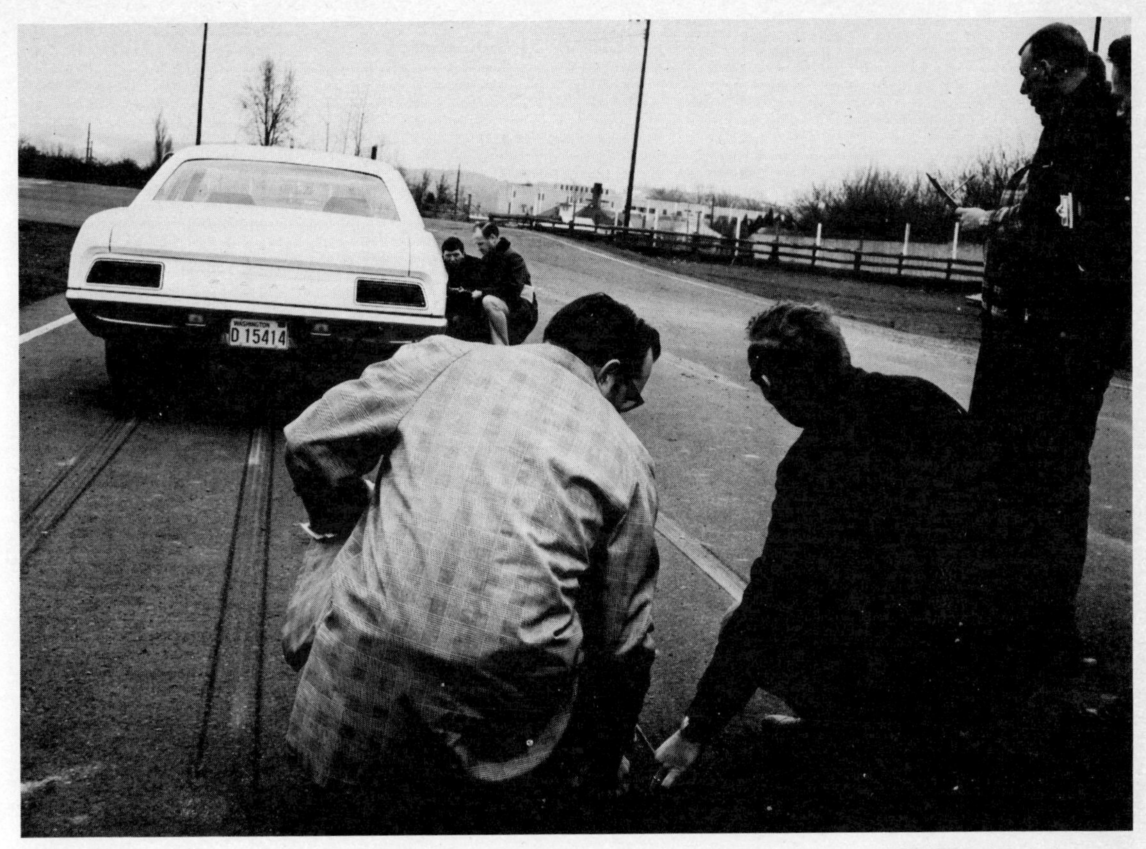

FIGURE 2.16 Police officers measure the length of skid marks. This is one way to determine the speed of the car that made them. *(Courtesy The Traffic Institute.)*

distances for 32 kilometers/hour, 64 kilometers/hour, and 128 kilometers/hour. Is there a simple relationship between speed and stopping distance? Perhaps it will be easier if you ask the following question. If the speed is doubled, what happens to the stopping distance? What happens if the speed is tripled? When you look at things this way, you find that doubling the speed gives four times the distance, and tripling the speed gives nine times the distance. Since $4 = 2 \times 2$, or 2 squared, and $9 = 3 \times 3$, or 3 squared, the relationship is that the stopping distance is proportional to the square of the speed.

It is found experimentally that the stopping distance, with the wheels locked and skidding, is about the same for all cars. What this means is that the acceleration is about the same. (Strictly speaking, an acceleration which slows something down is called a deceleration.) We see from the table that the stopping distance on wet pavement is greater than on dry pavement. This means that the acceleration is less on wet pavement (the velocity decreases more slowly).

When all the details are worked out (either mathematically or by

Velocity, kilometers/hour	Stopping distance, meters	
	Wet surface	Dry surface
8	0.6	0.45
16	2.4	1.8
32	9.6	7.2
48	21.6	16.2
64	38.4	28.8
80	60.0	45.0
96	86.4	64.8
112	117.6	88.2
128	153.6	115.2
160	240.0	180.0

TABLE 2.4 Velocity and stopping distance. This table gives the stopping distances for a car at various velocities and under different highway conditions. These distances are the stopping distances *after* the brakes have been applied and do not include reaction time. (Note that velocity in miles per hour is approximately two-thirds the velocity in kilometers per hour)

doing experiments), the following relationship is found for the stopping distance:

$$\text{Distance} = \frac{(\text{velocity})^2}{2 \times \text{acceleration}}$$

$$d = \frac{v^2}{2a} \tag{2.8}$$

A similar relationship holds for the acceleration of a drag racer or a falling body. The difference is that this second case involves the velocity at the end of the period of acceleration, whereas for the stopping car we are talking about the velocity just as the car starts to slow down.

Example 2.5

The measured value for the acceleration of a car with its brakes on for a certain piece of highway is 6 meters per second each second. Skid marks from a car are measured to be 75 meters long. How fast was the car moving at the beginning of the skid marks?

$$d = \frac{v^2}{2a}$$

$$75 \text{ meters} = \frac{v^2}{2(6 \text{ meters/second}^2)}$$

$$v^2 = 2(6 \text{ meters/second}^2) \times 75 \text{ meters}$$

$$v^2 = 900 \frac{\text{meters}^2}{\text{second}^2}$$

To find v we must find the number which, multiplied by itself, gives 900. This is termed taking the square root of 900.

$$v = 30\ \frac{\text{meters}}{\text{second}}$$

SUMMARY

We have found a number of relations among velocity, acceleration, distance, and time for moving bodies. It is well at this point to summarize these relationships and point out the restrictions on their use.

The defining relationships for velocity and acceleration are

$$\text{Velocity} = \frac{\text{distance traveled}}{\text{elapsed time}}$$

$$= \frac{\text{change in position}}{\text{elapsed time}}$$

$$\text{Acceleration} = \frac{\text{change in velocity}}{\text{elapsed time}}$$

In this entire discussion we have restricted ourselves to motion in a straight line. We have said nothing about motion which changes direction. The discussion has also been restricted to the case of constant acceleration (but not constant velocity). With these restrictions in mind, the relationships can be summarized as follows.

1. Distance = velocity $\times$ time. Useful either for constant velocity, or average velocity when the velocity is changing.
2. Velocity = acceleration $\times$ time. Allows calculation of the final velocity achieved, starting from rest, at constant acceleration. A consequence of the definition of acceleration.
3. Distance = $\frac{1}{2}$ acceleration $\times$ (time)^2. Can be used for all cases of constant acceleration, starting at rest. Falling bodies are a specific example.
4. (Velocity)^2 = 2 $\times$ acceleration $\times$ distance. Can be used for constant acceleration, starting at rest, or for constant deceleration, ending at rest, as with stopping cars.

One example of constant acceleration found in nature is freely falling objects. The acceleration of a falling object is given the symbol g. g varies slightly from point to point on the earth's surface, and is different on the moon or on other planets. The acceleration due to gravity g can be used in any of the equations involving acceleration.

SELECTED READING

Harold Lyon: "Atomic Clocks," *Scientific American*, February 1957, p. 71.

QUESTIONS

1 The earth is moving around the sun with a velocity of about 50 miles/second. Why do we not include this velocity in all our calculations?

2 You are colonizing a new planet and desire to establish a length standard. Describe how you would go about it.

3 Give some reason for and against the United States converting to the metric system. What is your personal opinion?

4 On the Apollo-12 moon flight an astronaut dropped a piece of paper and a hammer on the moon. What was the result of this experiment, and what conclusion can be drawn from it?

5 The velocity of objects thrown upward is zero at the peak of their motion. What would the consequences be if the *acceleration* of objects thrown upward were also zero at the highest point of their motion?

6 Velocity is the concept used to discuss how fast an object is moving and in what direction it is moving. Does acceleration also include the concept of direction?

7 Consider Fig. 2.15. What is the acceleration of the ball on its upward flight? What is the acceleration at its highest point? What is the acceleration of the ball on its downward path?

SELF-TEST

__________ speed
__________ average velocity
__________ instantaneous velocity
__________ acceleration
__________ meter
__________ kilometer
__________ millimeter
__________ Aristotle
__________ Galileo

1 Change in velocity divided by elapsed time
2 1,000 m

3 Standard of length in the metric system
4 Average velocity for a very short interval of time
5 0.001 meter
6 Proposed the correct result for falling bodies
7 Greek philosopher who proposed that objects fall with constant velocity, proportional to their weight
8 Distance traveled divided by elapsed time, including direction
9 Distance traveled divided by elapsed time

PROBLEMS

(Note: Numerical constants for use in the problems are to be found inside the front cover of the book. Answers in the back are calculated using these numerical constants which are rounded off in some cases for computational simplification.)

1 The diameter of the earth is 8,000 miles. Convert this to meters and to kilometers.
2 The speedometer of a bullet train in Japan reads 210 km/hour. What speed is this in miles per hour?
3 The 1,500-m track event is often called the metric mile. How near to 1 mile in length is this race?
4 The speed limit on a local freeway is 55 miles/hour. What is the speed limit in kilometers per hour?
5 A hockey puck is 3 in. in diameter. What is this in millimeters?
6 What is your height in meters?
7 A football field is 100 yd long. Convert this to meters.
8 How long does it take a rifle bullet to travel to a target 450 yd away if its velocity is 1,500 ft/second?
9 The graph below shows the position of a man walking along a highway. (*a*) Label the part of the graph which shows the greatest speed *A*. (*b*) Label the part of the graph during which he was not moving *B*. (*c*) Calculate his average speed for the first 2 hours. (*d*) Calculate his average speed for the first 4 hours.

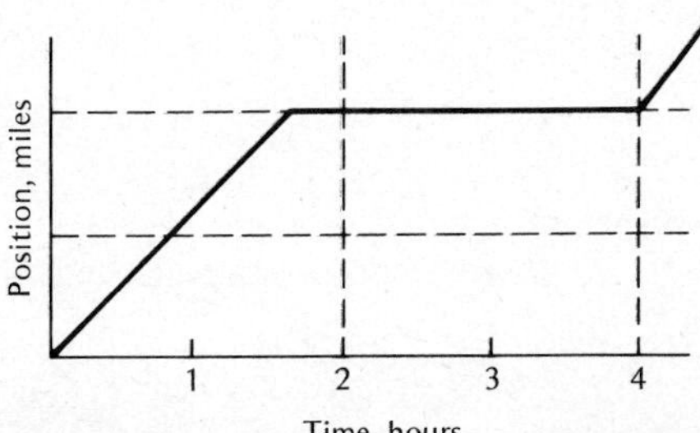

10 The maximum speed of a Boeing 747 is 640 miles/hour. The air distance from New York to Los Angeles is 2,450 miles. What is the shortest time this plane would take in going from New York to Los Angeles?

11 Draw a rough graph with position on the vertical axis and time on the horizontal axis. (*a*) Draw a line on this graph indicating the position of a person who is not moving as time goes on. (*b*) Draw a line on this graph indicating how the position of a person moving with constant velocity changes with time. (*c*) Draw a line on this graph indicating how the position of a person moving with a larger constant velocity than that of the person in part (*b*) changes with time.

12 The world record for the 1-mile run is (to the nearest second) 3 minutes 51 seconds, set by Jim Ryun in 1967. The world record for the 1,500-m run (the "metric mile") is 3 minutes, 33 seconds, also set by Jim Ryun in 1967. (*a*) What was Ryun's average speed in feet per second in the 1-mile run? (*b*) What was Ryun's average speed in meters per second in the 1,500-m run? (*c*) In which race did Ryun have the best average speed?

13 In the first event of the 1974 World Cross Country Cup series in skiing, Dieter Klause of East Germany took first place, covering the 30-km course in 1 hour, 28 minutes, 40 seconds. What was his average speed in meters per second?

14 The International Olympic Committee recently ruled that the 50-km walk would no longer be an Olympic event. The Olympic record was set by Abdon Pamich of Italy in 1964 with a time of 4 hours 11 minutes 12.4 seconds. What was his average speed in meters per second?

15 Our sun is part of a galaxy of stars and partakes in the rotational motion of these stars about the center of the galaxy. Thus the sun moves in a circle about the center of the galaxy. The radius of this circle is equal to the sun's distance from the galactic center, about 3.3×10^4 light years (3.1×10^{20} m). The sun, in one trip about the center, moves a distance equal to the circumference of this circle, 1.95×10^{21} m. The speed of the sun in its orbit is about 2.5×10^5 m/second. (*a*) How long does it take the sun to complete one full trip about the galactic center? (*b*) If the sun is about 5×10^9 years old, how many trips has it made? (These power-of-ten numbers are discussed in the appendix.)

16 The annual Housewives' Pancake Race at Olney, Buckinghamshire, England, was first mentioned in 1445. The record for the winding 415-yd course is 63 seconds, set by Janet Bunker, aged 17, on February 7, 1967. What was her average speed in feet per second?

17 A man drives a car 40 miles in 1 hour, and then 30 miles in the next $\frac{1}{2}$ hour in the same direction: (*a*) How far has he traveled? (*b*) What is his average speed for the whole trip?

18 Assume that thunder and lightning occur simultaneously, and that the light travels virtually instantaneously. If you hear the thunder 10 seconds after seeing the lightning, how far away is the thunderstorm? Assume the velocity of sound is 1,100 ft/second.

19 The world record in the 20,000-m race is held by Gaston Roelants of Belgium, with a time of 58 minutes 6 seconds. What was his average speed in meters per second for the race?

20 In one complete year the earth travels a distance of 5.8×10^8 miles in its orbit around the sun. In 1 year there are approximately 3.15×10^7 seconds. What is the average velocity of the earth in its orbit around the sun?

21 The trip to the moon in the flight of Apollo 11 took approximately 66.6 hours. The distance from the earth to the moon is approximately 238,000 miles. Compute the average speed of the Apollo-11 space vehicle for this trip in miles per hour and in feet per second.

22 In 1932 Hubert Opperman of Australia rode a bicycle 860.2 miles in 24 hours, a record which still stands. What was his average speed in miles per hour?

23 The total distance traveled on an automobile trip is 510 miles. If the total driving time is 17 hours, calculate the average speed.

24 A car is slowed from 88 to 66 ft/second in a time of 11 seconds. What is the acceleration?

25 A falling rock on earth achieves a velocity of 49.0 m/second. How long has it been falling?

26 A rocket accelerates with an average acceleration of 10 m/second^2 for a time of 5 minutes. (*a*) What is its final velocity in meters per second? (*b*) Express this also in feet per second.

27 An object on an unknown planet acquires a velocity of 25 m/second after falling a distance of 25 m. What is the value of g on this planet?

28 The velocity of a bullet leaving the barrel of a rifle is 1,000 ft/second. (*a*) If the length of the barrel is 2.5 ft, what is the acceleration of the bullet in the barrel? (*b*) How long does it take for the bullet to travel the length of the barrel?

29 A ball is dropped from a bridge into the river below. A splash is heard 4 seconds later. (*a*) What is the height of the bridge above the water? (*b*) What is the velocity of the ball the instant before it strikes the water?

30 A bicycle rider starts from rest and reaches a speed of 29 ft/second in a distance of 200 ft. (*a*) What is his final velocity? (*b*) How long does it take him to reach this final velocity? (*c*) What is his acceleration?

31 A 3,200-lb car is pushed from rest down the block by a second car. It is found that the first car moves a distance of 400 ft in 20 seconds. (*a*) What is the acceleration of the first car? (*b*) What is the final velocity of the first car?

32 A plane lands at 130 miles/hour on a 10,000-ft runway. (*a*) How long does it take the plane to come to rest, if it uses 7,000 ft of the runway? (*b*) What was the acceleration of the plane?

33 The driver of a car traveling 90 ft/second slams on the brakes. If the deceleration achieved by braking is 20 ft/second^2, how far does the car travel before coming to a complete stop?

34 It is desired to increase the speed of a rocket from 18,000 to 25,000 miles/hour. The engine can achieve an average acceleration of 60 ft/second^2. How long must the engine be operated?

35 Artis Gilmore of the American Basketball Association is about 7 ft 2 in. tall. (*a*) How long would it take a basketball released from rest just at the level of the top of his head to reach the ground? (*b*) Would this time change if the ball were dropped on the moon?

36 A man, standing at the base of an 80-ft tower, wishes to throw a ball to a friend who is at the top of the tower. He throws the ball and his friend catches it just as it reaches the peak of its motion, 80 ft above the ground. (*a*) How long does it take the ball to reach the top of the tower? (*b*) With what initial velocity was the ball thrown?

FORCE, MASS, AND NEWTON'S SECOND LAW

We have spent some time describing motion, without saying anything about the causes of motion. We have learned the definitions of velocity and acceleration, and how they can be used to describe motion. But what causes things to move? Why do objects fall? Under what circumstances do constant velocities such as we have discussed occur? Under what circumstances do constant accelerations occur?

Each of us has had, since we were small, a qualitative understanding that, if we want to move something, we give it a push. If the object is small, it requires "less push" to make it move than if it is large. This push or shove is what we mean by a force. But to understand quantitatively what we mean by a force, we must introduce some precise definitions, just as we did for velocity and acceleration. The first of these will be an attempt to state precisely what we mean by force.

Aristotle maintained that an object would keep moving only if one kept applying a force to it. This is consistent with most of our everyday experience. If we push a heavy box across a floor, it stops when we stop pushing. If we push a car, it moves only while we push it. There are other situations that do not seem to fit Aristotle's description exactly. A light box on a smooth floor will slide some distance after a push starts it moving; a hockey player can send a puck the length of the rink with a rather modest push; a marble rolling on a level floor can roll for yards before it stops, without anyone laying a hand on it. So we see that a careful look at the situation is in order.

3.1 GALILEO'S LAW OF INERTIA

Aristotle's influence on those who followed was such that many of his statements were not challenged for nearly 2,000 years. In the early seventeenth century Galileo made the first major step forward in our modern understanding of how bodies move. We have already seen that Galileo showed that an object rolling down an inclined plane is uniformly accelerated. He also observed that an object rolling up an inclined plane is decelerated, because it slows to a stop. A deceleration is a negative acceleration, or a slowing down. The acceleration of a body on an inclined plane is greater when the slope is greater. Similarly, the deceleration experienced by an object going uphill is greater, the steeper the hill. Now suppose we consider a ball rolling on a horizontal plane, that is, one with no slope either up or down. We can infer that an object moving in a horizontal direction will neither slow down nor speed up. That is, its acceleration will be zero; it will move with a constant velocity.

If this is true, how do we account for the observation that most objects do come to a stop, even on the level? We can make measurements which show that the rate of slowing down, the deceleration, depends on the nature of the surface of the object and the surface it is moving on. Everyone is familiar with the observation that an object will slide

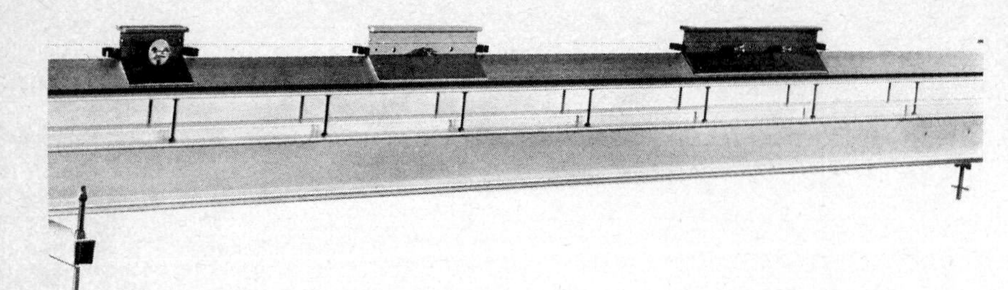

FIGURE 3.1 A photograph of an air track. *(Courtesy Ealing Corporation.)*

farther on a highly polished surface than on a rough area, and still farther on a smooth ice surface. Figures 3.1 and 3.2 show a piece of apparatus, called an air track, that has been developed to reduce surface effects as much as possible. A row of holes on the track allows air to be blown between the car and the track surface. The cars, in effect, float, and will move for a very long time before slowing down.

At this point we must take a flight of imagination. We must imagine an ideal experiment in which the interaction between the moving object and the surface has been completely eliminated, that is, the logical extension of the air track idea. Since the deceleration of an object is less and less as the surfaces become better and better, we would predict that, in the *ideal* experiment, the object *would not slow down at all!*

This type of reasoning is called a *Gedanken* experiment, or thought experiment. Very few experiments in science are ideal in the sense that they measure exactly what we wish to measure and nothing else. But if we can see the trend of the experiments as we make them closer and closer to ideal, we can extend the results in our mind to the ideal case. In this particular experiment of Galileo's, the undesired effect is the interaction between the moving object and the surface it moves on. This interaction is called *friction*. As the friction between surfaces is reduced, the object moves farther and farther without stopping. We therefore conclude that if we could do an ideal experiment, with no friction at all, the object would not slow down or stop.

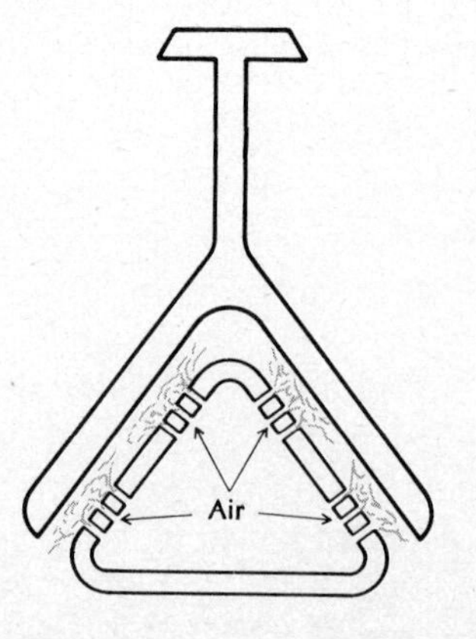

FIGURE 3.2 A diagram which shows how an air track works. Air is blown through the holes in the track so that the cars float and move almost without friction.

Observations such as this one, which seems to be applicable to all experience, are often summarized in a statement called a *principle* or *law of nature*. No legislature passes such a law. The law is simply a statement in words or mathematical symbols that generalizes a specific area of experience. The generalization of the experiments discussed above is called Galileo's law of inertia. It states:

Any object which is moving will continue moving with the same velocity unless some force acts upon it.

This law really contains two ideas. It states the property that bodies maintain their state of motion unless something is done to change it. This property is called *inertia*. Also, this law defines what we mean by force. Whatever changes the motion of an object, by changing its velocity, is called a *force*.

3.2 NEWTON'S FIRST LAW

Isaac Newton (1642–1727), who was born the year Galileo died, wrote what has been considered one of the two or three most important books on science ever written. *Principia Mathematica* is the complete exposition of Newton's work in mechanics. The starting point for Newton's work is three laws of motion. Two of these will be discussed in this chapter, and the third in Chapter 5. The first of these laws is a somewhat amplified restatement of Galileo's law of inertia.

Newton's first law states:

Any object which is at rest will remain at rest unless acted upon by a force. Any object in motion will continue in motion, with the same velocity and in the same direction, unless acted upon by a force.

The content of Newton's first law, either as stated by Galileo or by Newton, is to define what we mean by force. Newton's formulation includes the idea that to change the direction of motion requires a force, just as to increase or decrease the velocity requires a force.

FIGURE 3.3 Isaac Newton (1642–1727). *(Courtesy University of Pennsylvania Library.)*

Hidden in the statement of the first law is an implication which is not apparent at first glance. Most of us are familiar with the situation which occurs when we are riding in a car and the brakes are suddenly applied. A passenger is thrown forward as the car stops. This suggests that a force was applied to the passenger, since his state of motion was violently changed. Consider the same situation, however, as seen by an observer by the side of the road. The observer sees the car slow down, and the passenger keep moving. Her explanation, using the first law, is that the passenger continued with constant velocity while the car slowed down. Only when a force was applied by the car on some part of the passenger did he slow down also. Which observation is correct?

To answer this question we should carefully consider how each observer—the passenger in the car and the observer at the side of the road—measures the velocity of the passenger. The woman at the side of the road observes the number of fence posts the passenger and car pass per second. That is, she observes the velocity of the passenger and car with respect to objects fixed on the earth. The passenger, however, is most concerned about the fact that he travels the distance from his seat to the windshield, where he bumps his head, in a small fraction of a second. He is measuring distance, and thus velocity, with respect to objects fixed on the car. But we know the car is slowing down; that is, it is undergoing a deceleration. So the observation of the passenger is not that he has speeded up, but that his surroundings have slowed down. This leads to the appearance of his being propelled forward, because he continues to move at a constant velocity.

The conclusion from this discussion is a word of caution about the measurement of velocities. We assert that a force has been applied if we measure a change in velocity. But we must be careful to measure velocity changes with respect to objects that are not themselves accelerated. If we carefully adhere to this rule, we will not be led to conclude falsely that the passenger was propelled forward by a force when the car stopped.

3.3 NEWTON'S SECOND LAW

We have described motion in terms of velocities and accelerations, and have shown through Newton's first law that it is a force which causes a change in velocity—an acceleration. Now it is time to try to obtain a quantitative relationship between forces and accelerations.

Let us say, first of all, that there is only one proof of the relationships we are now going to discuss. Do these relationships agree with the results of experiments? The laws of physics are all like that. They are called laws only because they agree with the results of experiments. The court of last resort for a physicist is an experiment. It is the final arbiter of whether his theories stand or fall.

There is one other problem in thinking about the second law. In general its predictions do not agree closely with our everyday experience. This is because most of our experience involves events that are affected by friction, wind resistance, or other contaminating effects. In our discussion of the second law, as in that for the first law, we will discuss pure or idealized experiments. We will try to point out where the contaminating effects come in.

As an example of the above, we have previously stated that forces produce accelerations. In general, we would expect this to mean that big forces produce big accelerations, and small forces small accelerations. Suppose a husky student were asked to tie a rope on a 100,000-

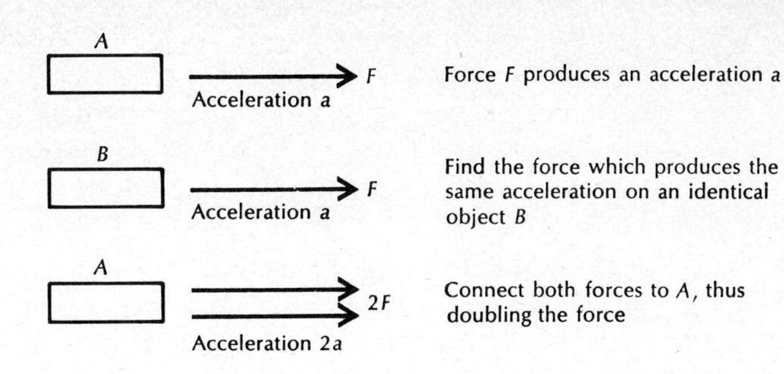

Figure 3.4 Doubling the force on object A doubles the observed acceleration.

pound freight car and start it moving. Most of you would deny that this is possible. Yet the second law, as we develop it, will say that, if the student exerted a 100-pound force for 100 seconds, the freight car would achieve a velocity of about 2 miles/hour! The difference between your experience and the prediction of the second law here is the enormous amount of friction in the wheels of the car. As we discuss the second law, we are talking about the effect of a single force on an object. Since friction does not behave in a simple way, we will ignore it in most cases.

How are accelerations measured? By measuring the rate at which velocity increases. Therefore one way to distinguish big and little accelerations is to wait a fixed interval of time and then see what the velocity is. In Chapter 2 we showed the definition of acceleration to be

$$\text{Acceleration} = \frac{\text{change in velocity}}{\text{elapsed time}} \tag{2.5}$$

Suppose we then see what happens if we double or triple the force on a single object. The way to find identical forces is to find forces that produce identical accelerations in identical objects. Then both forces are attached to the same object, and the acceleration calculated from the velocity, as shown in Fig. 3.4. Table 3.1 is an example of data collected in this manner. We see that the observed acceleration doubles when the force is doubled, triples when the force is tripled, etc. We stated in Chapter 2 that such a relationship is called a linear relationship, and the two quantities are said to be proportional to each other.

TABLE 3.1 Data for an experiment in which the force on a given object is doubled, tripled, etc., and the measured quantity is the velocity of the object after 10 seconds has elapsed. The acceleration is calculated from the relationship acceleration $= \dfrac{\text{change in velocity}}{\text{elapsed time}}$

Force	Velocity after 10 seconds	Acceleration
F	0.5 meter/second	0.05 meter/second^2
$2F$	1.0 meter/second	0.10 meter/second^2
$3F$	1.5 meters/second	0.15 meter/second^2
$4F$	2.0 meters/second	0.20 meter/second^2

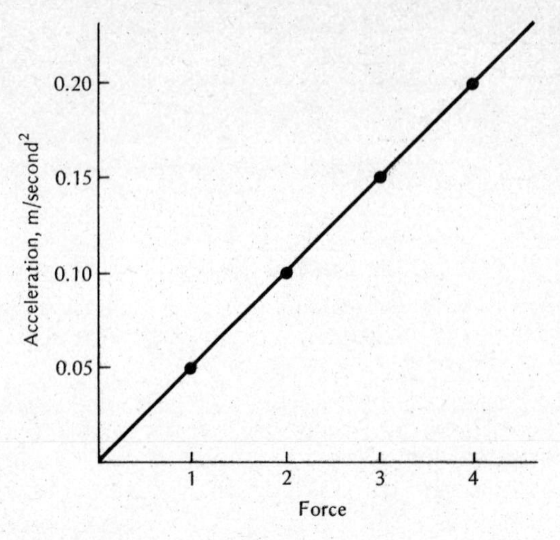

FIGURE 3.5 A graph of the data from Table 3.1 showing the linear relationship between force and acceleration.

This can be shown by making a graph of acceleration as a function of force. This is shown in Fig. 3.5. Note that the points fall on a straight line, hence the term linear relationship.

The discussion above leads to the conclusion that the observed acceleration of an object is proportional to the force applied to it. In symbolic terms this is

$$\text{Acceleration} \propto \text{force}$$
$$a \propto F$$

where we note again that $\propto$ means is proportional to. We cannot state a numerical relationship between force and acceleration, because as yet we have defined no unit in which force is to be measured.

But we know from experience that there must be more to it than this. All of us have experienced the fact that large objects are more difficult to move than small ones. Therefore there must be some property of large objects that causes them to be more difficult to accelerate than small ones. Let us return to the two identical objects used in the previous experiment, and by an analogous technique find out how the acceleration changes with the size of the object. This time we shall apply a fixed force to one of these objects and measure its acceleration as before. Then we can couple the two identical objects together and measure the acceleration again. The force is left unchanged in the two experiments. Table 3.2 gives data which might be obtained in such experiments. The result shows that the observed acceleration decreases as more and more identical objects are used. It decreases by one-half when two objects are used, by one-third when three are used, and by one-fourth when four are used. From this result we define a property of matter, called its *mass*. This property is defined such that doubling the

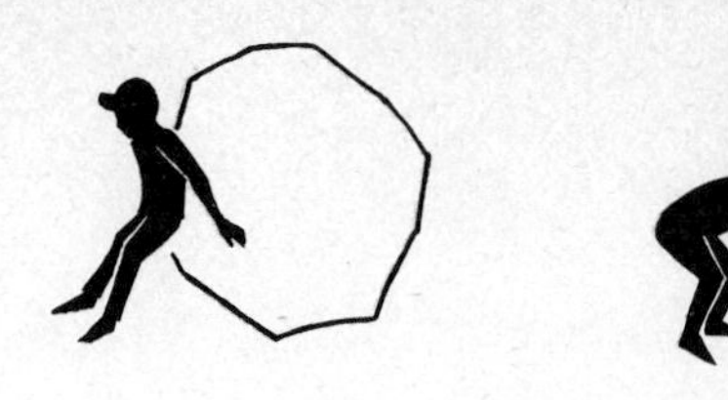

FIGURE 3.6 Large objects are more difficult to move than small ones.

mass of an object, while keeping the force constant, leads to one-half the acceleration. That is,

$$a \propto \frac{1}{m}$$

This statement is read "acceleration is inversely proportional to mass." Because of the way the experiment was done, we can see that mass is a measure of the amount of matter in an object. It is also a measure of the object's ability to resist acceleration under the action of a force. In the final analysis we will find that mass is determined by both the number of atoms in an object and by the kind of atom. Lead atoms have more mass than aluminum atoms, for example.

If we put together the two proportionalities we have obtained, we will have

$$\text{Acceleration} \propto \frac{\text{force}}{\text{mass}}$$

$$a \propto \frac{F}{m} \tag{3.1}$$

This last relationship is Newton's second law. In words it states:

The observed acceleration of any object is directly proportional to the net applied force, and inversely proportional to the mass of the object.

This statement contains two qualifications. Note the term net force. In Fig. 3.7*a* we see a boy who has taken two large dogs out on leashes. There is no doubt that he feels the forces, yet he is not accelerated,

Number of identical objects	Velocity after 10 seconds	Acceleration
1	0.5 meter/second	0.05 meter/second^2
2	0.25 meter/second	0.025 meter/second^2
3	0.17 meter/second	0.017 meter/second^2
4	0.12 meter/second	0.012 meter/second^2

TABLE 3.2 Data obtained when one, two, three, or four identical objects are acted on by a single force *F*. The quantity measured is again the velocity after 10 seconds. Acceleration is calculated using $a = v/t$. The numerical answers have been rounded off to two figures

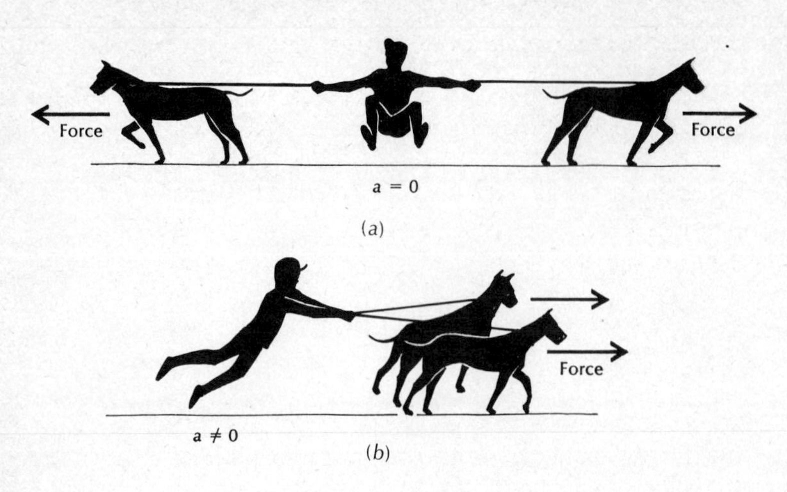

FIGURE 3.7 If the dogs pull in opposite directions, no acceleration results because the net force is zero. When they pull in the same direction, the net force and the acceleration are not zero.

because the forces are equal, *but in opposite directions*. In Fig. 3.7*b* the forces are in the same direction, and the boy is accelerated. Only the net force, or unbalanced force, produces an acceleration. The second qualification to the second law is the same as that discussed for the first law. Acceleration is a measure of the rate at which velocity is changing, but for the second law to hold we must be careful *not* to measure velocities with respect to some object that is itself accelerated.

UNITS OF MASS AND FORCE

If we wish to use the second law to calculate a numerical value for the acceleration, we need an equation, not a proportionality. Equation 3.1 shows the proportionality between acceleration and force, and the inverse proportionality between acceleration and mass, but it cannot be used for a numerical calculation. The reason is that as yet we have no way of assigning a numerical value to either mass or force. This is done by defining a unit of mass arbitrarily; then we can use the second law to define a unit of force.

In the metric system the unit of mass is the kilogram. It was originally defined as the mass of 1,000 cubic centimeters (centimeters^3 or cm^3) of water at a temperature just above the freezing point, where water has its maximum density. This system of units is called the mks system, because it is based on the meter, kilogram, and second as fundamental units. A permanent standard, the standard kilogram, has been prepared to have a mass as nearly as possible exactly equal to that of 1,000 centimeters^3 of water at this temperature. This international standard of mass is kept at the International Bureau of Weights and Measures, near Paris. All standards laboratories, such as the National

Bureau of Standards, have copies that are as nearly identical as possible with the original.

Once a standard, or unit, of mass has been chosen, we can use the second law to define a unit of force. The force that produces an acceleration of one meter per second each second on a one-kilogram mass is defined to be the unit of force and is called one *newton* (N).

In these units the second law is

$$a = \frac{F}{m} \tag{3.2}$$

$$1 \text{ meter per second each second} = \frac{1 \text{ newton}}{1 \text{ kilogram}}$$

where a is measured in meters per second each second, m in kilograms, and F in newtons. Once the standard mass has been chosen, the second law becomes the defining relation for the unit of force.

It is useful at this point to relate the unit of force you are most accustomed to—the pound—to the newton. The pound is the unit of force in the English system of units. Pounds and newtons are the same kind of thing, not pounds and kilograms. The relationship between pounds and newtons is

1 pound = 4.5 newtons

We will not have occasion to use the English unit of mass.

Example 3.1

We mentioned earlier the problem of a student trying to move a 100,000-pound freight car with a 100-pound force. In metric units, let us take a freight car with a mass of 50,000 kilograms (which weighs about 110,000 pounds). A 100-pound force is about 450 newtons, so let us use a 500-newton force. The acceleration produced is

$$\text{Acceleration} = \frac{\text{force}}{\text{mass}} = \frac{500 \text{ newtons}}{50{,}000 \text{ kilograms}}$$
$$= 0.01 \text{ meter per second each second}$$

Now calculate how much velocity this would yield if continued for 100 seconds.

$$\text{Velocity} = \text{acceleration} \times \text{time}$$
$$= 0.01 \text{ meter/second}^2 \times 100 \text{ seconds}$$
$$= 1 \text{ meter/second}$$

As we saw earlier, 1 meter/second is about 2 miles/hour.

FIGURE 3.8 A standard kilogram at the National Bureau of Standards. *(Courtesy National Bureau of Standards.)*

Do you believe that you could actually move a freight car? Probably not. The problem we have solved above is idealized, and ignores friction. Your experience always involves friction, which obscures your understanding of the second law.

3.4 WEIGHT AND MASS

We have discussed mass as resistance to being accelerated. Our experience indicates that mass is related to what we commonly call weight, since heavy objects are difficult to accelerate. But what is the relationship? The idea that heavy objects are hard to move is familiar. But the term heavy has not yet been defined. An object is commonly referred to as being heavy when it is difficult to lift. To lift an object requires a force upward. But we observe that the upward force we apply produces no acceleration when we hold an object in the air. Therefore the net force must be zero. For the net force to be zero on an object we hold, there must be a force downward and equal to the one we apply upward. This downward force is called the *weight* of the object. Figure 3.9 shows the relationship between the upward and downward forces. *Weight is therefore a force.*

We can now discuss again the experiment Galileo is said to have performed. He dropped two objects, one light and one heavy, from the Leaning Tower of Pisa. They struck the ground at almost the same time, indicating that both underwent the same acceleration. This observation can be generalized, as has already been stated, by more experiments to show that at a given point on earth all falling objects experience the same acceleration due to gravity. Even a feather and a lead ball will fall with the same acceleration if we remove all the air, so that air resistance does not slow down the feather.

In our discussion of the second law we stated that the observed acceleration of any object is proportional to the force applied to it. A freely falling object is observed to be accelerated. Therefore a force is acting on it. This force is called the *weight* of the object. Since all falling objects are observed to have the same acceleration, this acceleration has been given a special symbol, g. It is called the *acceleration due to gravity*. Let us write down the second law for this situation:

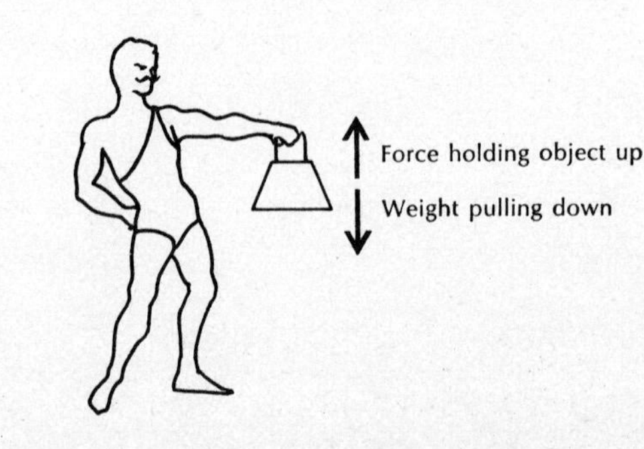

FIGURE 3.9 The weight of the object is a force pulling down. We must apply an equal force upward to keep the object from falling. The net force is then zero.

Place	g, meters/second^2
Reykjavik, Iceland	9.822
Copenhagen, Denmark	9.816
Brussels, Belgium	9.811
Seattle, Washington	9.807
New York, New York	9.803
Tokyo, Japan	9.798
Austin, Texas	9.793
Canal Zone, Panama	9.782

TABLE 3.3 Measured values of the acceleration due to gravity at sea level at different places on earth. Notice the variation with latitude

$$\text{Observed acceleration} = \frac{\text{net force}}{\text{mass}}$$

$$\text{Acceleration due to gravity} = g = \frac{\text{weight}}{\text{mass}} = \frac{W}{m} \qquad (3.3)$$

The acceleration due to gravity varies from point to point on the surface of the earth. So when we say that g is the same for all objects, this is true only if different objects are compared at the same location. Table 3.3 shows the value of g at several points on the surface of the earth. It is largest at the poles, and smallest at the equator. This variation occurs because the earth is not completely round, and also because of the earth's rotation.

Implicit in this discussion is the idea that mass is an intrinsic property of an object. Wherever in the universe the object may be, its mass is the same. Its weight is the force due to gravity on the object and depends on its location.

Because g is the same for all objects at a given point in space, we can compute weight if we know mass, or mass if we know weight, using

$$g = \frac{W}{m} \quad \text{or} \quad W = mg \qquad (3.3)$$

On the surface of the earth measurements give approximately $g = 9.8$ meters/second^2. Therefore the weight of a 1-kilogram object is

$$W = (1 \text{ kilogram})(9.8 \text{ meters/second}^2) = 9.8 \text{ newtons}$$

Since newtons are the unit of force in the metric systems, weight is measured in newtons.

On the moon, however, the acceleration due to gravity is less. The value of g on the moon is, approximately,

$$g_{\text{moon}} = 1.5 \text{ meters/second}^2$$

What would a 1-kilogram mass weigh on the moon?

$$W = mg = (1 \text{ kilogram})(1.5 \text{ meters/second}^2) = 1.5 \text{ newtons}$$

Weight is, by definition, a downward force, so that we do not usually discuss weight as a negative force. The direction of motion caused by weight is, of course, down. Because this is clearly understood, we do not try to use negative signs for this downward force or downward acceleration.

Although the mass of an object is the same no matter where in the universe it is, its weight depends strongly on where it is. The acceleration due to gravity, which could, in principle, be measured at any point in the universe, gives us the relationship between the mass of the object and its weight at that point.

Example 3.2

An object weighs 19.6 newtons on earth. What does it weigh on the moon? First calculate its mass:

$$g_{\text{earth}} = \frac{W_{\text{earth}}}{m}$$

$$9.8 \text{ meters/second}^2 = \frac{19.6 \text{ newtons}}{m}$$

$$m = \frac{19.6 \text{ newtons}}{9.8 \text{ meters/second}^2} = 2 \text{ kilograms}$$

The mass of an object is the same, no matter where it is. So we determine its weight from

$$g_{\text{moon}} = \frac{W_{\text{moon}}}{m}$$

$$1.5 \text{ meters/second}^2 = \frac{W_{\text{moon}}}{2 \text{ kilograms}}$$

$$W_{\text{moon}} = (1.5 \text{ meters/second}^2)(2 \text{ kilograms})$$

$$= 3 \text{ newtons}$$

An alternative method is to realize that weight is proportional to g. One can then set up a ratio

$$\frac{W_{\text{earth}}}{W_{\text{moon}}} = \frac{g_{\text{earth}}}{g_{\text{moon}}}$$

Solving this yields

$$W_{\text{moon}} = W_{\text{earth}} \frac{g_m}{g_e}$$

Of course this gives the same numerical result as before.

To summarize this entire discussion: *Weight is the force on an object due to gravity.* Weight is always a force, *never* a mass. Since all objects are observed to fall with the same acceleration at a given place, the weight of an object must be proportional to its mass. The mass of an object is the same at any place in the universe. The weight of an object depends on its location in the universe.

3.5 WEIGHTLESSNESS AND *g* FORCES

Much has been made in the press and TV about the state called weightlessness that occurs during space flight. One interpretation of this is that astronauts are so far from earth that the earth's gravity does not affect them. That statement is definitely *not* true.

Consider that you are standing on a scale in an elevator. The scale reads 150 pounds. Now imagine that the cable of the elevator breaks. The elevator begins to fall. It is accelerated downward. What does the scale read? To answer this we need to remember that when we stand on a scale it pushes upward on us with a force equal to our weight and opposite in direction. This upward force is what the dial on the scale reads.

When the cable breaks, both the elevator and the person inside are accelerated downward. If the scale were pushing upward on the person, he would have a different net force on him than his weight. His acceleration would be less than *g*, and less than that of the elevator. The scale must read zero if the elevator and person inside are to fall together.

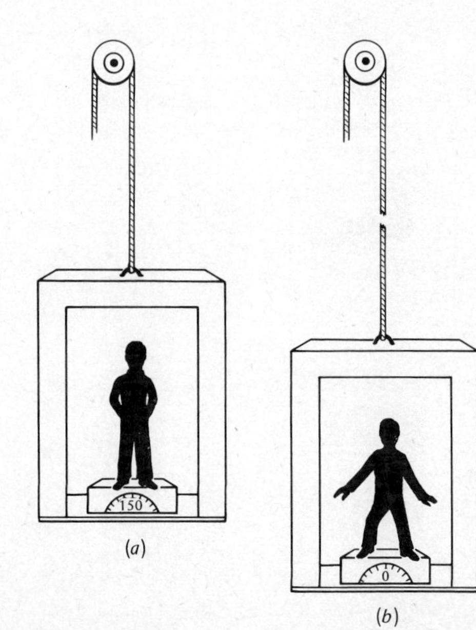

FIGURE 3.10 When the cable is broken, the elevator and boy are accelerated downward. The scale reads zero, because the only force on the boy is his weight, acting downward.

The paradoxes associated with weightlessness are a consequence of making measurements with respect to the elevator, which is itself accelerated. Gravity has not been removed. We are merely violating the caution with respect to the second law previously mentioned.

Astronauts are in the weightless state when their rocket engines are off, because their capsule is freely falling. Whatever velocity it may have is being changed because of the force of gravity on the capsule. All objects inside the capsule are also acted on by gravity. They all have the same acceleration and thus appear to be weightless.

We can understand the g forces experienced by an astronaut using a similar line of reasoning. When a rocket is accelerated off a pad by its engines, what force accelerates the astronaut? Clearly, it must be a force applied to him by the couch he lies on. This force must be such as to produce on him the same acceleration as the rocket. If the rocket is accelerated upward at a value equal to $2g$, there must be a net force upward on the astronaut given by twice his weight. Since gravity exerts a force downward equal to his weight, the couch must exert a force upward equal to three times his weight. This is shown in Fig. 3.11.

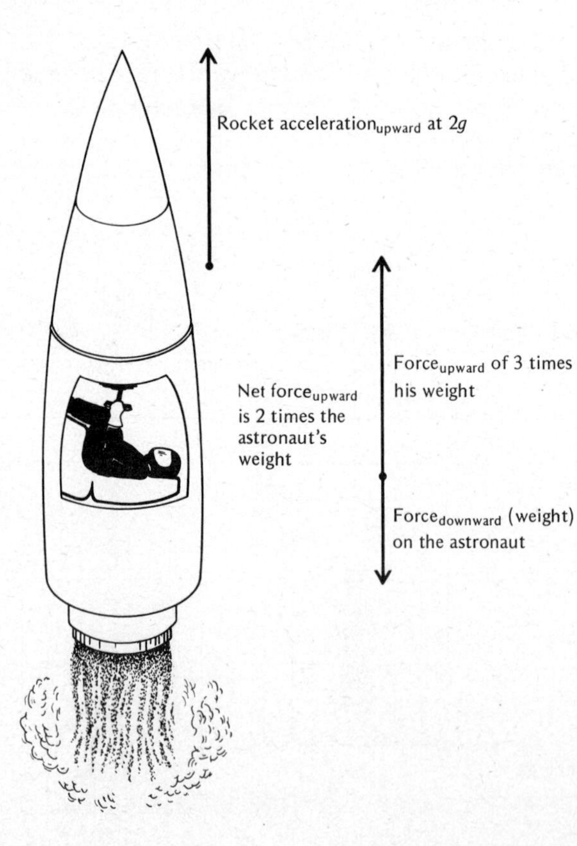

FIGURE 3.11 The couch pushes the astronaut upward with a force three times his weight, so that the net force on him is twice his weight, giving an acceleration of $2g$.

SUMMARY

The qualitative definition of what we mean by force is contained in Galileo's law of inertia and Newton's first law:

Any object which is at rest will remain at rest unless acted upon by a force. Any object in motion will remain in motion, with the same velocity and in the same direction, unless acted on by a force.

Newton's second law is written

$$\text{Acceleration} = \frac{\text{force (net)}}{\text{mass}}$$

The experimental fact that all objects fall with the same acceleration at a given point leads to the following relationship between g and the weight W.

$$g = \frac{\text{weight}}{\text{mass}} \quad \text{or} \quad \text{weight} = \text{mass} \times g$$

Weight is *always* a force, not a mass. The mass of an object is the same at any point in the universe. Weight is the force due to gravity on an object, and depends on where the measurements are made.

Weightlessness occurs when observations are made within a freely falling object, such as a space capsule.

SELECTED READING:

Manuel, Frank: "A Portrait of Isaac Newton," Harvard University Press, Cambridge, Mass., 1968. A psychological analysis of a very complex personality.

QUESTIONS

1 An object is observed to be at rest on a table top. Are we correct in concluding that no force acts on this object?

2 A magician deftly pulls a tablecloth from under a full set of dishes on a table. Explain why this is possible. What might be the result if the tablecloth had a hem?

3 In the United States potatoes are sold by the pound, and in Europe by the kilogram. What is the difference between the two?

4 A state of weightlessness is said to occur when observations are made in an object in free fall. Can a state of masslessness exist?

5 Is friction a force?
6 What is the purpose of a diet—to lose weight or mass?
7 Cars are often dislodged from holes by rocking. Discuss how this is done in terms of the laws of motion.
8 A car comes to an abrupt stop, and a camera on the back ledge moves forward and crashes into the windshield. Discuss the motion of the camera from the point of view of Newton's laws of motion.
9 What causes satellites to become hot when they approach the earth?

SELF-TEST

__________ force
__________ acceleration
__________ friction
__________ law of inertia
__________ Newton's first law
__________ Newton's second law
__________ mass
__________ net force
__________ kilogram
__________ newton
__________ pound

1 Amount of matter in 1,000 cm^3 of water at its maximum density
2 Objects continue moving at the same velocity *and* in the same direction unless acted on by a force
3 Force, opposing motion, which is eliminated or minimized in ideal experiments
4 Rate at which velocity is changing
5 Sum of all forces on an object
6 Unit of force in the English system
7 Unit of force in the metric system
8 The acceleration of an object is proportional to the net force on it, and inversely proportional to its mass
9 Causes a change in direction or velocity of an object
10 Objects continue to move at constant velocity unless acted on by a force
11 Quantity which determines the resistance to being accelerated

PROBLEMS

1 A European car weighs 14,700 newtons. What is its weight in pounds?

2 Express your own weight in newtons. Express your mass in kilograms.

3 The baggage limit for overseas flights is 44 lb. What is this in kilograms?

4 If beef in Japan is priced at 100 yen/100 g, what is its price per pound? (Use 300 yen = $1.00 and 1,000 grams = 1 kg.)

5 What is the weight, in newtons, of 5 lb of potatoes? What is its mass?

6 A girl's weight is 125 lb. Convert this to newtons.

7 A drag racer accelerates from a standing start to 60 m/second in 12 seconds. (*a*) What is the average acceleration? (*b*) If the mass of the car is 1,000 kg, what force was applied to it to give this acceleration?

8 A horizontal force of 10 newtons is applied to a 100-kg object on the moon. Find the acceleration.

9 An electron whose mass is 9×10^{-31} kg is acted on by a force of 10^{-28} newton. What acceleration is produced?

10 A golf ball of mass 0.06 kg driven by a golf club acquires a speed of 80 m/second during the impact. The impact lasts 2×10^{-4} second. Assume that a constant force acts on the ball. (*a*) What is the acceleration of the ball? (*b*) What force acts on the ball?

11 An object which weighs 64 lb on earth is found to weigh 256 lb on another planet. Calculate g on this other planet.

12 On another planet g is found to be 30 m/second^2. Calculate the weight of an object which weighs 100 newtons on earth.

13 An object weighs 32 lb on the moon. What is its weight on earth?

14 Calculate the weight of a 15-kg object on the moon.

15 An object weighs 450 lb on the moon. What does it weigh on earth?

16 Calculate the weight on the moon of an object which weighs 490 newtons on earth.

17 Find the weight of a 150-lb man on a planet where g is 200 ft/second^2.

18 A 3,000-kg car experiences a braking force of 12,000 newtons. It stops in a distance of 200 m. What was its velocity before the brakes were applied?

19 A force of 6 newtons acts on a 10-kg mass for 10 seconds. (*a*) Find the acceleration. (*b*) How far does the object move in this 10 seconds?

20 A bullet whose mass is 10 g strikes a tree and stops in 0.1 m. Its velocity was 300 m/second. Find the average force on the bullet.

21 A car whose mass is 2,000 kg is observed to stop from a speed of 30 m/second in a distance of 90 m. (*a*) Calculate the braking force, assuming that it is constant. (*b*) Calculate the acceleration. (*c*) What is the weight of the car?

22 A 10-newton force is applied to a 2-kg object. (*a*) How long does it take to achieve a velocity of 55 m/second? (*b*) How far does it travel in this period of time?

23 A cart of mass 30 kg is pushed from rest by a man for 30 ft. In that distance the cart accelerates to a speed of 8 ft/second. (*a*) What is the acceleration of the cart? (*b*) What time does it take the cart to go 30 ft? (*c*) What force acts on the cart?

24 A block of wood which weighs 19.6 newtons is initially observed to be sliding at 30 m/second along a table. Over a distance of 10 m it is seen to gradually slow down and finally come to rest. (*a*) How long did it take the block to come to rest? (*b*) What was the acceleration of the block? (*c*) What was the magnitude of the force acting on the block?

25 A car of mass 1,200 kg accelerates from rest to a speed of 30 m/second over a distance of 400 m. (*a*) What is the acceleration of the car? (*b*) How long does it take the car to cover 400 m? (*c*) What force acts on the car?

26 The brakes are suddenly applied to a 1,000-kg car going 25 m/second. The wheels lock, and the car skids to a stop in 5 seconds. (*a*) What is the acceleration of the car? (*b*) How far does the car travel while it is coming to rest? (*c*) What force acts on the car to bring it to rest?

27 A 1-kg wrench is dropped by an astronaut standing on the edge of a cliff on the moon. At the end of 3 seconds, how far has it fallen?

28 A rock on the moon weighs 300 newtons. A horizontal force of 10 newtons is applied. (*a*) What horizontal acceleration is achieved? (Assume a frictionless situation.) (*b*) If this horizontal force is applied for 10 seconds, what velocity is reached, assuming the rock started at rest?

29 In an effort to determine the acceleration due to gravity on a strange planet, a ball is dropped a distance of 200 m. It is found that it takes 5 seconds for the ball to drop this distance. The ball has a mass of 5 kg. (*a*) What is the acceleration due to gravity on this planet? (*b*) What is the velocity of the ball as it passes the point 200 m below the point where it was released? (*c*) What does the ball weigh on this planet? (*d*) What would the ball weigh on the earth?

30 A 2-kg mass is found to weigh 100 newtons on some other planet. (*a*) Find the value of g. (*b*) How far will an object fall in 5 seconds? (*c*) How fast will it be going at the end of 5 seconds? (*d*) How fast will it be going after falling 100 m?

CIRCULAR MOTION AND GRAVITY

Isaac Newton, in his *Principia Mathematica*, first proposed the idea of artificial satellites of the earth (Fig. 4.1). He considered how one might launch a projectile horizontally from the top of a high mountain so that it would not fall to earth. If the projectiles were given larger and larger initial horizontal velocities, they would fall to earth farther from the starting point. With an initial velocity high enough, the projectile would circle the earth and come to the starting point without ever striking the earth.

To discuss the relationships of circular motion we shall use as a primary example the motion of an artificial satellite, like those which now circle the earth. For simplicity we shall make one major idealization: We shall look only at the motion of a satellite in a circular orbit. Although not all satellites move in circular orbits, it is possible to achieve such an orbit. It is well known that satellites in orbits lower than 100 miles above the earth's surface quickly burn up or fall to earth. Therefore the orbit of this satellite must be higher than 100 to 150 miles.

4.1 THE INWARD ACCELERATION

To begin this discussion consider first a professor with a student asleep in class (a reasonably common occurrence). Our professor has on his lecture table a ball at the end of a string. He decides to swing the ball around in a circle above his head and then let it go so that it will strike

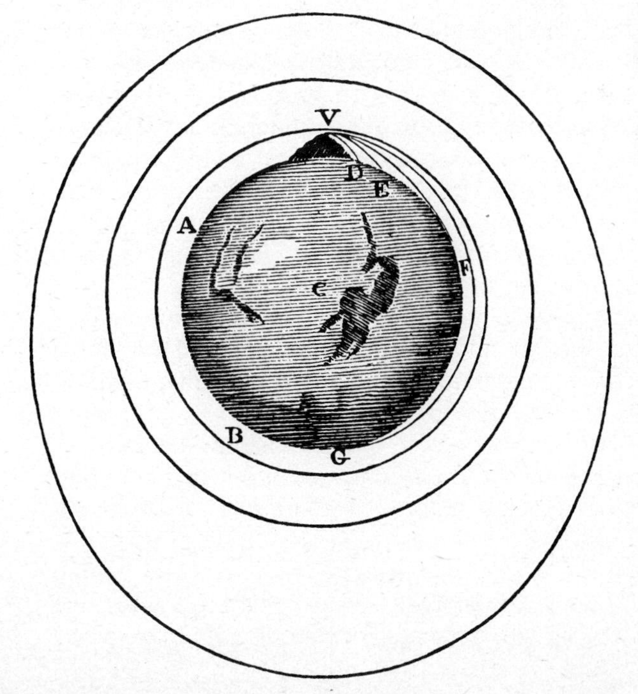

FIGURE 4.1 Newton's drawing showing how a projectile that is given sufficient horizontal velocity from a mountain top would become an artificial satellite of the earth.

FIGURE 4.2 At what position in its circular path should the professor release the ball so as to hit the sleeping student?

the sleeping student (Fig. 4.2). At what point in the motion of the ball, *A*, *B*, or *C*, should the instructor let go of the string so that it will fly toward the sleeping student, marked *X*?

If you do not believe that *B* is the correct answer, try the experiment with or without a sleeping student.

What does this experiment show? As long as our professor applies a force through the string directed toward the center, the ball moves in a circle. When he lets go of the string, the ball moves in a straight line with exactly the velocity it had when the string was released. This is simply another example of Newton's first law: In order to change the direction of motion of an object a force is required. Otherwise it moves in a straight line. The force required to keep an object in circular motion is directed toward the center of the circle. Newton's second law says that whenever there is a force there is also an acceleration. Therefore, for an object in circular motion, there is an acceleration directed toward the center of the circle.

You are probably aware that, if you swing a ball on a string, the faster you swing it, the greater the force you must exert on the string. In fact, if you swing it fast enough, the string may break. Therefore the force needed to keep the ball moving in a circle increases as its speed increases. This must also mean that the inward acceleration also increases as the speed increases. We will not attempt a mathematical derivation of this result. Rather we will state the result that can be verified experimentally.

Consider first our ball on a string, so that the size of the circle is fixed. The inward acceleration of the ball can be shown to be proportional to the square of the velocity:

$$\text{Acceleration}_{\text{inward}} \propto (\text{velocity})^2$$

If we keep the velocity fixed, but look at circles of different sizes, we find that the inward acceleration is *inversely* proportional to the radius of the circle.

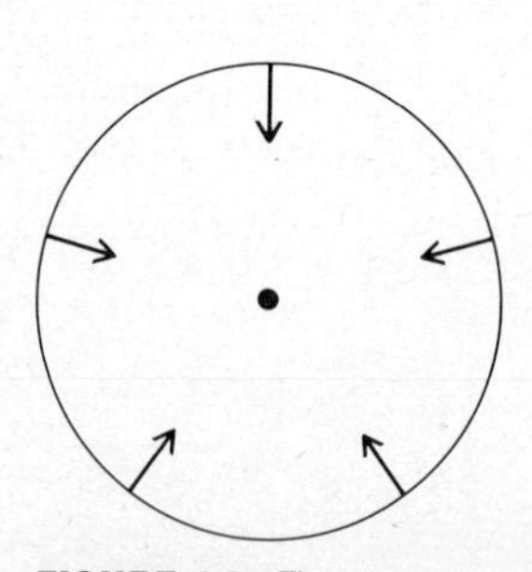

FIGURE 4.3 The arrows show the direction of both the inward force and the inward acceleration for circular motion.

$$a_{\text{inward}} \propto \frac{1}{\text{radius}}$$

Putting these two expressions together we have

$$a_{\text{inward}} = \frac{(\text{velocity})^2}{\text{radius}} = \frac{v^2}{R} \qquad (4.1)$$

This relationship always holds for circular motion. It is a geometric relationship which holds for any type of circular motion. If an object is to move in a circle of radius R with velocity v, it must have an acceleration toward the center of the circle, which is given by v^2/R. Conversely, if an object with velocity v experiences an acceleration a, at right angles to its motion, it *must* move in a circle whose radius is given by

$$\frac{v^2}{R} = a \qquad \text{or} \qquad R = \frac{v^2}{a} \qquad (4.2)$$

Equations 4.1 and 4.2 also imply that, if the velocity is fixed, the acceleration will be smaller the larger the size of the circle. Conversely, it requires a greater acceleration to move any object in a small circle with a given velocity.

One point should be made clear. Equations 4.1 and 4.2 are geometric results. They allow a calculation of the acceleration if the velocity and radius of the circle are known, just as for motion in a straight line we used

$$a = \frac{v}{t}$$

None of these relationships tells us anything about why an object *should* experience a particular acceleration. That information must come from a knowledge of the force applied in a particular situation and an application of Newton's second law.

Note that what has been calculated here is the magnitude of the inward acceleration. The fact that it is inward specifies its direction, and we do not need to attach a sign to signify this.

Let us do a simple example.

Example 4.1

A student whirls a ball on a string 2 meters long. The ball achieves a velocity of 2 meters/second. What is the acceleration toward the center?

$$a = \frac{v^2}{R} = \frac{(2 \text{ meters/second})^2}{2 \text{ meters}} = \frac{4}{2} \text{ meters/second}^2$$

$$a = 2 \text{ meters per second each second}$$

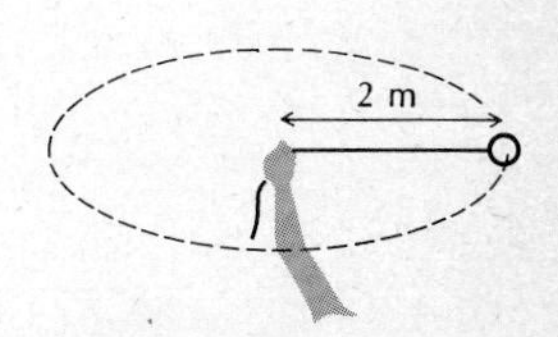

FIGURE 4.4 A ball whirled on a string has an inward acceleration toward the boy's hand.

What force must be supplied by the student to keep the ball in this example moving in the circle? The ball is known to have an inward acceleration of $a = 2$ meters/second². Newton's second law states that the observed acceleration is always given by the force divided by the mass; so

$$a = \frac{F}{m} \qquad \text{or} \qquad F = ma$$

We cannot go further until we know the mass of the ball. Assume that its mass is 0.1 kilogram. Then we can write

$$F = (0.1 \text{ kilogram})(2.0 \text{ meters/second}^2)$$
$$F = 0.2 \text{ newton}$$

4.2 THE INWARD FORCE

Following the procedure in Example 4.1 we can write down the force necessary to keep *any* object moving in a given circle. The acceleration toward the center is given by

$$a_{\text{inward}} = \frac{v^2}{R}$$

By the second law, this acceleration must always be produced by a force, and is given by

$$a = \frac{F}{m}$$

Combining these two equations gives the result

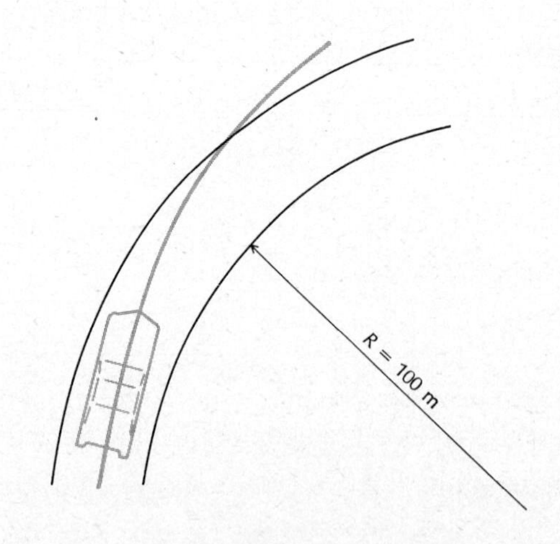

FIGURE 4.5 A car will leave the road on a curve if the inward force is not sufficient.

$$a = \frac{F}{m} = \frac{v^2}{R}$$

$$\text{Force}_{\text{inward}} = m\frac{v^2}{R} \qquad (4.3)$$

Equation 4.3 gives the force needed to keep an object of mass m moving in a circle of radius r with velocity v. If the applied force is less than this, the acceleration is smaller, and the object must move in a circle with a larger radius. If the force is greater than this, the acceleration is greater, and the object will move in a smaller circle.

One application of this result is to the problem of a car rounding a curve at a high speed. Will it stay on the road? Figure 4.5 shows a curve in a road. A curve can be looked at as part of a circle; the radius of this particular circle is 100 meters. This is called the *radius of curvature* of the curve. If the velocity of the car is 30 meters/second (about 65 miles/hour), will it stay on the curve? If the mass of the car is assumed to be 1,000 kilograms, the force necessary to keep it on the curve can be calculated:

$$F = m\frac{v^2}{R}$$

$$= (1{,}000 \text{ kilograms}) \frac{(30 \text{ meters/second})^2}{100 \text{ meters}}$$

$$F = 9{,}000 \text{ newtons}$$

The force necessary to keep the car on the road is, then, 9,000 newtons, or about 2,000 pounds. The source of this force is friction between the tires and the road, as shown in Fig. 4.6. The frictional force is best stated in terms of the maximum sideward force on the car before it begins to slide. The maximum frictional force between the tires and the road is given as a fraction of the weight of the car. Depending on the type of road surface, this can take various values, but is rarely more than 75 percent of the weight of the car. For the problem being discussed, this means that the maximum sideward force available to make the car move along the curve of the road is, since weight is given by $W = mg$,

$$F_{\text{max}} = 0.75mg$$

$$= 0.75(1{,}000 \text{ kilograms})(9.8 \text{ meters/second}^2)$$

$$F_{\text{max}} = 7{,}350 \text{ newtons}$$

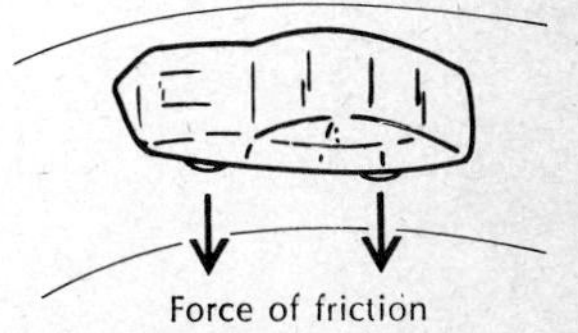

FIGURE 4.6 Friction between the tires and road prevents a car from sliding sideways.

If the maximum force available (7,350 newtons) is compared with that needed to negotiate the curve (9,000 newtons), it is clearly too little, and the car will leave the road, following the grey line in Fig. 4.5.

The reader should note that the force necessary to keep a car going along a specified curve, according to Equation 4.3, increases as the *square* of the velocity. Therefore doubling the velocity of the car does not double the danger of leaving the road, but increases it by four times.

BANKED CURVES

One method for creating a safer road, in terms of curves at least, is to bank the roadway. This means that the road surface is tilted toward the inside of the curve. The essential purpose of this is to provide additional force toward the center of the circle, so that there is less tendency for the car to slide off the road.

One way of seeing why this is true is to examine an extreme case of banking, as seen in some carnivals and circuses. Figure 4.7 shows the general idea. The motorcycle rider rides the inside of the wall, even though the wall is completely vertical. Here *all* the force to keep the motorcycle going in a circle is provided by the wall; none is provided by friction. Friction between the tires and wall keeps the motorcycle from falling down because of gravity.

A banked curve is somewhere between the extremes of the vertical wall on which the motorcycle runs and a level road. On a banked curve part of the force keeping the car on the curve comes from friction, and part comes from the road itself, as in the case of the motorcycle.

4.3 CENTRIPETAL AND CENTRIFUGAL FORCES

Some terminology associated with circular motion is frequently used in such a way as to create confusion regarding the true situation. We should clarify these terms at this point.

It has been stated that for motion in a circle to occur there must be an acceleration *toward* the center of the circle. To create this acceleration a force *toward* the center is also required. This acceleration and this force are called the *centripetal acceleration* and *centripetal force*, respectively. Centripetal means center-seeking.

Almost everyone has had the experience of riding in a car around a curve and feeling as if he were being thrown or pushed toward the outside. Frequently, it is stated that this occurs because of *centrifugal*

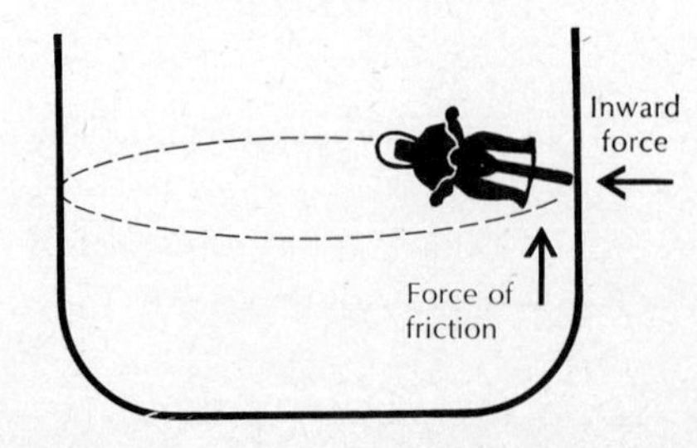

FIGURE 4.7 The inward force required to keep the motorcycle moving in a circle is provided by the wall. Friction keeps the motorcycle from sliding downward.

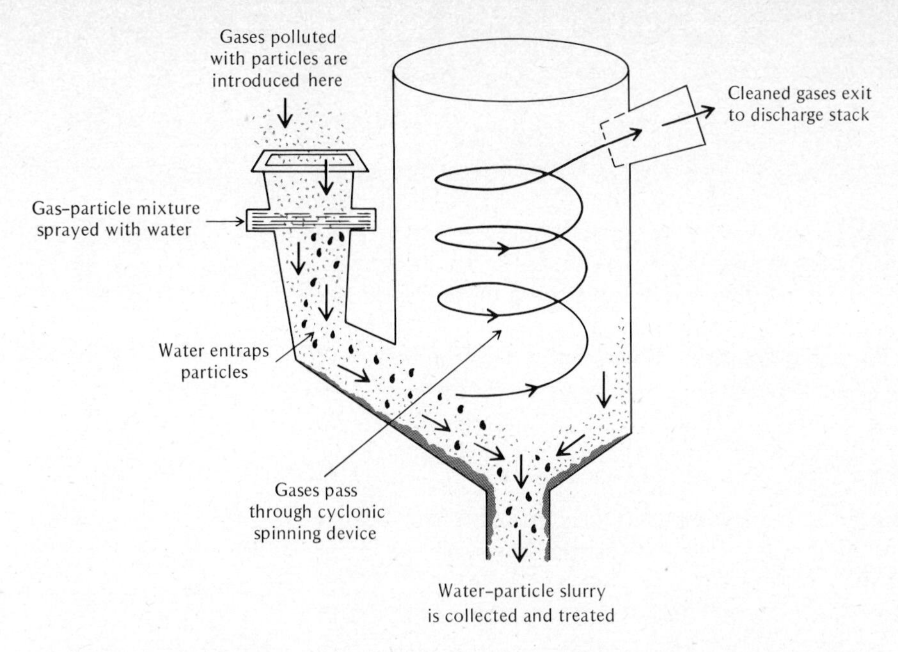

FIGURE 4.8 Schematic illustration of a cyclone particle collector.

(center-fleeing) force. Centrifugal force is stated to be the force pushing one toward the outside of the circle, but this is a misleading statement. The true state of affairs is that we slide toward the outside of the curve because there is not sufficient force to keep us moving in the same curve as the car. Centrifugal force is thus a fictitious force. Nothing is ever thrown toward the outside of a circle. If there is insufficient force toward the center, centripetal force, an object will move more nearly in a straight line. This makes it appear to be thrown to the outside.

There are some very practical consequences of these ideas. Suppose we want to design an automobile tire. What keeps the outer parts of the tire moving in a circle? Obviously it is the strength of the tire itself. To go 100 miles/hour requires four times the strength needed to go 50 miles/hour. For this reason recaps and retreads are not recommended for high-speed driving, even though they may work perfectly well around town. Similar considerations are involved in the design of any piece of rotating equipment.

Another practical use of rotational motion is in a cyclone device for separating particulate matter from a stream of gas. The gas stream is whirled about, and the particulates move to the outside, because there is no force holding them in. There they can be separated and collected, as shown in Fig. 4.8.

4.4 PRECOPERNICAN ASTRONOMY

Now that we have had a brief encounter with motion in a circle, we are in a position to understand the developments and arguments which led to the law of universal gravitation, developed by Isaac Newton. As background for this we shall need a brief excursion into the astronomy of his day.

We can give only a small glimpse into the fascinating history of astronomy and the solar system before the time of Copernicus. Aristotle's name, among others, occurs prominently in the development of geocentric astronomy. By geocentric we mean an astronomy which viewed the earth as the center of the universe and all celestial bodies as revolving about it. There are many difficulties with such a theory because, although the moon and sun move regularly, the motion of the other planets relative to the earth is not simple. Viewed against a background of fixed stars, a planet's path consists of a series of loops (Fig. 4.9). These observed planetary motions require, in the geocentric theory, a system of circular motions within circular motions, called epicycles, that are as complicated to describe mathematically as they are to explain. The most elaborate geocentric description of motion in the solar system was constructed by the astronomer Ptolemy of Alexandria in the second century (Fig. 4.10).

Despite its complexity, the ptolemaic system was sufficiently good that tables useful to navigators for many centuries were constructed. To do so, however, required a large number of arbitrary assumptions (many wheels within wheels). The ptolemaic view of the universe was almost universally accepted until after 1600.

There existed an opposing theory even in the time of Ptolemy, called the heliocentric (sun-centered) theory. Associated with this idea is the name of Aristarchus of Samos, who was also the first to obtain a relatively good idea of the size of the earth, assuming it was round. Despite Aristarchus' suggestion of a heliocentric universe in the third century B.C., it was not until almost 2,000 years later that the theory, as stated by Copernicus, gained many followers.

4.5 COPERNICUS

A Polish churchman, Nicolaus Copernicus (1473–1543), published in 1543 a book entitled *De Revolutionibus Orbium Caelestium*. He de-

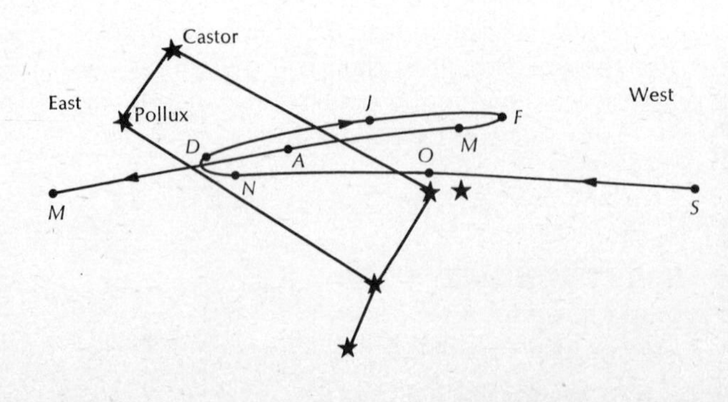

FIGURE 4.9 Position and motion of the planet Mars in the constellation Gemini from September 1960 through May 1961. The position in September 1960 is labeled *S*. The backward motion occurring from December through February in this example caused the most difficulty for the ptolemaic system.

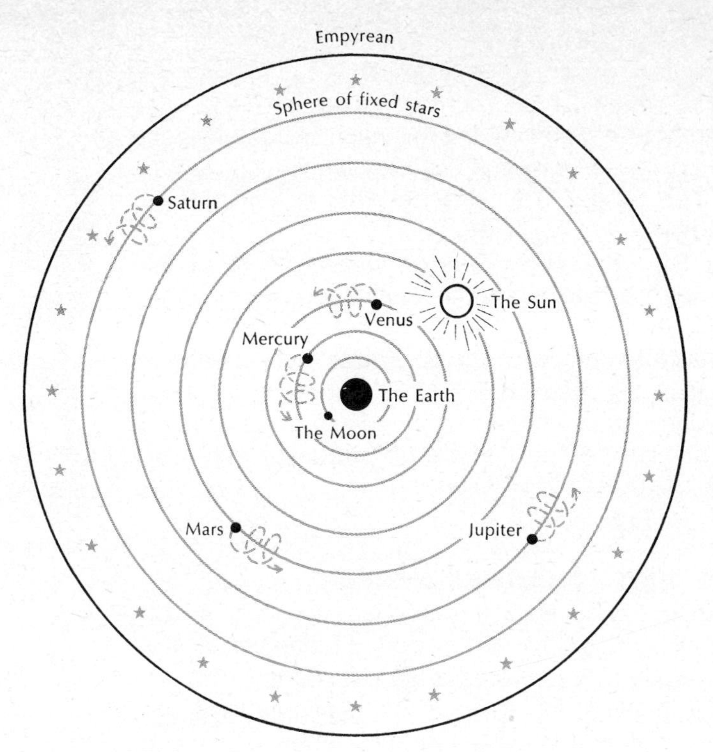

FIGURE 4.10 The ptolemaic system. Each of the planets is shown with its epicycle.

scribed an astronomical system in which the sun is at rest and the planets move about it in circular orbits. Although many of the ideas he presented previously had been suggested by one or another writer as far back as the early Greeks, Copernicus joined these ideas with some of his own into a logical and coherent system. Because Copernicus was unwilling to admit the existence of orbits other than circles, his system became in practice nearly as unwieldy as the ptolemaic, and was little better for making navigational calculations, one of the prime uses of astronomy in his day. In addition to this complexity, the copernican system left unexplained why, as the earth rushed through space around the sun, everything not nailed down was not left behind. One

FIGURE 4.11 The four largest, or galilean, moons of Jupiter at three different times. The planes of their orbits are nearly the same as the orbit of Jupiter around the sun, so we see them edge on. They appear to move back and forth across the planet. *(Courtesy Yerkes Observatory.)*

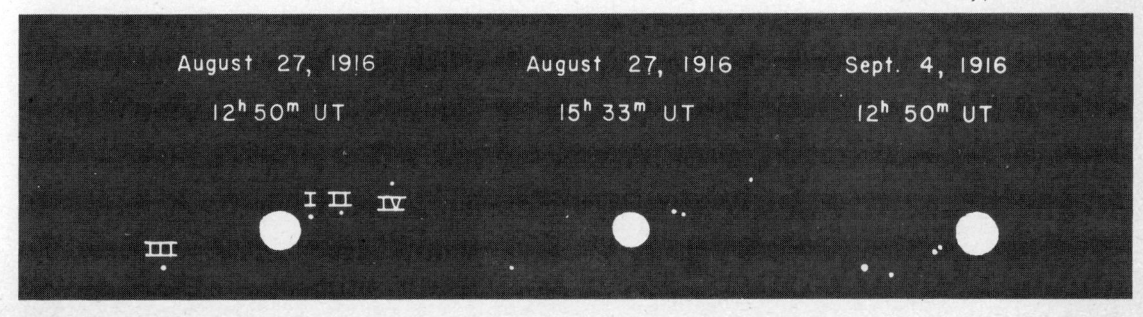

of the virtues of an astronomical system in which the earth is at rest is that there is no difficulty of this sort. Before Galileo and Newton it was not at all clear why objects should not be left behind if the earth were moving through space. In fact, until the invention of the telescope (1608), there was little to choose between the copernican and ptolemaic systems in terms of their relative ability to describe or interpret the observations of astronomy.

Galileo used the telescope (invented in Holland) to show that Jupiter had four moons which revolved about it in a fashion very similar to that which Copernicus had proposed for the planets around the sun. Even though Galileo was later forced by the Church to recant his belief in the copernican system, his observations gave credence to the copernican heliocentric view. Further events and discoveries made it more and more clear as time passed that the planets revolve in some sort of orbit with the sun at or near the center.

FIGURE 4.12 Tycho Brahe (1546–1601). *(Courtesy Burndy Library.)*

4.6 TYCHO BRAHE AND JOHANNES KEPLER

Tycho Brahe (1546–1601) was a Danish astronomer whose life work was careful recording of the positions and motion of planets and other celestial bodies. In his lifetime he amassed a large amount of observational data, perhaps 20 times as accurate as that available to Copernicus, without the use of a telescope, which had not yet been invented. Brahe believed in the geocentric system, but his assistant, Johannes Kepler (1571–1630), was a copernican and, after Brahe's death, used the unexcelled data of Brahe to develop three laws that describe the motion of the planets. Kepler's laws are *empirical laws*. This means that they represent a synthesis, or summary, of a set of observations. Interestingly, Kepler had to hide Brahe's data so that Brahe's heirs would not make off with it. Kepler is surely one of the all-time most interesting characters in science.

Brahe's data were sufficiently accurate that Kepler was able to show that, if the earth moves around the sun, it does *not* move in a circle. The geometric simplicity of a circle had always exerted a magical fascination on all who studied the heavens. Even Copernicus rejected the idea that the motions of the heavenly bodies could be anything other than circles. After several years of calculation and study of Brahe's data, Kepler discovered the first of his three laws:

> *All planets move in elliptical paths, with the sun at one focus of the ellipse.*

FIGURE 4.13 Johannes Kepler (1571–1630). *(Courtesy Burndy Library.)*

An ellipse is a well-known egg-shaped geometric figure which is closely related to a circle. A circle is a line all of whose points are equally distant from a point such as A in Fig. 4.14. An ellipse is a figure constructed so that the line AOB in Fig. 4.14 is the same length for any

FIGURE 4.14 *(a)* A circle: All parts are equidistant from the center *A*. *(b)* An ellipse: The length of the time *AOB* is the same for every point on the ellipse.

point on the curve. The points *A* and *B* are called the foci of the ellipse. As *A* and *B* approach each other, the ellipse becomes more and more nearly circular. When they coincide, the figure is a circle.

Of course, the statement that the shape of the orbit of a planet is an ellipse does not tell an astronomer where to look for the planet. In addition, some information is needed about the velocity of the planet in its orbit. This Kepler also discovered by studying the data of Brahe. The basic idea is shown in Fig. 4.15. Kepler showed that, during equal intervals of time, the area swept out by a line from the sun to a planet is the same, no matter where the planet is in its orbit. This is Kepler's second law. The shaded areas in Fig. 4.15 correspond to the area swept by the line from the sun to the planet in equal time intervals, such as 1 day. The areas A_1, A_2, and A_3 were found by Kepler to be the same. His second law is:

The line from the sun to a planet sweeps out equal areas in equal periods of time, at any point in the planet's orbit.

4.7 KEPLER'S THIRD LAW

After the discovery of his first two laws, Kepler spent many years searching in Brahe's data for a relationship between the distance of the planets from the sun and their period, the time required for one revolution about the sun. Finally, in 1619, he found the result he sought. For each of the planets which revolve about the sun, he found that the square of its period is proportional to the cube of its distance from the sun ($T^2 \propto R^3$). Thus

$$\frac{(\text{Period})^2}{(\text{Radius})^3} = \frac{T^2}{R^3} = \text{constant for all the sun's planets} \qquad (4.4)$$

Table 4.1 lists the observational data for the planets known to Kepler in terms of the astronomical unit (AU) for distance. One astronomical unit is the average distance from the earth to the sun. With this choice of units, the ratio T^2/R^3 is 1.00 if the period is measured in years.

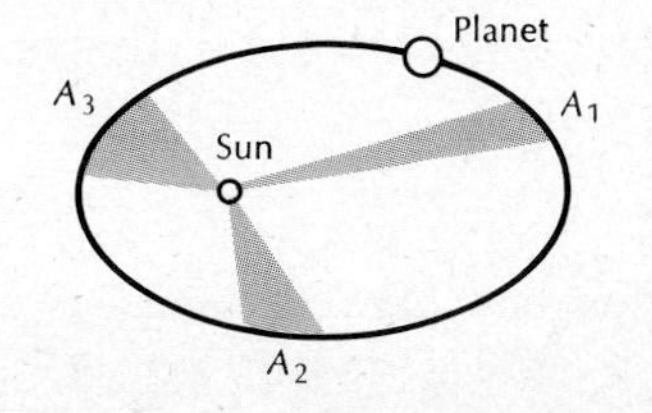

FIGURE 4.15 Kepler's second law: The shaded areas are the same, and represent the distance swept out by the line from the sun to planet in equal intervals of time.

Planet	Period T, years	Distance from sun R, astronomical units	T^2	R^3	T^2/R^3
Mercury	0.2408	0.3871	0.05798	0.05800	0.9996
Venus	0.6152	0.7233	0.3785	0.3784	1.0000
Earth	0.1000	1.0000	1.0000	1.0000	1.0000
Mars	1.8808	1.5237	3.537	3.5373	0.99997
Jupiter	11.862	5.2028	140.71	140.84	0.9991
Saturn	29.457	9.5388	867.72	867.92	0.9998

TABLE 4.1 Relationships between the period of the planets and their distance from the sun, showing Kepler's third law. The more figures used in the expressions for T and R, the more accurately the ratio T^2/R^3 is equal to 1.000. The astronomical unit (AU) has a value of approximately 93 million miles, or 9.3×10^7 miles.

4.8 NEWTON AND THE LAW OF UNIVERSAL GRAVITATION

We are now in a position to understand and appreciate the work of Isaac Newton in explaining the source of Kepler's laws. Obviously, since the distance from the sun to the planets appears in all three laws, the sun occupies a key position in the solar system. Newton was also aware that the moons of the planets obey Kepler's laws in their motion about the planets. In particular, the several moons of Jupiter and Saturn obey Kepler's third law, but with a different value for the quantity T^2/R^3 for each planet. That is, the third-law relationship between distance and period is found to hold if the distance from the moon to its planet is taken in the third law. The value of the ratio T^2/R^3 is not the same as found for the planets around the sun, and is different for the moons of each planet.

Newton was led by these, and other reasons, to believe that there was a force acting between the sun and each of the planets, and between Jupiter and each of its moons. He believed that this force was the same force which made an apple fall, that is, the force of gravity.

The basic problem, then, was to find a description of the force of gravity that would account for the observational data. That is, Newton had to find a force law that predicted each of Kepler's laws. This assumes that the observed motions of the planets and their regularities as evidenced in Kepler's laws are a reflection of a force acting between the planets and the sun. Newton's contemporaries, Edmund Halley, Robert Hooke, and Christopher Wren, were convinced that the force between the sun and a planet must be one which decreases as the square of the distance between the sun and the planet increases, but they were unable to show that this led to Kepler's laws. A force between two objects that decreases as the square of the distance between them increases is an inverse-square-law force. This is written

$$F \propto \frac{1}{R^2} \tag{4.5}$$

Newton was asked by Halley what the shape of the orbit would be for a planet under such a force. He replied that he had already calculated it to be an ellipse, showing that this form for the force agrees with Kepler's first law.

In order to show that Equation 4.5 is the correct form of the force law, Newton compared the acceleration of falling objects near the surface of the earth with the acceleration of the moon toward the earth. We know that all objects on the surface of the earth are accelerated toward the earth, and this acceleration is what we have called g, where $g = 9.8$ meters/second^2. Since the moon moves in a nearly circular orbit around the earth, previous discussion shows that it must be accelerated toward the earth. We shall approximate the moon's orbit as a circle, for simplicity.

Now let us calculate the acceleration shown by the moon in its orbit. The moon is approximately 238,000 miles, or (since 1 mile $\cong$ 1,600 meters) 3.8×10^8 meters away from the earth. It travels around the earth once every $27\frac{1}{3}$ days. The distance once around its orbit is $2\pi R$, where R is the earth-moon distance. By using these numbers it is possible to calculate the velocity of the moon in its orbit, and then its acceleration toward the earth, using $a = v^2/R$. The result of this calculation is, approximately, $a = 2.7 \times 10^{-3}$ meter/second^2 or 0.0027 meter/second^2. This acceleration must be proportional to whatever force is acting on the moon to keep it in its orbit around the earth.

POWER-OF-TEN NUMBERS

We introduce at this point a method frequently used in science for handling very large and very small numbers. We will have occasion to use this technique frequently. Further discussion of this is included in the mathematical appendix, Section A.3.

Rather than write a large number as 238,000, in power-of-ten notation this is written as 2.38×10^5, since $10^5 = 10 \times 10 \times 10 \times 10 \times 10 = 100{,}000$. If 2.38×10^5 is the earth-moon distance in miles and there are $1{,}600 \cong 1.6 \times 10^3$ meters/mile, the earth-moon distance in meters is

$$(2.38 \times 10^5) \times (1.6 \times 10^3) =$$
$$(2.38 \times 1.6) \times (10^5 \times 10^3) = 3.8 \times 10^8$$

Notice that we multiply separately the numerical part and the power of-ten part.

In addition to its compactness, this method allows us to indicate clearly how accurately we are expressing numbers. When we write 1.6×10^3 meters/mile, we are claiming only two-digit accuracy. If we write 1.60×10^3, we are claiming three-digit accuracy. The number of figures used this way are called significant figures. 1.6×10^3 is expressed to two significant figures.

We have previously seen that all objects on the surface of the earth fall with the same acceleration. If we presume that the source of this acceleration extends its influence as far as the moon, we would expect the same result to hold. Therefore the acceleration of the moon toward the earth must be the same as that of any other object that distance from the earth.

Thus we can compare the acceleration of objects at the distance of the moon from the earth with the acceleration shown by objects at the surface of the earth. Since accelerations are proportional to forces, this should tell us something about the force law. The ratio of the observed accelerations is approximately

$$\frac{g \text{ at surface of earth}}{a \text{ at distance of moon}} = \frac{9.8 \text{ meters/second}^2}{2.7 \times 10^{-3} \text{ meter/second}^2} \cong 3{,}600$$

What does this ratio of 3,600 mean? The distance from earth to moon is about 3.8×10^8 meters. The distance from the center of the earth to its surface is 4,000 miles or about 6.4×10^6 meters. The ratio of these two numbers is

$$\frac{\text{Distance from moon to center of earth}}{\text{Distance from earth surface to center}} = \frac{3.8 \times 10^8}{6.4 \times 10^6} = 59$$

Now 59^2 is just slightly less than 3,600. Therefore the ratio of the forces is 3,600 which is 60^2, and the ratio of the distances is 59; so it seems as though the acceleration decreases as the square of the distance, to the accuracy with which we have done the arithmetic.

We infer that if the force which holds the moon in its orbit is the same as that which causes objects to be accelerated on the surface of the earth, then this force must decrease as the square of the distance from the center of the earth, since $60^2 = 3{,}600$ and 59 is, approximately, 60. We conclude this because the acceleration decreases as the square of the distance, and we associate acceleration with force. Isaac Newton made the calculation we have just done, and found it to fit "pretty nearly" with the numbers he used.

Newton was bothered by the fact that in this calculation the force on an object at the surface of the earth seemed to depend on its distance from the center of the earth, rather than in some complicated way on the distance to each part of the earth. Later he was able to show, using the calculus, a branch of mathematics he had invented, that the force of gravity of a spherical body such as the earth is exactly that which would be calculated if all its mass were concentrated at its center.

We have shown that the gravitational force between two objects depends on $1/R^2$, where R is the distance between them. It should also depend on the masses of the two objects. We have already seen that the force of gravity on an object at the surface of the earth, which is its weight, is given by mg, where m is its mass and g is the observed acceleration due to gravity. Therefore the force of gravity is proportional to the mass of the object. We can also infer by symmetry that the force upward on the earth should be proportional to the mass of the earth as well as the mass of the object. Therefore the force of gravity should be proportional to the mass of both objects involved.

Newton was led by these arguments to propose the *law of universal gravitation*:

> *Every mass in the universe attracts every other mass. The force of attraction is proportional to each of the masses involved, and inversely proportional to the square of the distance between their centers.*

The mathematical expression of this law is

$$F \propto \frac{m_1 m_2}{R^2}$$

Here m_1 and m_2 are the masses of the two objects attracting each other. We know from our experience that two masses of 1 kilogram 1 meter apart do not exert a very large gravitational force on each other. In fact, careful experiments show that this force is 6.67×10^{-11} newton, a very tiny force. So the law of gravity must be written

$$F = G\frac{m_1 m_2}{R^2} \tag{4.6}$$

where G has the numerical value 6.67×10^{-11}. We will discuss shortly how G is measured. The units of G are such as to make the force come out in newtons.

Using calculus, Newton was able to show that, in general, the form of the orbit of a planet attracted to the sun by a force which falls off as the square of the distance will be an ellipse. This agrees with the first of Kepler's laws. Therefore the law of gravity predicts Kepler's first law.

Kepler's second law is a result which can be shown to hold no matter what the form of the force law, so long as the force on the planets is directed toward the sun. Therefore it cannot be used as a test of the law of gravity.

Kepler's third law is a direct consequence of the $1/R^2$ dependence of the gravitational force on distance. We can say that Newton's law of universal gravity "predicts" Kepler's laws. This is one of the main ways in which scientific theories are tested: Do they predict results that are in agreeent with experiment? The constant in Kepler's third law is found to depend only on G and the mass of the sun.

4.9 ARTIFICIAL SATELLITES

Since Sputnik in 1957, many hundreds of artificial satellites of the earth have been launched by the United States, the U.S.S.R., and several other nations. Many of them are still in orbit around the earth. In what way is an artificial satellite like our professor's ball on a string? In what way is it different?

In the original discussion of circular motion we said that there must be a force toward the center of the circle, and that this force must obey the relation

$$F = m\frac{v^2}{R} \tag{4.3}$$

In the case of an artificial satellite this force is provided by gravity. Gravity is different from the professor's string, because the force provided by gravity decreases as the satellite orbits further from the earth. But before discussing that, consider a satellite in a circular orbit at a distance of, say, 100 miles from the earth's surface. The mass of the satellite is fixed, and so is the size of its orbit. The force is also constant, since the distance from the center of the earth is unchanged. What about the velocity? Clearly, if all other quantities in the equation are fixed, the velocity must also be. What this means is that only one particular velocity will keep a satellite in a circular orbit at that height. This is shown in Fig. 4.16. If there is too little velocity, the satellite will fall to earth. If there is too much, it will go into an elliptical orbit, longer than the circular orbit. This is the reason that to go to the moon the Apollo astronauts increased the velocity of their rocket at a point on the side of the earth opposite the moon.

What happens if we look at satellites in orbits further and further from the earth? Because the force of gravity decreases as the distance

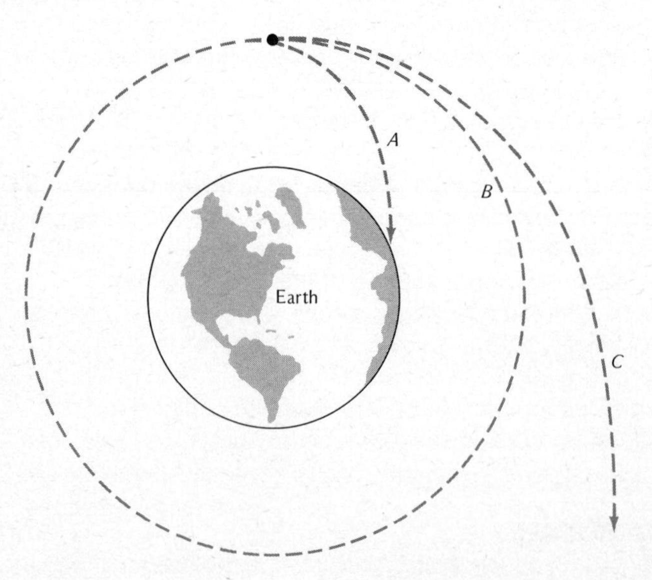

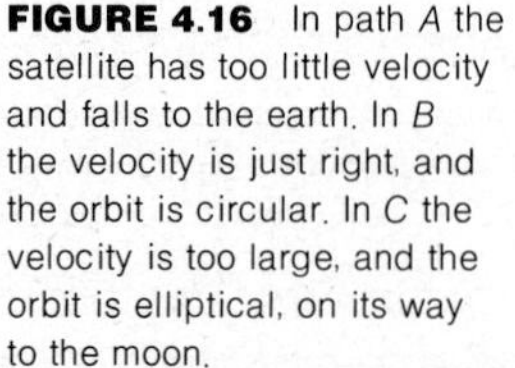
FIGURE 4.16 In path *A* the satellite has too little velocity and falls to the earth. In *B* the velocity is just right, and the orbit is circular. In *C* the velocity is too large, and the orbit is elliptical, on its way to the moon.

increases, the velocity necessary for the circular orbit also decreases. Kepler's third law involves the period, which is the time required to go once around. The periods become longer for orbits of larger radius, both because the distance around is longer and also because the velocity is less. The detailed result is Kepler's third law.

One interesting application of this is a synchronous satellite, which is a satellite at a distance above the earth such that its period is 24 hours. It revolves around the earth at the same speed as the earth's rotation. Therefore it stays always above one spot on the earth's surface. This type of satellite is extremely useful for communication between continents. Several are now in orbit over both the Atlantic and Pacific Oceans. The altitude of a synchronous satellite is about 22,000 miles above the surface of the earth.

4.10 MEASUREMENT OF G

If one investigates the gravitational force between laboratory-sized objects, using Equation 4.6, the force comes out to be incredibly small. The gravitational force between a 100-kilogram boy and a 50-kilogram girl who are 0.1 meter apart is

$$F = G\frac{m_1 m_2}{R^2}$$

$$= (6.67 \times 10^{-11}\text{ newton-meters}^2/\text{kilogram}^2) \times \frac{(100\text{ kilograms})(50\text{ kilograms})}{(0.1\text{ meter})^2}$$

$$= 6.67 \times 10^{-11} \times 5 \times 10^5\text{ newtons}$$

$$F = 3.33 \times 10^{-5}\text{ newton}$$

This is approximately 1/100,000 pound. (Obviously, in the situation discussed, other forces may operate.)

To determine experimentally the value for G, the force between two known masses must be measured. To do this requires the measurement of these very tiny forces. The first reliable measurement was performed by Henry Cavendish in 1798. The forces in gravitational experiments are very small, and therefore the experiments are technically very difficult. Direct measurement of the force between two known objects yields the value of G given above. Figure 4.18 indicates the general arrangement of the Cavendish experiment.

FIGURE 4.17 Henry Cavendish (1731–1810). *(Courtesy University of Pennsylvania Library.)*

If G is known, we can "weigh the earth." The force on a 1-kilogram mass at the earth's surface is known to be 9.8 newtons, and the radius of the earth is 6.4×10^6 meters. If G is known, the only unknown in the gravitational-force equation between this 1-kilogram object and the earth is the mass of the earth itself.

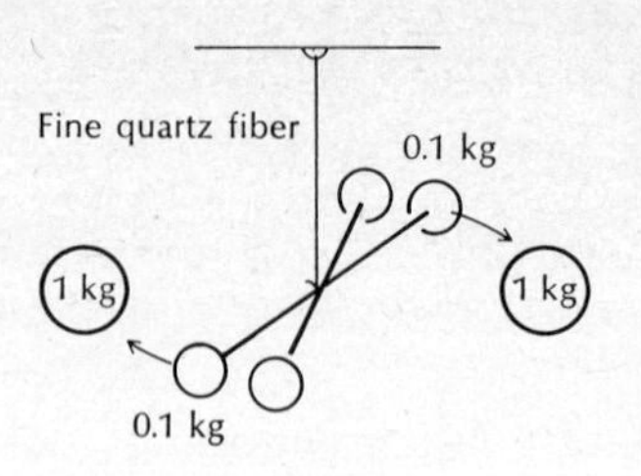

FIGURE 4.18 The Cavendish experiment. The small gravitational force between the 1-kilogram and 0.1-kilogram masses causes a small twist in the quartz fiber.

$$F = G\frac{M_{\text{earth}}m}{R_{\text{earth}}^2}$$

$$9.8 \text{ newtons} = G\frac{M_{\text{earth}}(1 \text{ kilogram})}{R_{\text{earth}}^2}$$

$$M_{\text{earth}} = \frac{(9.8 \text{ newtons})\, R_{\text{earth}}^2}{G(1 \text{ kilogram})}$$

$$= \frac{9.8 \text{ newtons } (6.4 \times 10^6 \text{ meters})^2}{(6.67 \times 10^{-11} \text{ newton-meters}^2/\text{kilogram}^2)(1 \text{ kilogram})}$$

$$M_{\text{earth}} = \frac{9.8 \times 6.4 \times 6.4}{6.67} \times 10^{6+6+11}$$

$$M_{\text{earth}} = 6.0 \times 10^{24} \text{ kilograms}$$

SUMMARY

The inward acceleration of an object in circular motion is given by

$$(\text{Acceleration})_{\text{inward}} = \frac{v^2}{R}$$

The force needed to keep an object of mass m moving in a circle of radius R is

$$(\text{Force})_{\text{inward}} = m\frac{v^2}{R}$$

If less than this force is available, the object will move off the circular path to the outside.

Early astronomy considered that the earth was the center of the universe, and that all celestial objects rotated around it. Copernicus proposed that the sun was the center of the solar system, and that all planets revolved about it. Kepler, using Tycho Brahe's data, developed three laws that described the motions of the planets. They are:

All planets move in elliptical paths, with the sun at one focus.

The line from the sun to a planet sweeps out equal areas in equal periods of time, at any point in the planet's orbit.

For all of the planets, the quantity $\frac{(period)^2}{(radius)^3}$ *is the same number.*

Newton proposed the law of universal gravity that accounts for all of Kepler's laws: The force between any two objects in the universe is given by

$$F = G\,\frac{m_1 m_2}{R^2}$$

where $G = 6.67 \times 10^{-11}$ newton-m²/kg², and is a constant whose value is determined by direct measurement. R is the distance between the two objects.

SELECTED READING:

Koestler, Arthur: "The Watershed: A Biography of Johannes Kepler," Science Study Series S-16, Doubleday and Company, Garden City, New York, 1960. A fascinating biography of one of the truly extraordinary characters in the history of science. This book is a portion of a larger work, "The Sleepwalkers," that covers science from the Greeks to Newton.

QUESTIONS

1 Science fiction movies have shown space stations slowly rotating. Why is this, and how does it work?
2 The masses of planets with moons are better known than the masses of those that have no moons. Why?
3 In recent years, race cars have used "wings" to improve their cornering ability. How do they work?
4 Why is the constant in Kepler's third law the same for all planets?
5 Why is the acceleration due to the earth's gravity at the position of the moon not 9.8 m/second^2?
6 Both Kepler's and Newton's discoveries are called laws. How do their origins differ?
7 Why are there seasons?
8 Does the distance from the earth to the sun vary during the year? Why?
9 As a planet moves about the sun, its orbital velocity changes so that equal areas are swept out in equal times by a line joining it and the sun. In what orbital position does the planet have the largest orbital velocity? Where does it have the smallest orbital velocity?
10 If you sit with your body against a car door and the car rounds a

corner, what is the direction of the force the car door exerts on you?

11 An adult holds a small child by the hands and spins him in a circle. What is the direction of the force the adult must exert on the child? Is this consistent with your experience?

12 Could a synchronous satellite be placed in orbit above your hometown?

13 If we know the mean distance of a planet from the sun, we can calculate its period. How would we do this?

14 From Newton's law of universal gravitation one can obtain all three of Kepler's laws of planetary motion. Can one obtain Newton's law of universal gravitation from any one of Kepler's laws?

SELF-TEST

__________ period of circular motion
__________ centripetal acceleration
__________ centripetal force
__________ centrifugal force
__________ Nicolas Copernicus
__________ Tycho Brahe
__________ Ptolemy of Alexandria
__________ Aristarchus of Samos
__________ Johannes Kepler
__________ universal law of gravity
__________ Henry Cavendish

1 Early proponent of heliocentric astronomy
2 Early proponent of geocentric astronomy
3 Developed three laws describing planetary motion
4 Developed, in 1543, a heliocentric astronomy
5 Made very accurate measurement of planetary positions
6 Measured G
7 Time to travel around a circle
8 Inward acceleration
9 Spurious outward force
10 Inward force
11 Describes the force holding planets in their orbits

PROBLEMS

1 What is the acceleration of a man standing 20 ft from the axis of rotation of a merry-go-round making four turns a minute?

2 A boy is whirling a ball weighing 8 lb at the end of a rope of length 4 ft. The boy exerts a force of 100 lb on the rope, and the ball moves in a horizontal circle. (*a*) Find the velocity of the ball. (*b*) Draw a picture of the ball and the circle it moves in, and with an arrow show the direction of the force on the ball due to the rope. (*c*) Draw another picture as in part (*b*), and with another arrow show the direction the ball would move should the rope break.

3 A ball is tied to the end of a string 2 ft long and is whirled in a horizontal circle. The ball weighs 1 lb, and the string will break if the tension on it exceeds 16 lb. (*a*) What is the maximum velocity with which the ball can be whirled? (*b*) What is the force on the ball when the velocity is one-half that found in part (*a*)? (*c*) Draw a picture of the ball and the circle in which it moves, and show with an arrow the direction of the force on the ball.

4 The planet Mercury travels around the sun once in 88 days. Its distance from the sun is approximately 6×10^{10} m. What is its acceleration toward the sun (in meters per second per second)?

5 The acceleration toward earth of a satellite 1,000 miles above its surface is 20.5 ft/second^2. Calculate the velocity of such a satellite.

6 Calculate the inward acceleration of an electron moving in a circle 10^{-10} m in diameter at a velocity of 10^6 m/second.

7 A boy swings a ball whose mass is 0.1 kg around his head on a string 3 m long. If the force required to break the string is 20 newtons, how fast will the ball be going when the string breaks?

8 A car travels on a curve with a radius of curvature of 100 m. The maximum sideward force the tires can sustain is 10^4 newtons. The mass of the car is 2,000 kg. What is the maximum speed the car can travel on this curve without going off the road?

9 Calculate the inward acceleration for a car traveling on a curve of radius 300 m at a velocity of 40 m/second. If the car has a mass of 2,000 kg, how much force is needed to keep it on the road?

10 A fly of mass 1 g lands on the outer edge of a phonograph record and remains there, a distance of 15 cm from the center. The phonograph record is spinning at a rate of $33\frac{1}{3}$ revolutions per minute, so that the fly has a period of 1.8 seconds. (*a*) What is the velocity of the fly? (*b*) What is the acceleration of the fly? (*c*) What is the magnitude and direction of the force which gives rise to the acceleration found in part (*b*)?

11 The asteroids (minor planets) travel in orbits approximately 2.8 astronomical units from the sun. (The earth-sun distance is by definition 1.00 astronomical unit.) What is the period of these minor planets about the sun? (*Hint:* Use Table 4.1.)

12 Suppose NASA orbits an earth satellite that is four times as far from the center of the earth as a satellite whose period is 2 hours. Using Kepler's third law, calculate the period of this new satellite.

13 Suppose someone told you another planet circled the sun within the orbit of Mercury at a distance of $\frac{1}{4}$ astronomical unit from the sun. What would be its period? (At one time there was speculation that such a planet did exist, and it was called Vulcan. This notion has now been generally discredited.)

14 An asteroid is observed to be in a circular orbit about the sun at a distance of 5.6 astronomical units. The mass of the asteroid is 10^4 kg. (*a*) What is the period of this asteroid? (*b*) How far does it move in one trip around the sun? (*c*) What is the velocity of this asteroid in its orbit?

15 Calculate the gravitational force between two objects whose masses are 2 kg and which are 10 m apart.

16 Calculate the gravitational force between a 50,000-ton ship and its captain, standing on the dock alongside. The captain's mass is 100 kg, and he is 50 m from the center of the ship. (Assume all the ship's mass is concentrated at this point, which is only a fair approximation.)

17 Calculate the gravitational force between two objects that are 10 m apart whose masses are 1.5 and 4 kg.

18 Calculate the gravitational force between two objects each of mass 10 kg that are 3 m apart.

19 At an altitude of 4,000 miles above the surface of the earth, g is approximately 8.0 ft/second^2. Calculate the velocity of a satellite in orbit at this height.

20 Given that the radius of the earth is 4,000 miles, the radius of the moon is 1,000 miles, g on earth is 32 ft/second^2, and g on the moon is 5 ft/second^2, calculate the ratio R of the mass of the moon to the mass of the earth.

$$R = \frac{\text{mass of moon}}{\text{mass of earth}} = ?$$

21 (*a*) Calculate the velocity of a satellite in orbit above the earth at a distance of 5×10^7 m (about 30,000 miles) from the center of the earth. (*b*) Calculate the period of such a satellite.

22 The earth moves about the sun with a velocity of 3.0×10^4 m/second. The mean distance from the sun to the earth is 93 million miles. (*a*) What is the acceleration of the earth toward the sun? (*b*) If the mass of the earth is 6×10^{24} kg, what is the force of gravity between the earth and the sun? (*c*) What is the mass of the sun?

23 A synchronous satellite is 36,000 km above the earth's surface. (*a*) What is its period? (*b*) How far does it go in 1 day? (*c*) What is its orbital speed? (*d*) What is its acceleration in orbit? (*e*) At what radius would one have to place a satellite so that its period is 1 hour? Would this be possible? If not, why?

24 The mass of the moon is 7.4×10^{22} kg. A moon satellite of mass 4,000 kg is launched and orbits the moon at a distance of 7,000 km from its center. (*a*) What is the force of attraction between the moon and the satellite? (*b*) What is the velocity of the object in its orbit (assumed circular) around the moon? (*c*) What is the circumference of the orbit, that is, how far does the satellite travel in one trip around the moon? (*d*) What is the period of this satellite?

25 The moon is a satellite of the planet Earth. The radius of its orbit, measured from the center of the earth, is 3.84×10^8 m. The mass of the moon is 7.4×10^{22} kg. The mass of the earth is 6×10^{24} kg. (*a*) What is the velocity of the moon in its orbit? (*b*) What is the total distance traveled by the moon in one orbit about the earth? (*c*) What is the period of the moon? (*d*) If the radius of the moon is 1.6×10^6 m, what is the acceleration due to gravity at its surface?

26 The planet Neptune has two moons, Triton and Nereid. Triton moves in an orbit of radius 3.5×10^8 m and has a mass of 1.37×10^{23} kg. Neptune has a mass of approximately 1.1×10^{26} kg. (*a*) What is the orbital velocity of Triton? (*b*) What is the distance Triton moves in one period? (*c*) What is the period of Triton? (*d*) If the radius of Triton is 2×10^6 m, what is the acceleration due to gravity at its surface? Is it necessary to include gravitational effects due to Neptune in this calculation?

27 The planet Mars has a mass of 6.2×10^{23} kg. A satellite of mass 1,000 kg is put into a circular orbit about Mars whose radius is 6.8×10^6 m. (*a*) What is the magnitude of the force acting on this satellite? (*b*) What is its acceleration in its orbit? (*c*) What is the velocity of this satellite in its orbit? (*d*) What is the period of this satellite? (*e*) What is the acceleration due to gravity at a distance of 6.8×10^6 m from the center of Mars?

28 Ganymede is a moon of the planet Jupiter. The radius of its orbit measured from the center of the planet is 1.1×10^9 m. The mass of Ganymede is 1.52×10^{22} kg. The mass of Jupiter is 1.9×10^{27} kg. Assume Ganymede moves in a circular orbit. (*a*) What is the gravitational force acting on Ganymede due to Jupiter? (*b*) What is the acceleration of Ganymede in its orbit? (*c*) What is the velocity of Ganymede in its orbit? (*d*) What is the period of Ganymede?

29 The earth rotates on its axis such that a point on the equator makes one complete rotation in 1 day. The average radius of the earth is 6.38×10^6 m. (*a*) How far does a person on the equator travel in 1 day because of this rotation of the earth on its axis? (*b*) What is the velocity of the person on the equator due to this rotation? (*c*) What is the acceleration of a person on the equator due to this rotation?

30 One of the largest asteroids is called Ceres. Its radius is about 150 km. If the rock of Ceres has the same average density as Earth, what would a 1,000-newton man weigh on Ceres?

31 The acceleration due to gravity on the moon is 1.5 m/second^2.

The radius of the moon is 1,000 miles. (*a*) Find the mass of the moon. (*b*) Find the velocity of a satellite in orbit 1,000 miles above the surface of the moon.

32 Some stars, at the end of their evolution, become neutron stars. These stars have masses very nearly the mass of our sun (2×10^{30} kg) but radii of only (approximately) 20 km. (*a*) What is the acceleration due to gravity at the surface of a neutron star? Consider an object orbiting a neutron star at its surface. (*b*) What distance does this object travel in one period? (*c*) What is the orbital velocity of this object? (*d*) What is the period of this object?

MOMENTUM AND ENERGY

We began the discussion of mechanics by studying concepts which are, in a general way at least, familiar to most of us. They are velocity, acceleration, force, weight, and mass. In this section we shall introduce two concepts, momentum and energy, which are less direct and less intuitively obvious. It is possible to use these ideas in mechanical calculations in a way that greatly simplifies matters. That is, it is possible, using energy and momentum considerations, to do calculations when much less information is given about the system than would be necessary using only the tools of Part 1.

Momentum and energy are the first examples we shall discuss of a kind of quantity which is extremely important in physics today. These are quantities which are *conserved*. That is, these quantities do not change in the course of some process. According to present-day thinking, conservation laws are the most fundamental statements of physical laws. Many conservation laws can be shown (we shall discuss this in Chapter 23) to be indirect consequences of the symmetry or isotropy of space. For instance, the fact that the laws of physics are the same everywhere leads directly to the law of momentum conservation.

In Part 2 we shall introduce momentum and energy and demonstrate how useful they can be in various types of calculation. Chapter 5 is a discussion of Newton's third law of motion and its relationship to momentum. Chapter 6 develops the ideas of work, kinetic energy, and potential energy. Chapter 7 is a discussion of the relationship between mechanical and heat energy and a development of the general energy principle. Chapter 8 is a discussion of some aspects of the "energy crisis" we face in our society today, based on the concepts of Chapters 6 and 7.

Very early in the study of motion it was believed that there must be something which was a measure of what can be called the "quantity of motion." As early as the sixth century A.D., John Philoponus suggested that an object possesses a property, which he called *impetus*, that tends to keep it in motion. Galileo used this term in the same manner. This "quantity-of-motion" idea has evolved into what we now call momentum.

NEWTON'S THIRD LAW AND MOMENTUM

5.1 NEWTON'S THIRD LAW

The third of Newton's laws of motion is a more original contribution than the other two. Both of the first two laws existed in some form in the writings of others before him, principally those of Galileo, but the third law appears to be uniquely Newton's.

No object is able to exert a force on itself. Therefore, if any object changes its velocity, it must be acted on by some outside agent. The agents we shall consider include gravity, electric forces, nuclear forces, strings, ropes, and so on. Any of these can produce a force on an object.

Suppose we take as a particular example an apple falling to the earth, which is supposed to have sparked Newton's thinking on gravity. We know, because the apple is accelerated toward the earth, that a force is acting on it. This force is what we have called the weight and, as we have discussed, is due to the gravitational attraction between the apple and the earth. Newton's third law says that, if we observe a force on the apple which is due to the gravitational attraction between the apple and the earth, we must also observe a force of exactly the same magnitude but opposite in direction acting on the earth:

> *In nature, forces always come in pairs. These pairs are equal in magnitude but exactly opposite in direction. If one member of the pair is caused by object A and acts on object B, the other member of that pair will be the force due to object B acting on object A.*

For example, consider the sprinter about to start a race in Fig. 5.2. When the starter's gun goes off, he wants to accelerate his body as much as possible. This means there must be a force acting on his body to

FIGURE 5.1 Newton's third law says that the downward force on the apple is matched by an equal and opposite force upward on the earth.

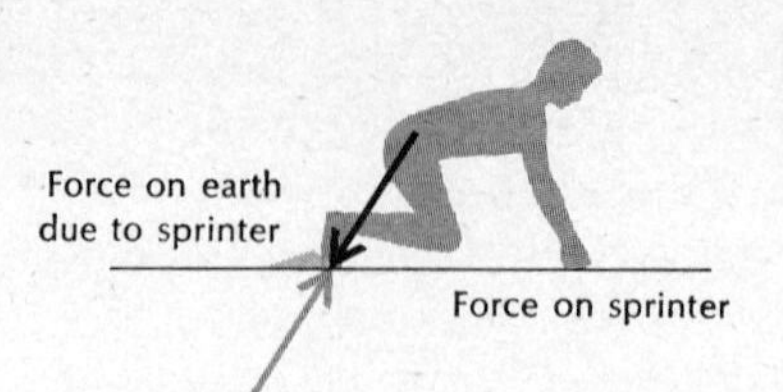

FIGURE 5.2 The sprinter exerts a force on the starting blocks anchored to the earth. The force of the earth on the runner propels him forward.

accelerate it. When the gun goes off, the runner exerts as much force as possible against the starting blocks fixed in the earth. The other force of the pair, the force of the earth against him, causes his body to be accelerated. Why is the earth not accelerated by the force on it? In fact it *is* accelerated by the push of the runner against it. But the mass of the earth is so great that the magnitude of the acceleration caused by this force is impossibly small to measure. The figure shows the sprinter pushing against starting blocks. However, if the starting blocks are not carefully anchored to the earth, they will accelerate backward when he tries to start, and the start will be spoiled.

Another example occurs in the situation shown in Fig. 5.3. The boy is trying to step from the canoe to the dock. When he tries to step forward, the boat moves backward, leaving him in the water. The explanation is simple. To walk forward, we push *backward* on the floor. The third law says, then, that there is an equal force *forward* acting on our foot due to the floor. This forward force propels our body forward. But the whole process depends on the fact that the floor is attached to the earth, which is very massive. Therefore, the backward push on the floor produces no measurable acceleration. But a canoe is light, and the backward force produces a very real acceleration. The body of the boy is then propelled forward with a smaller velocity than if the boat were fixed, and he lands in the lake.

We have previously encountered pairs of forces that are equal and opposite. If a mass sits on a table there are two forces on it. One is the force of gravity downward, the second is the force upward on it, due

FIGURE 5.3 When the boy pushes backward on the canoe in order to step out onto the dock, he does not have as much forward motion as he would if the canoe were a very massive object.

to the table. This pair of forces is *not* a third-law pair, because both act on the same object. Third-law pairs always act on different objects.

5.2 ROCKETS AND JETS

Newton's third law is basic to an understanding of how a rocket works. To see this, consider a rather different sort of example. A fisherman has lost the oars to his boat while still out on a lake. Now, how is he to start moving toward the shore? Figure 5.4 shows the procedure. It is to throw a fish as hard as possible away from the shore. A force equal, but opposite in direction, to that on the fish is applied to the man and the boat through the third law. The boat will then be given some velocity toward the shore.

A rocket works on the same principle. Instead of a fish, the rocket ejects exhaust gases to the rear. According to the third law, this then gives a forward force to the rocket. The exhaust gases are given the maximum possible velocity by choosing appropriate fuel mixtures. The greater the velocity and mass of the exhaust gases, the greater forward thrust given to the rocket. Of course, as more and more fuel is burned, the remaining mass of the rocket becomes less and less. Therefore the acceleration of the rocket will steadily increase while the engine is burning at a steady rate and producing a constant force. The maximum acceleration will occur just before the rocket runs out of fuel.

5.3 COMPARISON OF MASSES

The third law gives us a way of setting up a scale of masses that is both elegant and logically superior to the way it was done using the second law. Let us consider two cars on a frictionless track, with a compressed spring between them, tied together with a string preventing them from moving apart, as shown in Fig. 5.6.

What do we expect to happen if we suddenly cut the string? The third law says the force on car 1 is exactly the same as the force on car 2, but in the opposite direction. While the two cars are in contact through the spring, they experience the same force, and are accelerated. Once there is no contact between the cars, they will move at constant velocity. The velocity each car achieves is proportional to its accelera-

FIGURE 5.4 Throwing fish in one direction exerts a force on the boat in the opposite direction, so that the boat acquires a velocity in the direction opposite that in which the fish are thrown.

FIGURE 5.5 The upward force of a rocket is achieved by the ejection of the exhaust gases with the maximum possible velocity. The rocket shown is a Saturn V, launching Apollo 6, an unmanned flight. *(Courtesy of NASA.)*

tion. But since each car experiences the *same* force at *every* instant of time as the other, the only difference between the accelerations experienced by the two cars will occur because of the difference in their masses. If one of the cars is exactly twice as massive as the other, its average acceleration will be exactly one-half as much as the other, because the forces are equal at every instant.

Thus the ratio of the final velocities of the two cars is directly indicative of the ratio of their masses. The *heavier* car will have the *smaller* velocity. We can express this ratio thus:

$$\frac{\text{Velocity of 1}}{\text{Velocity of 2}} = \frac{\text{mass of 2}}{\text{mass of 1}}$$

$$\frac{v_1}{v_2} = \frac{m_2}{m_1} \quad \text{or} \quad m_1 v_1 = m_2 v_2 \tag{5.1}$$

That is, the quantity mass times velocity is the same in magnitude for each car. This quantity, mass times velocity, turns out to be a sufficiently useful and important quantity that is given a name of its own, *momentum*.

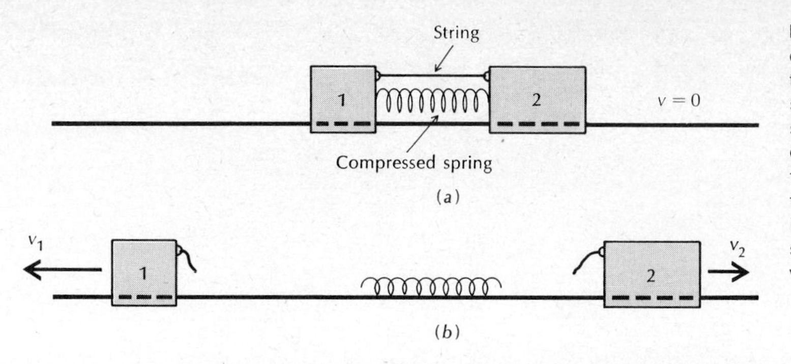

FIGURE 5.6 *(a)* Two cars on a frictionless track are tied together with a compressed spring between them. *(b)* The string is gently cut, and the cars move apart owing to the force of the spring. Since the force on each car is the same in magnitude, the more massive car will have the smaller velocity.

The momentum of an object is defined as the mass of the object times its velocity.

Ernst Mach (1838–1916) claimed that the preceding experiment is the proper way to develop a scale for mass. First, one defines an arbitrary unit of mass, such as the kilogram, as before. Then, every other mass is compared with this basic unit by measuring velocities in the experiment above. This method is considered logically superior, because it does not depend in any way on the introduction of a scale of forces, but only on the third law.

5.4 MOMENTUM

In the experiment discussed in Section 5.3, we came to the conclusion that

$$m_1v_1 = m_2v_2 \tag{5.1}$$

Momentum of 1 = momentum of 2

In drawing this conclusion, we did not pay attention to the direction of the velocities, only to their magnitude. If velocities in one direction are chosen as positive, velocities in the opposite direction are negative. Therefore it is possible to have both positive and negative momenta. In the experiment described in Section 5.3, it should be clear that the momenta of the two cars are equal and opposite. That is, one is positive, with a given value, and the other is negative, with the same numerical value. Therefore Equation 5.1 should be properly written as

$$m_1v_1 + m_2v_2 = 0 \tag{5.2}$$

Momentum of 1 + momentum of 2 = 0

Equation 5.1 was written using only the magnitude of the two velocities. Equation 5.2 includes their sign; that is, v_1 and v_2 have opposite signs, since the two cars go in opposite directions. Equation 5.2 is the more general result. The sum of the momenta of the two cars is zero. But *before* the string was cut, when both cars were standing still, the

FIGURE 5.7 In the drawing the momentum given to a fish is equal in magnitude, and opposite in direction, the momentum given to the boat.

velocity was also zero; so the momentum of each car was zero before the string was cut. The total momentum of the two cars, which is the sum of the momenta of each, was zero both before and after the string was cut.

This experiment is a particular case of an important and general principle. Before and after we cut the string, the total momentum of the system was the same. In other words, the momentum did not change. This is a very general principle, called the *Law of conservation of momentum:*

> *If no outside force is applied to a system, the momentum of the system does not change.*

When we cut the string between the two cars, it must be done carefully, so that we do not exert any outside force on the system of the two cars and spring. Any time a system is isolated from outside forces, the total momentum of the system will remain constant. The total momentum is the algebraic sum of the momenta of all parts of the system.

We can now look at the man in the rowboat with the fish, or at a rocket, from the point of view of the momentum of the system. In each case there are no outside forces acting. In order for the rocket or the rowboat to acquire momentum in a forward direction, it is necessary for some part of the system to be given momentum in a backward direction. The total momentum of the entire system will always remain zero, if it was initially zero. The forward momentum of the rocket will then exactly equal the backward momentum of the fuel exhausted. (This statement applies to a rocket in space, so that gravity does not have to be overcome.)

The conservation-of-momentum principle is particularly useful in collision problems. If two cars collide, we can say that the sum of the momenta just before the collision is equal to the sum of the momenta after the collision. The only forces are between the two cars. No outside

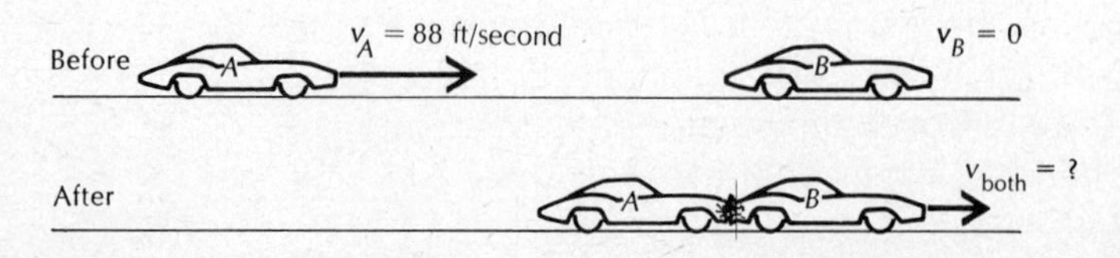

FIGURE 5.8 The moving car *A* collides with the stationary car *B*, and the combined wreckage moves at a velocity determined by momentum conservation.

forces act. Therefore momentum is conserved. This is convenient, because we do not have to inquire about the complex details of the forces which act during the collision.

Example 5.1

A 2,000-kilogram Mercedes strikes a 1,000-kilogram Volkswagen. The Volkswagen is not moving when struck. From the distance the two cars slide after the collision it is calculated that after the collision the velocity of the wreckage is 20 meters/second. How fast was the Mercedes going before the collision?

The basic principle is that total momentum before the collision is the same as total momentum afterward:

$$\text{Momentum before} = \text{momentum after}$$

$$(2{,}000 \text{ kilograms})(\text{velocity}) + (1{,}000 \text{ kilograms}) \times 0 = (2{,}000 + 1{,}000 \text{ kilograms})(20 \text{ meters/second})$$

We add the two masses on the right because both cars are moving at 20 meters/second after the collision.

$$2{,}000 \text{ kilograms} \times v_{\text{before}} = 3{,}000 \text{ kilograms} \times 20 \text{ meters/second}$$

$$v_{\text{before}} = \frac{3{,}000 \text{ kilograms}}{2{,}000 \text{ kilograms}} \times 20 \text{ meters/second}$$

$$v_{\text{before}} = 30 \text{ meters/second or} \sim 60 \text{ miles/hour}$$

5.5 THE CHANGE IN MOMENTUM

The last several sections have discussed situations in which no outside force acts and therefore the momentum is constant. Now let us consider a situation in which there is an outside force, and relate this force to the change in the momentum it produces.

You will remember that we wrote Newton's second law as

$$\text{Observed acceleration} = \frac{\text{applied force}}{\text{mass}}$$

We also wrote as the definition of acceleration

$$\text{Acceleration} = \frac{\text{change in velocity}}{\text{elapsed time}}$$

If we set these equal to each other, we have

$$\frac{\text{Applied force}}{\text{Mass}} = \frac{\text{change in velocity}}{\text{elapsed time}} \tag{5.3}$$

or

$$\text{Applied force} \times \text{elapsed time} = \text{mass} \times \text{change in velocity}$$

But mass times change in velocity is the change in momentum. Therefore we conclude that the change in momentum of an object is a result of the force applied to that object and the time during which that force acts. Thus the same change in momentum can be produced by a small force acting for a long time, or a large force acting for a short time.

A spectacular example of this is shown in Fig. 5.9. This is a high-speed photograph of a tennis racket hitting a ball. Notice the degree to which the ball is squashed. This indicates the tremendous force between the racket head and the ball at the moment of impact. The time during which this force acts is very small. This large force does *not* come from the player's arm and hand. He applies a much smaller force for a much longer time in order to accelerate and give momentum to the racket head. Then the collision between racket head and ball imparts momentum to the ball.

The ideas in the discussion above apply to all sports in which one object is struck by another: golf, tennis, football (kicking), hockey, etc. In order to give the maximum momentum to the object being struck, the force applied during the blow must be continued as long as possible. The time of impact is not instantaneous, although very short. Thus the importance of the follow-through which all coaches emphasize. In tennis, in particular, the ball is squashed on the racket for a considerable time.

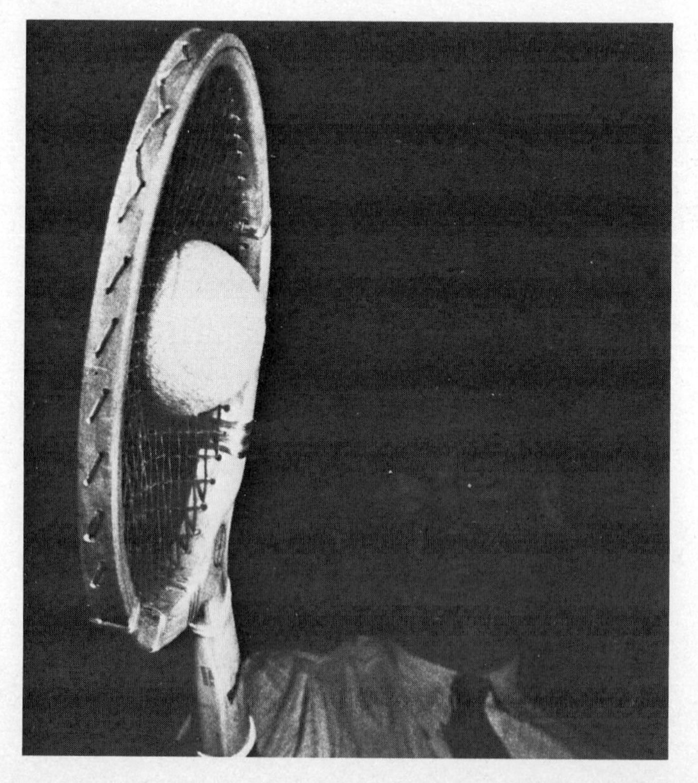

FIGURE 5.9 The momentum imparted to the tennis ball depends on the force, and the time the force acts. We can see from the shape of the ball that the force is large. *(Courtesy Dr. Harold E. Edgerton, MIT.)*

Only if force is applied in the proper direction all during this time will the shot come out right.

Example 5.2

We can make an estimate of the force between a golf club and golf ball using these ideas. A golf ball has a mass of about 0.1 kilogram and can achieve a velocity of over 100 miles/hour, or about 50 meters/second, when well hit. The most difficult question is to estimate how long the impact between club and ball lasts. A good estimate might be 0.01 second or less.

From Equation 5.3 we have

$$\frac{\text{Force}}{\text{Mass}} = \frac{\text{change in velocity}}{\text{time}}$$

$$\text{Force} = \frac{(50 \text{ meters/second})(0.1 \text{ kilogram})}{0.01 \text{ second}}$$

$$= 500 \text{ newtons or} \sim 120 \text{ pounds}$$

If a better value for the time of impact is 0.001 second, the force is 5,000 newtons or about 1,200 pounds. Note that this is the force between the club and the ball. Momentum is given to the club by a much smaller force applied by the player for a much longer time.

5.6 ANGULAR MOMENTUM

So far, what we have discussed is called *linear momentum*, because it relates to motion in a straight line. There is another form of momentum called *angular momentum*, or rotational momentum.

One of the most striking demonstrations of angular momentum is a frequently used classroom demonstration. The professor stands on a

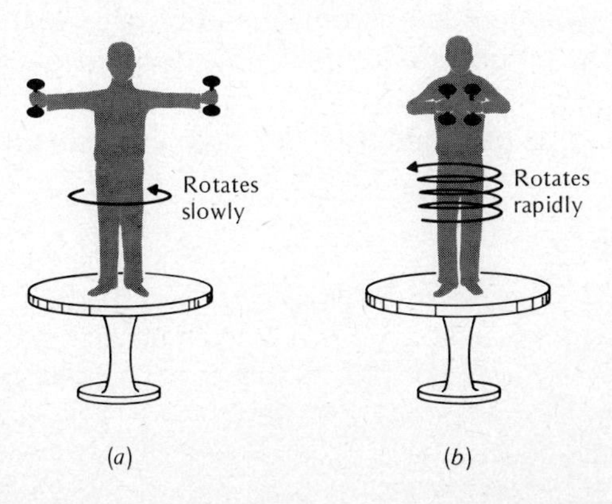

FIGURE 5.10 When the professor brings the dumbbells close to his body, his rotational speed increases.

turntable which rotates with very little friction. He holds two weights in his hands and starts himself rotating slowly (Fig. 5.10*a*). Then he pulls the weights in toward his body and is observed to spin more and more rapidly (Fig. 5.10*b*). We have arranged this demonstration so that no outside forces are acting. The quantity that is conserved or constant is the angular momentum.

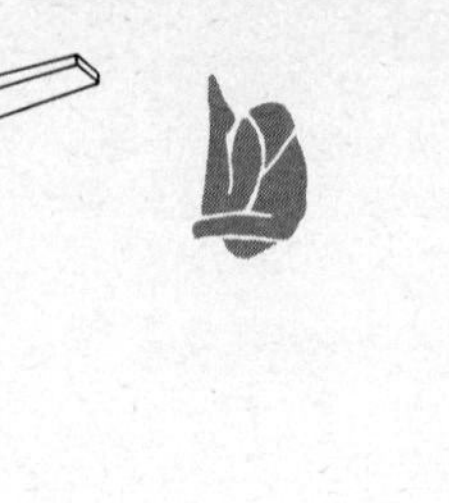

FIGURE 5.11 When a diver tucks tightly, he achieves a higher rotational velocity due to angular momentum conservation.

How can we say that angular momentum is conserved in this experiment if the rotational speed is increased? This occurs because the proper form for angular momentum is mass times velocity times distance from the axis of rotation. When the professor pulled the weights in close to his body he reduced their distance from the axis of rotation. Therefore if the angular momentum was to remain constant, the rotational velocity had to increase. We write

$$\text{Angular momentum} = mvR \tag{5.4}$$

(This works for circular motion. For more complex motions, a more complex formula is needed.)

Springboard divers and figure skaters use the principle illustrated above. When a diver goes into a tight tuck, he is able to increase his rotational speed and achieve more somersaults (Fig. 5.11). When an ice skater is spinning, she gradually brings her arms close to her body to increase the rotational speed.

In astronomy, Kepler's second law is a direct consequence of angular momentum conservation. The planets move about the sun in elliptical orbits, which means they are sometimes nearer, and sometimes further, from the sun. Because of angular momentum conservation, they move fastest when nearest the sun, and slowest when furthest away (see Fig. 4.15).

SUMMARY

Newton's third law states that forces come in pairs, equal to each other but opposite in direction. If there is a force caused by *A* on *B*, the other force is caused by *B* on *A*. Rockets and jets move forward because of third-law forces.

Ratios of masses can be obtained from the third law, using the ratio of the velocities that two masses achieve, as

$$\frac{m_2}{m_1} = \frac{v_1}{v_2}$$

Momentum is defined as mass times velocity. If no external forces act on a system, the momentum is constant. This principle can be used to calculate the results of collisions, even though the details of the collision process are not known.

The change in momentum is given by the applied force multiplied by the time it is applied.

Angular momentum, or rotational momentum, is defined, for motion in a circle, as

Mass × velocity × distance from the center

If no outside forces act, angular momentum is constant.

QUESTIONS

1 What is a conservation law?
2 A man stands on a scale and the scale reads his weight. Discuss all the forces that act on the man and the scale and identify all the third-law pairs of forces which act in this case.
3 A blown-up balloon is released and darts about the room as it deflates. Describe the forces that act during this motion.
4 Would you expect it to be easy or hard to hold a high-pressure water hose? Is it easier if the hose is straight or coiled?
5 Does a rocket depend on the surrounding atmosphere to work?
6 Why is an automobile dashboard padded?
7 A person stands on the periphery of a freely spinning turntable and begins to run toward the center. Will the turntable speed up or slow down?
8 If no object can exert a force on itself, what holds a watermelon together?
9 An apple of mass $\frac{1}{2}$ kg falls to the earth with an acceleration of 9.8 m/second^2. What is the acceleration of the earth toward the apple?
10 If there were truly an immovable object, would its existence violate Newton's laws of motion?
11 Discuss in terms of Newton's laws of motion what force causes a car to move down a road.
12 In the case of the man in the rowboat throwing fish to attain the shore, does gravity have to be taken into account?
13 Do the forces an object *A* exerts on an object *B* have any effect on object *A*'s motion?

SELF-TEST

__________ Newton's third law
__________ momentum

__________ conservation of momentum
__________ angular momentum
__________ change of momentum

1 In the absence of external forces, no change is observed
2 Mass times velocity
3 Force times the time during which the force acts
4 Mass times velocity times distance from the center
5 Forces come in pairs, equal and opposite to each other. The force on *A* due to *B* is equal and opposite to the force on *B* due to *A*.

PROBLEMS

1 A car of mass 500 kg moves with a speed of 30 m/second. What is the momentum of the car?

2 A typical car may have a mass of 2,000 kg. What is the momentum of this car when it is traveling at a speed of 25 m/second (about 50 miles/hour)?

3 The earth has a mass of about 6×10^{24} kg and moves with an orbital speed of 3×10^4 m/second. How much momentum does it have?

4 What is the speed of an electron if its mass is 9.1×10^{-31} kg and its momentum is 1.82×10^{-27} kg-m/second?

5 When in operation the Queen Elizabeth was the largest passenger vessel ever built. The ship had a mass of about 8×10^7 kg and cruised at a speed of 32.8 miles/hour. When cruising, what was the momentum of this ship?

6 Mercury has a mass of 3.4×10^{23} kg and moves in its orbit around the sun with a speed of 4.8×10^4 m/second. What is the momentum of Mercury?

7 The moon has a mass of about 7×10^{22} kg and orbits the earth with a speed of about 10^3 m/second. What is the momentum of the moon?

8 A baseball whose mass is about 250 g is thrown with a speed of 40 m/second (about 90 miles/hour). What is its momentum?

9 The slowest moving chelonian is the giant tortoise found in Mauritius. Even when hungry and enticed by a cabbage, it cannot cover more than 5 m in a minute. A typical male of this species may have a mass of 200 kg. What is the momentum of this cabbage-crazed tortoise?

10 Calculate the momentum of a car whose mass is 1,500 kg traveling at a speed of 30 m/second.

11 The experiment shown in Fig. 5.6 is performed using two masses.

Mass 1 is 1 kg. Mass 2 is unknown. The velocities after separation are mass 1, 5 m/second, and mass 2, 3 m/second. What is the mass of 2?

12 A ball whose mass is 0.1 kg is struck with a bat. The velocity of the ball initially was 20 m/second, and after being struck is 40 m/second in the opposite direction. If the impact between the ball and the bat lasts 0.01 second, calculate the force between ball and bat.

13 The fisherman in Fig. 5.4 begins with his boat at rest. He throws 10 fish, each with mass 1 kg, with a velocity of 5 m/second. If the boat and fisherman together have a mass of 200 kg, what velocity will the boat achieve? (In this problem the change in the mass of the boat as additional fish are thrown can be neglected. This is not true for a rocket.)

14 (*a*) How fast would a car have to be going in order to bring a truck weighing 10 times as much to a dead stop in a head-on collision, if the truck is going 20 miles/hour and both are coasting? (*b*) What is the momentum of the system (car plus truck) before the collision? (*c*) What is the momentum of the car-truck system after the collision?

15 A 1,000-newton man stands in a 1,500-newton boat at rest on the surface of a calm lake. The man dives from the boat with an initial horizontal velocity of 2 m/second westward. What is the initial recoil of the boat? Neglect any momentum given to the water.

16 Two automobiles collide in a completely inelastic collision. Initially car *B*, whose mass is 1,500 kg, was at rest. Car *A*, whose mass is 2,000 kg, was moving 25 m/second before the collision. Find the velocity of the wreckage.

17 Two automobiles collide in a head-on collision. Car *A* whose mass is 2,000 kg, was traveling 30 m/second north before the collision. Car *B*, whose mass is 2,500 kg, was traveling 20 m/second south before the collision. If the collision is completely inelastic, what is the velocity of the wreckage after the collision, including its direction?

18 A cannon on wheels whose mass is 200 kg shoots a shell of mass 5 kg. If the shell leaves the muzzle with a horizontal velocity of 500 m/second, what is the backward velocity of the cannon?

19 A rifle whose mass is 2 kg is suspended on strings. It shoots a bullet whose mass is 15 g at a velocity of 200 m/second. What velocity does the gun achieve just as the bullet leaves the barrel?

20 A bullet whose mass is 15 g is fired into a block of wood suspended on strings. The mass of the wood block is 1 kg. If the velocity of the block of wood at the impact of the bullet is measured to be 5.0 m/second, what was the velocity of the bullet?

21 A golf ball whose mass is 0.2 kg is struck and given a velocity of 60 m/second. If the impact between club and ball lasts 10^{-3} second, what is the average force on the ball during this time?

22 A man who weighs 160 lb is running at a speed of 8 ft/second. He comes upon a child's wagon at rest which weighs 4 lb, flops into it, and rolls away. What is the velocity of the man and the wagon after the encounter?

23 A car which weighs 3,200 lb is moving north at a speed of 30 miles/hour. It collides with a loaded truck weighing 6,400 lb which is stopped at a stop sign. If the collision is totally inelastic, what is the final velocity and in what direction is the motion?

24 A 160-lb man stands on ice skates and throws a 16-lb rock with a velocity of 30 miles/hour toward the south. What is the final velocity of the man and in what direction is he moving?

25 A comet is observed to have a velocity in its orbit of 1,000 miles/hour when it is at its maximum distance of 10^9 miles from the sun (10^9 miles = 1 billion miles, which is somewhere near the orbit of Saturn). What is its velocity when it is 10^7 (10 million) miles (its minimum distance) from the sun?

6

WORK AND ENERGY

This chapter and the next are devoted to developing some of the relationships involving energy, a quantity which is even more useful than momentum in the solution of a wide variety of physical problems. Historically, the major development of ideas about energy was made between 1665 and 1840. As recently as 1905 the concept of energy was modified by Einstein. And, of course, almost no one is unaware that the world faces an "energy crisis," or at least an "energy problem."

In 1665 Christian Huygens (1629–1695), a Dutch physicist, reported to the Royal Society the results of a study of the collisions of rigid, or elastic, balls (billiard balls are a good example). He found that the mass times the square of the velocity of the balls seemed to have an interesting property. In the collision of two balls, which we label 1 and 2, Huygens found that the quantity $m_1v_1^2 + m_2v_2^2$ before the collision was the same as $m_1v_1^2 + m_2v_2^2$ after the collision. Thus the quantity mass times velocity squared is a constant in this type of collision, or is *conserved.* Notice that this is an experimental result. The German mathematician Gottfried Wilhelm Leibniz (1646–1716) also discovered the usefulness of the quantity mv^2. A major theme in the growth of science for the next 250 years was the development of the energy concept from this small beginning. We will devote the next three chapters to this subject, and it will recur frequently later in the book.

6.1 WORK AND SIMPLE MACHINES

One of the oldest mechanical principles known to man is that of the lever (Fig. 6.2). A long rod or crowbar is placed over the fulcrum. If the distance ℓ_1 between the man and the fulcrum is greater than ℓ_2, the distance between the fulcrum and the rock, the man can move a rock heavier than he could lift if unaided. The greater the difference between

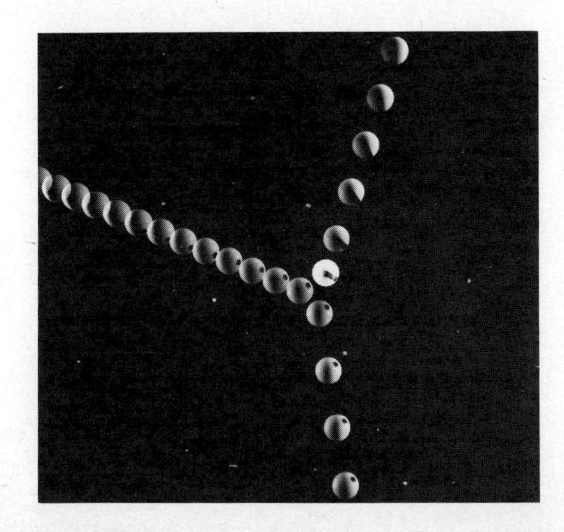

FIGURE 6.1 This photograph shows a ball approaching from the bottom and striking a stationary ball near the center of the photograph. The ball originally in motion moves up and to the left in the photograph. The ball originally at rest moves up and to the right. The apparatus was arranged so that the light flashed 30 times per second, so that there was $\frac{1}{30}$ second between pictures. The two balls have the same mass. The kinetic energy can be measured to be the same before and after the collision. *(Courtesy Educational Development Center.)*

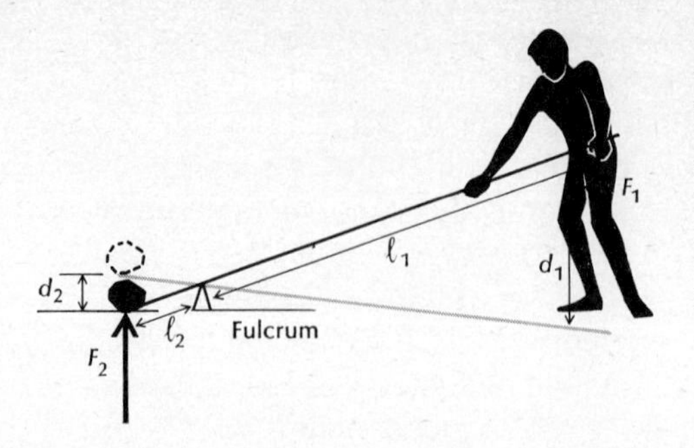

FIGURE 6.2 The principle of the lever. The lever acts as a force multiplier. A larger force is applied to the rock than the man is applying at his end, but the rock is moved a smaller distance than the man moves his end of the lever.

ℓ_1 and ℓ_2, the more force the man can apply to the rock. The lever thus acts as a force multiplier. There is, however, a penalty extracted for this. It is easy to see that the distance d_1 that the man moves his end is greater than the distance d_2 that the rock moves. Therefore the lever allows a man to apply to the rock a large force for a small distance, while he is applying to his end of the lever a smaller force for a larger distance.

This relationship can be made quantitative by direct experiment. Measure the force applied at one end of the lever to lift a rock of a known weight, and then measure the distances d_1 and d_2. The result is, using the notation of Fig. 6.2,

$$F_1 \times d_1 = F_2 \times d_2 \tag{6.1}$$

This result states that the product of force times distance is the same for each end of the lever. You may have discovered the same result as a child when you found that a fat kid had to sit on the short end of a teeter-totter.

The quantity force times distance is given a special name. It is called *work*. In Fig. 6.2 a certain amount of work is done by the man at one end of the crowbar. The same amount of work is done at the other end on the rock. All machines operate in a similar fashion. The work put in is the same as the work put out, but a machine can *amplify forces* by making the distances different. As we will see later, only a perfect machine will deliver out as much work as is put in. No such perfect machine can be built.

You will encounter levers in many situations in everyday life. Many examples of levers exist in the human body. Figure 6.4 is a drawing of a human arm holding a 50-newton weight, and shows the attachment of the biceps muscle to the forearm bone. Here the fulcrum is at the

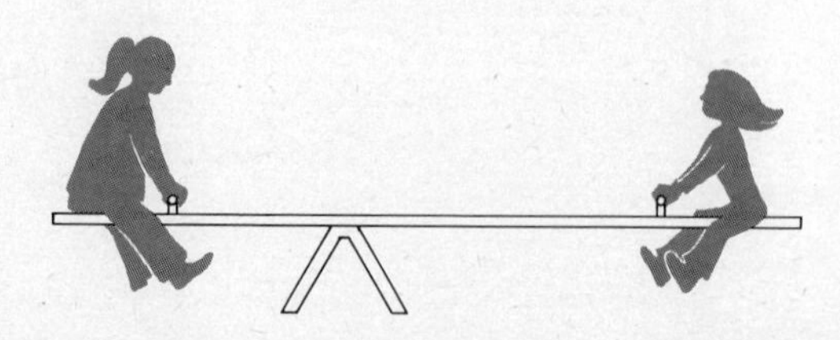

FIGURE 6.3 A heavier kid must sit on the short side of the teeter-totter if it is to balance.

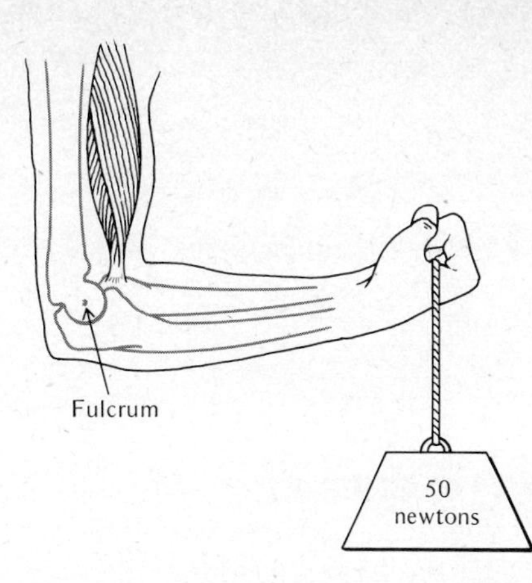

FIGURE 6.4 In the human arm, the force applied by the biceps muscle is upward and near the fulcrum. The force of 50 newtons is downward, and further from the fulcrum. The muscle applies a force of between 350 and 400 newtons.

elbow, and both forces are applied to the right of the fulcrum. As you can see from their distances from the fulcrum, the muscle must apply a force larger than 50 newtons to the bone. (About 7 or 8 times larger, in fact.) To compensate for this the total distance the muscle must contract is much less (by the same factor of 7 or 8) than the distance the weight moves.

One qualification must be made to the definition of work. This qualification distinguishes the scientific term work from the everyday use of this word. To be considered work in the technical sense used here, the force and the distance must be parallel. If the force and the distance moved are at right angles, the work is zero, no matter how difficult it may seem to the person doing it. In Fig. 6.5 are shown two situations. In Fig. 6.5*a* the man is pushing a cart. The force and distance are parallel, and the work is force times distance. In Fig. 6.5*b* the man is carrying a suitcase; the force is vertical, and the distance it is carried is horizontal. The work is zero, even though the man may become tired.

Example 6.1

A teeter-totter is 5 meters long. Where must the fulcrum be placed if a boy weighing 500 newtons is to balance one weighing 750 newtons?

To achieve balance the force (the boy's weight) times the distance to the fulcrum must be the same on both sides. That is,

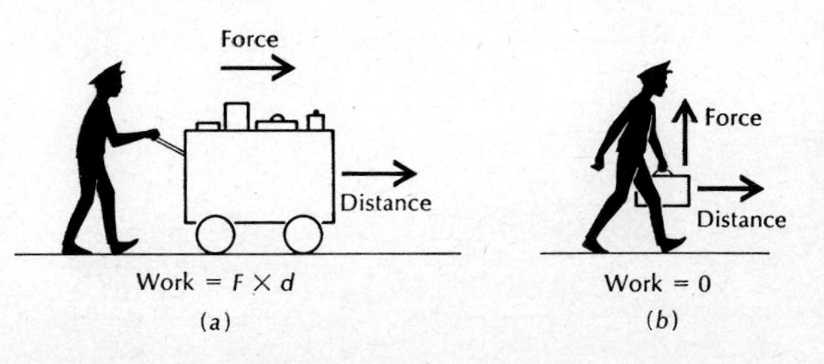

FIGURE 6.5 In *(a)* the work done is given by the force times the distance, because the force and the distance the object is moved are parallel. In *(b)* the force and the distance the object is moved are perpendicular, and the work is zero.

$$500 \text{ newtons} \times \ell_1 = 750 \text{ newtons} \times \ell_2$$

or

$$\frac{500 \text{ newtons}}{750 \text{ newtons}} = \frac{\ell_2}{\ell_1}$$

The ratio $\frac{500}{750} = \frac{2}{3}$, so the ratio ℓ_2/ℓ_1 must be $\frac{2}{3}$. This can be achieved if the 500-newton boy is 3 meters from the fulcrum and the 750-newton boy is 2 meters from the fulcrum.

6.2 WORK AND KINETIC ENERGY

It is clear that there is a difference between the motion of a large truck and a small car, even if they are traveling at the same speed. We have already seen in Section 5.3 that the momentum of two such cars is different. Another useful way of describing the difference is in terms of their energy of motion, called *kinetic energy*.

The kinetic energy of an object is defined as

$$\begin{aligned} KE &= \tfrac{1}{2} \text{ mass} \times (\text{velocity})^2 \\ &= \tfrac{1}{2} mv^2 \end{aligned} \tag{6.2}$$

Why do we define this quantity in this way?

In Chapter 2 we developed a relationship from studying the stopping distance of cars. This relationship was

$$(\text{Velocity})^2 = 2 \times \text{acceleration} \times \text{distance}$$

or

$$\text{Acceleration} = \frac{(\text{velocity})^2}{2 \times \text{distance}}$$

We also have Newton's second law:

$$\text{Acceleration} = \frac{\text{force}}{\text{mass}}$$

If we set these two expressions equal to each other, we have

$$\frac{\text{Force}}{\text{Mass}} = \frac{(\text{velocity})^2}{2 \times \text{distance}}$$

or

$$\text{Force} \times \text{distance} = \frac{\text{mass} \times (\text{velocity})^2}{2}$$

$$F \times d = \tfrac{1}{2}mv^2 \tag{6.3}$$

FIGURE 6.6 A large truck has more kinetic energy than a small car.

What does this say? The work done on an object is equal to its kinetic energy. A better way to say this is to say that work done on an object changes its kinetic energy. The amount of change is equal to the work done. For Equation 6.3 the kinetic energy is zero either at the beginning or at the end. If this is not true, the full equation would be

$$\text{Work} = \text{change in kinetic energy produced} \qquad (6.4)$$
$$F \times d = \left(\tfrac{1}{2}mv^2\right)_{\text{after}} - \left(\tfrac{1}{2}mv^2\right)_{\text{before}}$$

Here before and after refer to before and after the work is done. Work can either increase or decrease the kinetic energy, depending on how the force is applied. Which happens is determined by whether the force is directed so as to increase or decrease the velocity.

UNITS OF WORK, ENERGY, AND POWER

The unit we use for work and energy can be obtained quite easily from the definition of work. *Work* is defined as force times distance, so we can define a unit of work in terms of our units of force and distance. The unit of work is 1 newton-meter. It is the work done by one newton of force acting through a distance of one meter. This unit is given a name of its own, the *joule* (J) (for James Joule, an English scientist, whose work is discussed in Chapter 7).

$$1 \text{ joule} = 1 \text{ newton} \times 1 \text{ meter} \qquad (6.5)$$

The joule is a unit of either work or kinetic energy. Both are forms of energy.

Example 6.2

A 1,000-kilogram Volkswagen accelerates from 0 to 30 meters/second (about 60 meters/hour) in a distance of 200 meters. What force must be applied to achieve this result?

To handle this problem we use the result

$$\text{Kinetic energy achieved} = \text{work done}$$
$$\tfrac{1}{2}mv^2 = F \times d$$
$$\tfrac{1}{2}(1{,}000 \text{ kilograms})\left(30\ \frac{\text{meters}}{\text{second}}\right)^2 = F \times 200 \text{ meters}$$
$$450{,}000 \text{ joules} = F \times 200 \text{ meters}$$
$$2{,}250 \text{ newtons} = F$$

The total work done is 450,000 joules, and the force is 2,250 newtons (about 500 pounds).

It might appear that the units of work and kinetic energy should be different, since work is force times distance and kinetic energy is mass times velocity squared. But if you remember how we defined a newton force, in Chapter 3, as mass times acceleration, you can see that it all works out.

$$\text{Force} \times \text{distance} = \text{mass} \times \text{acceleration} \times \text{distance}$$
$$= \text{kilograms} \times \frac{\text{meters}}{\text{second}^2} \times \text{meters}$$
$$= \text{kilograms} \frac{\text{meters}^2}{\text{second}^2}$$

The result is mass times velocity squared, the same as for kinetic energy.

Frequently one is interested in the rate at which work is being done. This quantity is called *power*. The unit of power is called the watt (W), and is defined as 1 joule/second.

$$1 \text{ watt} = \frac{1 \text{ joule}}{1 \text{ second}} \tag{6.6}$$

One watt of power used for one second is then one joule of energy. You pay your electric bill in terms of the number of kilowatthours used. A kilowatt is 1,000 watts. Therefore, a kilowatthour (kWh) is

$$1{,}000 \text{ watts} \times 3{,}600 \text{ seconds} = 3{,}600{,}000 \text{ watt-seconds}$$
$$= 3{,}600{,}000 \text{ joules}$$

Occasionally in discussions of energy one encounters the English unit of energy. This is the foot-pound, distance times force as before. One has to be careful in dealing with kinetic energy because pounds are a unit of weight, not mass. Mass is given by pounds weight divided by g = 32 feet/second^2. The English unit of mass is called the slug.

6.3 POTENTIAL ENERGY

We have stated that a force continuously applied to an object for a given distance leads to a change in the kinetic energy of the object. But there are two kinds of situations. If we apply a horizontal force to a car on our frictionless air track, a change in kinetic energy occurs. If we apply a vertical force to an object to lift it slowly off the floor, we find no change in the kinetic energy of the object. We start with an object at rest on the floor and end with an object held steady above the floor; it has no velocity and therefore no kinetic energy. In the first situation we do work on the car, and it undergoes a change in kinetic energy. In the second we also do work, because the force is parallel to the distance the object moves, but there is no change in kinetic energy.

There is a clear difference between the two cases. For any object to change its kinetic energy, which means that its velocity must be changed, the *net* force on the object must be different from zero, so that some acceleration is produced. When we push an object horizontally, the net force, neglecting friction, is the force we apply. But when we lift an object, there are two forces acting. One is the pull upward exerted on it, and the other is the downward force of gravity we have called its weight. If we lift an object very slowly, these two forces will be almost exactly equal. The force upward need only be a trace larger than the weight of the object in order that it be lifted. So the *net* force on the object is zero, and it is not accelerated.

What happens, then, to the work we have done on the object when we lift it? As an example, take the lifting of a 10-kilogram object 2 meters from the floor. To do this requires an amount of work equal to the weight times the distance it is raised:

$$\begin{aligned}\text{Work} &= F \times d = \text{weight} \times d\\ &= mgd\\ &= (10\ \text{kilograms})(9.8\ \text{meters/second}^2)(2\ \text{meters})\\ \text{Work} &= 196\ \text{joules}\end{aligned}$$

If we now release the object, it drops and acquires kinetic energy as it falls toward the floor. What is the value of the kinetic energy just as it reaches the floor? To find this we need to calculate the velocity the object acquires in falling 2 meters, using the results of Chapter 2:

$$\begin{aligned}v^2 &= 2 \times \text{acceleration} \times \text{distance}\\ &= 2(9.8\ \text{meters/second}^2)(2\ \text{meters})\\ &= 39.2\ \text{meters}^2/\text{second}^2\\ \text{Kinetic energy} &= \tfrac{1}{2}mv^2 = \tfrac{1}{2}(10\ \text{kilograms})(39.2\ \text{meters}^2/\text{second}^2)\\ &= 196\ \text{joules}\end{aligned}$$

The kinetic energy the object acquires in falling 2 meters is exactly

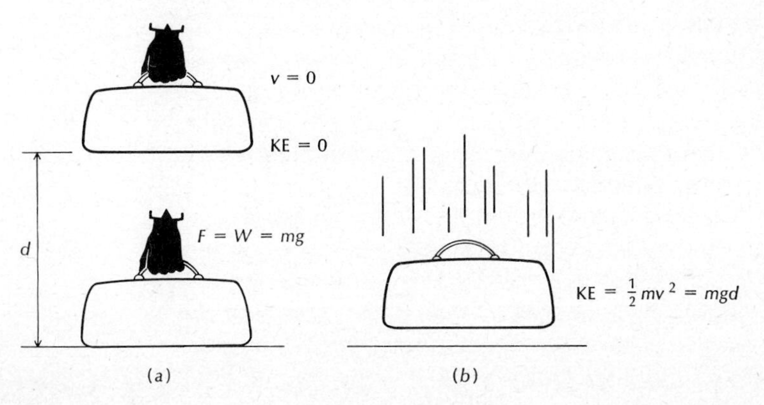

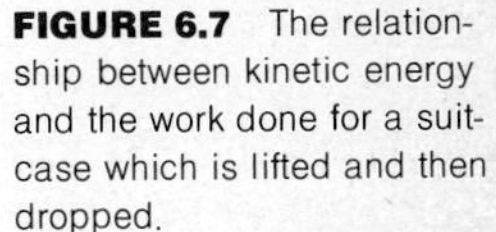

FIGURE 6.7 The relationship between kinetic energy and the work done for a suitcase which is lifted and then dropped.

FIGURE 6.8 When work is done to compress a spring, it goes into potential energy which is released when the spring is released.

equal to the work required to raise it 2 meters. So it seems that, in this situation, the work done on the object turns into kinetic energy, but not until later. That is, we do work on the object to lift it up, and then later, when we release it, it acquires a kinetic energy exactly equal to the work originally done to lift it up, if it falls the same distance. This leads to the definition of another quantity, called *potential energy*. Potential energy can be called energy of position, or stored energy. We say that, in doing 196 joules of work to raise a 10-kilogram object 2 meters, we have given it 196 joules of potential energy. When we drop the object, this potential energy is converted to kinetic energy.

The change in potential energy of an object is exactly equal to the work done on the object to lift it. This work is the force times the distance. Since the force is the weight, which is given by mg, the potential-energy change in raising an object a distance d is given by potential energy $= mgd$.

If an object falls a distance d, it loses some potential energy. It acquires a kinetic energy exactly equal to the potential-energy loss. We say that potential energy has been converted to kinetic energy. We have shown that the potential-energy change in raising an object a distance d is mgd. When it falls a distance d, the potential-energy change is also mgd. For a falling object, starting at rest, the kinetic energy after falling a distance d is given by

$$\tfrac{1}{2}mv^2 = mgd \tag{6.7}$$

We see that potential energy is, in a sense, stored energy. We do work on an object to lift it, and store an amount of energy equal to the work done. When we release the object, the stored energy, or potential energy, is turned to kinetic energy as it falls. The concept of potential energy is, in fact, more general than this discussion has made it appear. Any time work is done against any force, we can store energy for future release. This is always called potential energy.

An example of this is the work done to compress a spring. Here also, energy is stored which will be converted to kinetic energy when the spring is released (Fig. 6.8). Other examples—most of which we will discuss—are: the energy of the sun stored chemically in oil, which we release by burning (chemical potential energy); the energy stored in nuclei at the origin of the universe, which we release in nuclear reactors (nuclear potential energy); the energy stored in water behind dams, which is released when the water turns turbines to make electricity.

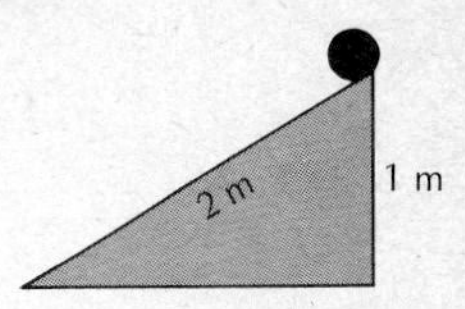

FIGURE 6.9

Example 6.3

Earlier in this chapter we discussed the lever as a simple machine. Another simple machine or tool is useful in discussing the ideas of potential energy. This is the *inclined plane*. The essential idea is that the work needed to make a change in potential energy is the same, no matter how the change is achieved (if friction can be ignored).

Figure 6.9 shows a large object of mass m at the top of an incline. What force was necessary to push it up the hill?

The principle here is that the work required to push it to the top is the same no matter how it is done. If it were lifted straight up, the work would be

$$mgd = mg \times 1 \text{ meter}$$

The distance up the incline is 2 meters, so therefore the force must be half as great, or $\frac{1}{2}\, mg$. That is, the force needed to push it up the incline, neglecting friction, is one-half of its weight, since the distance is twice as much as the vertical distance, and the work must be the same.

6.4 THE ENERGY GAME

So far we have considered three forms of energy: work, kinetic energy, and potential energy. We have found that these three can each be converted into one of the others. That is, work can be converted into kinetic energy, and kinetic energy into work; work can be converted into potential energy, and potential energy into work; kinetic energy can be converted into potential energy, and potential energy into kinetic energy. In all these conversions no energy is lost. The same amount is around afterward as before. [This last statement assumes perfect (frictionless) machines or devices.] We have here the beginning of a completely general and powerful idea: Energy seems to be convertible from one form to another, but not lost. In Chapter 7 we will find that, when it appears to be lost, it has in fact merely turned up in a new form, heat.

The three forms of energy discussed so far are only three out of many. The interconversion of energy among these three is one part of the larger theorem that we develop in Chapter 7.

6.5 WORK AGAINST FRICTION

There are many situations in which work is done as defined in this

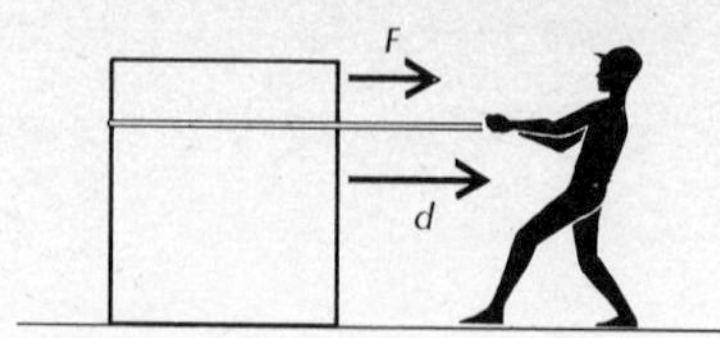

FIGURE 6.10 The boy is certainly doing work, since the force and distance are parallel. In this case the work goes to overcome friction.

chapter, but no kinetic energy or potential energy seems to be acquired by the system. An example is a boy dragging a box, as shown in Fig. 6.10. We know from experience that this box will stop moving the instant he stops pulling on the rope. Yet it is also clear that work will be done on the box in this situation. What has happened to it? The work has been done, and no kinetic or potential energy has appeared. In Chapter 7 we shall show that the work has appeared as another form of energy, heat. The force which retards the motion of the box is friction. In this case work is done to overcome friction.

One example of this is the stopping distance of cars, which was discussed in Chapter 2 and at the beginning of this chapter. We find that the stopping distance is proportional to the velocity squared. That means that stopping distance is proportional to kinetic energy. The work done on the car is done by the force of friction on the tires. An amount of work equal to the original kinetic energy of the car must be done before the car stops. Heavier cars do not take a longer distance to stop, because the frictional force is proportional to the weight of the car. It is larger for larger cars, but the kinetic energy is also larger by exactly the same amount.

Example 6.4

A 2,000-kilogram car is traveling at 30 meters/second. If the car is to be brought to a stop in 75 meters, what must the braking force be?

Here again, the work done must equal the change in kinetic energy. The braking force is the force between the tires and the road, or the brake shoes and the drum. Since the car stops, the work is equal to the original kinetic energy

$$\frac{1}{2}mv^2 = F \times d$$

$$\frac{1}{2}\,(2{,}000 \text{ kilograms})\left(30\,\frac{\text{meters}}{\text{second}}\right)^2 = F \times 75 \text{ meters}$$

$$900{,}000 = 75F$$

$$12{,}000 \text{ newtons} = F$$

The braking force is about 60 percent of the weight of the car, a typical result for dry roads.

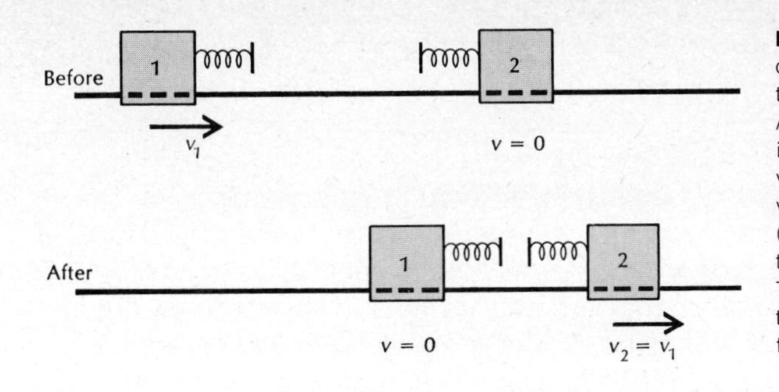

FIGURE 6.11 An elastic collision between two air-track cars of equal mass. After the collision, car 2 (initially at rest) moves with a velocity equal to the initial velocity of car 1; and car 1 (initially in motion) stops at the point of the collision. Thus the kinetic energy after is equal to the energy before.

6.6 COLLISIONS

In Section 5.4 we discussed the idea that the momentum of a system is unchanged if no outside force acts on the system. In a collision between two objects, no outside force acts. There may be very large forces between the two objects. But if the system is the two objects in collision, the effect of outside forces is very small during the short time of the collision. Therefore, we can say that momentum is a constant in a collision—of two cars, for example. This statement is true for all collisions.

Energy in collisions is a more complex matter. We can consider two classes of collisions. The first we will call "bouncy" or elastic. Billiard balls give a good example of this sort of collision. Huygens' result was that, in billiard ball collisions, the total kinetic energy is the same before and after the collision. This turns out to be true in all "bouncy" collisions.

The second kind of collision we will call "sticky," or inelastic. Automobile collisions are representative of this group. In a sticky collision, some of the kinetic energy is used up. In auto collisions it is used smashing up the cars. In the end, most of the energy used up in such a collision turns up as heat energy. A collision is called *completely inelastic* if the two objects stick together after the collision. A collision is called *completely elastic* if no kinetic energy at all is used up. These are, in fact, only the two extreme cases. All gradations of partly inelastic collisions also exist.

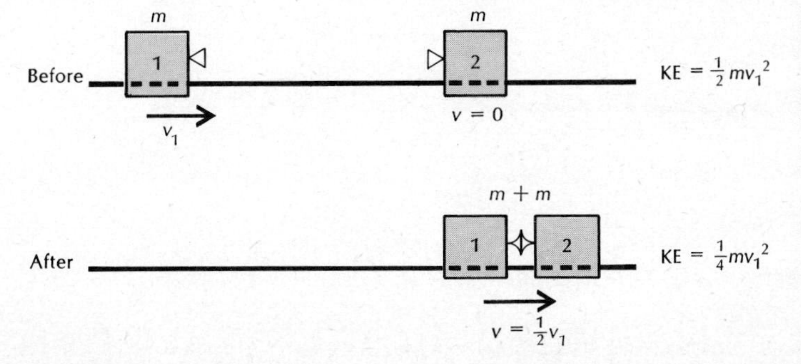

FIGURE 6.12 In a completely inelastic collision between two air-track cars the cars stick together after the collision. The velocity after is one-half the velocity of car 1 before. The kinetic energy after is one-half as great as before. (For unequal masses the factor would not be $\frac{1}{2}$.)

SUMMARY

Work is defined as the force multiplied by the distance through which that force acts. A lever is a device whereby a large force may be applied for a short distance by exerting a smaller force for a larger distance.

Kinetic energy is defined as $\frac{1}{2}$ mass times (velocity)^2. The work done on an object is equal to the change in its kinetic energy.

The joule is a unit of energy. It is defined as a force of one newton acting for a distance of one meter. A watt is a unit of power. One watt is defined as a rate of doing work of one joule/second.

Potential energy is stored energy that can be released at a later time. It can be calculated by calculating the work done to store the energy. Potential energy can be called energy of position. The change in potential energy of objects due to gravity is given by PE $= mgd$, where d is the vertical displacement.

All three forms of energy—work, kinetic energy, and potential energy, are convertible exactly into each other (if friction is ignored).

In collisions energy may or may not be constant. A "bouncy" or elastic collision is one that leaves the kinetic energy unchanged. A "sticky" or inelastic collision leaves some of the energy converted to heat.

QUESTIONS

1. Give an example in which a force acts to decrease the kinetic energy of something.
2. A weight lifter holds a 200-lb weight over his head. Is he doing work? Does he use any energy in doing this?
3. Does the kinetic energy of an object depend on the direction in which it is moving?
4. One object moves twice as fast as a second object, but their kinetic energies are equal. How are their masses related?
5. Which requires more work, running up a flight of stairs or walking up a flight of stairs?
6. Which requires more work, running at 8 miles/hour on a level field or running uphill? Why?
7. How many joules of energy are required to light a 100-watt light bulb for 1 year?
8. Is it true that whenever work is done on an object its energy changes?
9. If I lift an object 10 m vertically, and I lift a second object 10 m vertically and carry it 10 m horizontally, which has the greater potential energy?
10. Does a screen door attached to the door jamb by a spring acquire potential energy when it is opened?

11 Two cars collide in a totally inelastic collision. Since the forces acting between them are equal (they are third-law pairs), they will both have the same amount of work done on them. True or false?

12 Is it correct to conclude that, if no net force acts on an object, we cannot change its energy? Is it correct to conclude that, if no net force acts on an object, we cannot change its kinetic energy?

SELF-TEST

__________ Christian Huygens
__________ work
__________ kinetic energy
__________ joule
__________ power
__________ watt
__________ potential energy
__________ friction
__________ elastic collision
__________ inelastic collision

1 Force retarding motion that leads to a dissipation of energy
2 1 joule per second
3 1 newton force through a distance of 1 m
4 Force times distance
5 The rate of doing work or using energy
6 Energy of motion
7 Energy of position, stored energy
8 A collision in which some kinetic energy is turned to heat
9 A collision in which no kinetic energy is lost
10 Discovered the importance of mv^2

PROBLEMS

1 Calculate the kinetic energy of a 3-kg object moving with a speed of 5 m/second.

2 Calculate the kinetic energy of an electron ($m = 9 \times 10^{-31}$ kg) moving with a speed of 2×10^6 m/second.

3 Calculate the kinetic energy of a proton ($m = 1.65 \times 10^{-27}$ kg) moving with a speed of 3×10^5 m/second.

4 Calculate the kinetic energy of an 0.02-kg bullet moving with a speed of 10^3 m/second.

5 A typical car may have a mass of 2,000 kg: What is the kinetic energy of the car at 60 miles/hour?

6 The earth has a mass of 5.98×10^{24} kg and moves with an orbital speed of 2.98×10^4 m/second. What is the kinetic energy of the earth?

7 (*a*) What is the speed of an electron if its mass is 9.1×10^{-31} kg and the magnitude of its momentum is 1.82×10^{-27} kg-m/second? (*b*) What is the kinetic energy of this electron?

8 Mercury has a mass of 3.4×10^{23} kg and moves in its orbit about the sun with a speed of 4.8×10^4 m/second. What is the kinetic energy of Mercury?

9 The moon has a mass of about 7×10^{22} kg and orbits the earth with a speed of 10^3 m/second. What is the kinetic energy of the moon in its orbit?

10 A baseball, which has a mass of about 250 g, is thrown at a velocity of 90 miles/hour (about 40 m/second). What is the kinetic energy of the ball?

11 (*a*) Calculate the potential-energy change in lifting a 10-lb weight a distance of 7 ft. (*b*) Calculate the potential-energy change in lifting this same object 7 ft on the moon.

12 (*a*) Calculate the potential-energy change in lifting a 10-kg mass a distance of 6 m. (*b*) Calculate the potential-energy change in lifting the same object the same distance on the moon.

13 A 10-kg mass passes my window falling at a rate of 50 m/second. (*a*) Neglecting air resistance, from what height did this object fall? (*b*) What was the potential energy of this object before it began to fall?

14 A rock is hurled upward at an initial velocity of 40 m/second. Ignoring air resistance and using conservation of mechanical energy, find how high the rock will rise.

15 Using the law of conservation of energy, find how high a ball thrown upward with a velocity of 5 m/second will rise.

16 A girl throws a 0.2-kg ball a distance of 6 m straight up in the air. (*a*) What is the potential energy of the ball at its highest point, relative to the point of release? (*b*) What was the kinetic energy of the ball as it left the girl's hand? (*c*) How much work does the girl do in throwing the ball? (*d*) If the girl's arm muscle contracted a distance of 0.5 m while throwing the ball, what was the average force exerted by the muscle?

17 The champion jumper among fleas is the common flea (*Pulex irritans*). In an experiment carried out in 1910, a flea high-jumped a distance of 130 times its own height, a distance of 19.7 cm. The mass of the flea was 0.005 g. Assume the flea jumped straight up. (*a*) What was the kinetic energy of the flea the instant it left the ground? [Ignore air resistance.] (*b*) What work did the flea do in launching itself? (*c*) Assuming the work was done by a force acting a distance of one-half the height of the flea, what was the size of this force?

18 A block rests on a smooth track (no friction) and is released and slides around the track as in the diagram. The block has a mass of 15 kg. (*a*) How far above the level surface does the block rise on the right-hand side of the track before coming to rest? (*b*) At the lowest point on the track, what is the kinetic energy? (*c*) What is the velocity at the lowest point on the track? (*d*) Where is the block moving fastest?

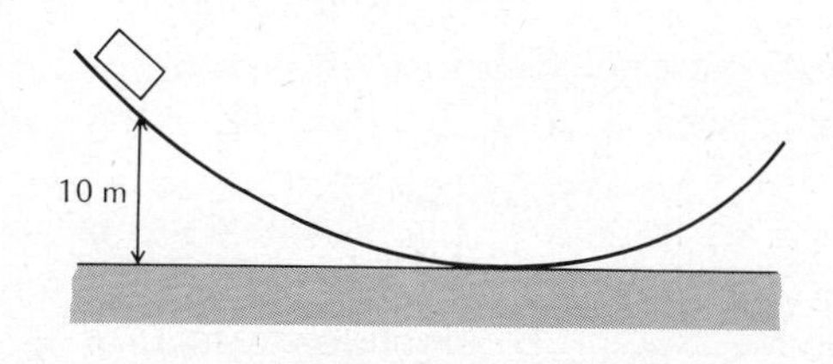

19 A block weighing 6 lb slides down a frictionless chute and then strikes a stationary 10-lb block on a horizontal table. After the collision the blocks stick together. (*a*) What is the velocity of the first block when it reaches the bottom of the chute? (*b*) After the collision what is the velocity of the two blocks?

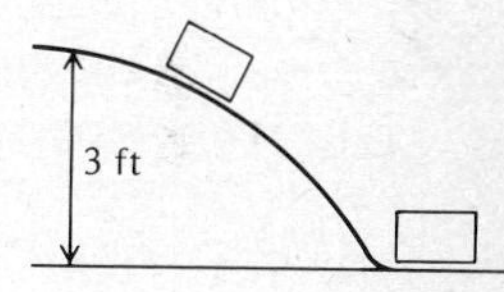

20 A body of mass 5 kg coasts up an incline. As it starts up the incline, its speed is observed to be 4 m/second. At the top of the incline it comes momentarily to rest and then slides back down the incline. When it reaches the bottom of the incline, its speed is seen to be 3 m/second. (*a*) How much mechanical energy is converted to thermal energy during the roundtrip? (*b*) If one-half of this energy conversion occurs during the trip up, how high above the lowest point on the incline does the block rise?

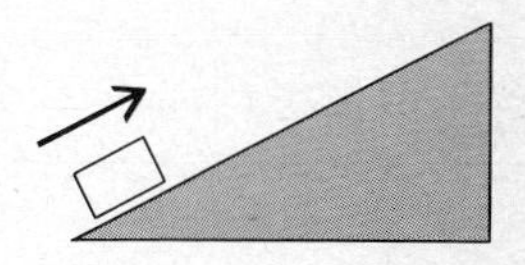

21 A 4-kg block rests at the top of a frictionless inclined plane as shown in the picture. It is propelled downward with an initial velocity of 1 m/second. (*a*) What is its initial kinetic energy? (*b*) What is its initial potential energy relative to the bottom of the plane? (*c*) What is its kinetic energy at the bottom of the plane? (*d*) What is its velocity at the bottom of the plane?

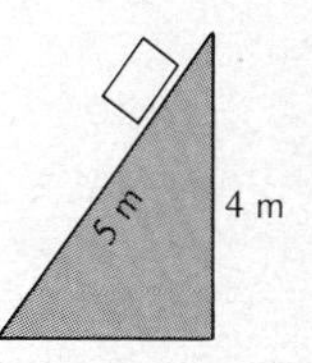

22 A roller coaster of mass 500 kg reached point *A* on its run, moving at a rate of 2 m/second. It moves from point *A* to point *B*, and at point *B* it is moving at a rate of 9 m/second. At point *C* the roller

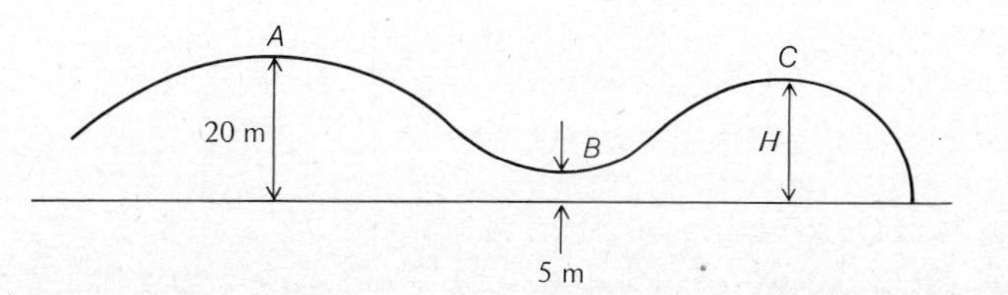

coaster is moving at a speed of 4 m/second. (*a*) How much mechanical energy is dissipated in going from point *A* to point *B*? (*b*) Assume that in going from *B* to *C* one-tenth of the energy dissipated in going from *A* to *B* is dissipated. What is the height of *H*?

23 A 150-kg handcart coasts on a straight level track at 10 m/second. A 75-kg man drops onto the cart from a bridge as the cart passes underneath him. (Neglect friction.) (*a*) Find the new speed of the cart. (*b*) Find the momentum of the cart and the man after the "collision." (*c*) Find the kinetic energy of the cart and the man after the "collision." (*d*) The man jumps sideways off the cart after he has ridden a short distance. What is the speed of the cart after he jumps?

24 A block of mass 5 kg is held in place at the top of an inclined plane as shown in the drawing. The block is released and allowed to slide to the bottom. As the block slides, a total of 15 joules of mechanical energy is converted to thermal energy. Take the zero of potential energy to be the bottom of the plane. (*a*) What is the potential energy of the block before it is released? (*b*) What is the kinetic energy of the block at the bottom of the plane? (*c*) What is the velocity of the block at the bottom of the plane? (*d*) What force would be required to push the block up the plane from the bottom if the same frictional force acts?

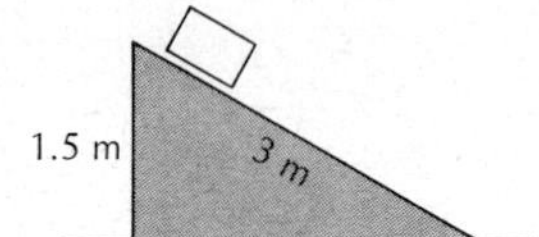

25 A 3-kg mass slides down the incline shown. (*a*) If the incline is frictionless, what is its kinetic energy at the bottom? (*b*) What is its velocity?

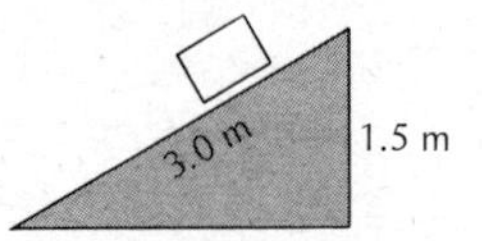

26 (*a*) If the incline in Problem 25 is frictionless, how much work is required to move a 5-kg object up the incline? (*b*) What force must be applied to the object?

27 A railroad car of mass 10^5 kg moving at 10 m/second catches and couples to two other cars, also of mass 10^5 kg, which were moving at 5 m/second. (*a*) What is the speed of the three cars after they couple together? (*b*) What was the total kinetic energy before the collision? (*c*) What is the total kinetic energy after the collision? (*d*) What is the energy lost in the collision?

28 A novice skater, mass 50 kg, traveling with a speed of 3 m/second, overtakes and collides with another skater (mass 80 kg) whose speed is 2 m/second in the same direction. In a frantic effort to avoid falling, the novice clutches the other skater, and they both fall and go sliding across the ice together. (*a*) What is the final velocity of the two skaters? (*b*) What is the initial kinetic energy of the system? (*c*) What is the final kinetic energy? (*d*) What is the change in energy? (*e*) Is this an elastic collision?

29 A 20-lb bomb is seen to be sliding north on the ice at a speed of 10 ft/second. The bomb explodes into two equal pieces, and after the explosion one piece is seen to be at rest on the ice. (*a*) What is the final speed of the other half, and in what direction is it moving? (*b*) What was the initial kinetic energy of the system? (*c*) What is the final kinetic energy of the system? (*d*) What is the change in kinetic energy of the system and to what do you attribute this change?

30 A 2,900-kg truck moving at a speed of 15 m/second collides with a stationary car of mass 1,000 kg in a completely inelastic collision. (*a*) What is the final velocity of the wreckage? (*b*) What was the initial kinetic energy of the system? (*c*) What is the final kinetic energy of the system? (*d*) How much kinetic energy is lost in the collision?

31 Two cars collide in a completely inelastic collision. Car *A* has a mass of 2,000 kg, and car *B*, 1,000 kg. Initially car *B* was at rest, and car *A* had a velocity of 10 m/second. How much kinetic energy is lost in the collision?

32 An automobile of mass 2,000 kg collides with another of mass 1,500 kg. Before the collision the first car had a velocity of 30 m/second, and the second a velocity of 10 m/second. If, after the collision, both cars have a velocity of 20 m/second, how much kinetic energy is lost in the collision?

33 Car *A* whose mass is 2,000 kg collides with a car *B* whose mass is 1,500 kg. Initially, *B* was at rest, and the velocity of *A* was 20 m/second. (*a*) If they stick together after the collision, what is the velocity of the wreckage? (*b*) How much kinetic energy is lost in the collision?

34 A 2-kg block slides on a curved track, as shown in the diagram. It is released from rest, allowed to slide down the track to the bottom and up the other side, where it is found to rise to a height of 2.0 m. (*a*) How much mechanical energy is converted to thermal energy? (*b*) If 60 percent of this change takes place while the block slides to the lowest point on the curve, what is the velocity of the block at the lowest point?

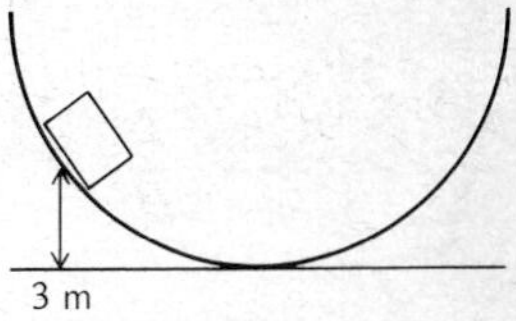

3 m

35 Two cars collide head-on and stick together, and immediately after the collision the wreckage is seen to be at rest. The first car weighs 3,200 lb and before the collision was heading north at 60 miles/hour. The second car was heading south before the collision at a speed of 50 miles/hour. (*a*) What is the weight of the second car? (*b*) What was the total kinetic energy before the collision? (*c*) What is the total kinetic energy after the collision? (*d*) How much kinetic energy is lost and where might it possibly have gone?

36 A truck which weighs 3,200 lb collides head-on with a motorcycle which weighs 640 lb, including its driver. The truck and motorcycle, with its driver, end up stuck together. The truck was originally headed north at a speed of 40 ft/second, and the motorcycle was headed south at a speed of 40 ft/second. (*a*) After the collision, what is the speed of the wreckage? (*b*) What was the initial kinetic energy? (*c*) What is the final kinetic energy? (*d*) How much energy is lost in the collision?

37 A woman throws a 1.0-slug ball upward with an initial velocity of 8 ft/second. (*a*) What is the initial kinetic energy of the ball? (*b*) What is the kinetic energy of the ball at its highest point? (*c*) What is the potential energy of the ball relative to the point at which it was released when it reaches its highest point? (*d*) How high does the ball rise?

TEMPERATURE, HEAT, AND ENERGY

In this chapter we will develop the concepts of temperature and heat and relate them to energy. Then we will discuss the law of energy conservation. We will try to give some idea of the importance and usefulness of this law. In Chapter 8 we will use these ideas to discuss the present "energy crisis."

7.1 TEMPERATURE AND HEAT

We are all familiar with heat and the concept of relative "hotness," and almost everyone is accustomed to assigning a numerical value that we call *temperature* to indicate the hotness of an object. But what is really meant by heat? By temperature? And how are they related?

Consider a rather simple experiment involving two bricks. Leave one outside on a winter day and put the other in the oven for awhile. One brick will be cold, and the other hot. Put the hot brick on top of the cold brick, as in Fig. 7.2*a*. After a short time the two bricks will no longer be hot and cold, but both will be warm, as in Fig. 7.2*b*. That is, the hot one has become cooler than it was before, and the cold one warmer than it was before. The sensation of hotness or coldness corresponds qualitatively to what we call temperature. This experiment allows us to say more precisely what we mean by temperature. Two objects are said to have different temperatures if, when placed together, one becomes warmer and the other colder. Conversely, we say that two objects have the same temperature when there is no change in their hotness or coldness when they are placed together. Two objects in contact with one another which have the same temperature are said to be in *thermal equilibrium*.

We can use the experiment with the bricks to indicate also what is meant by heat. When the two bricks are put together, one becomes

FIGURE 7.1 Our senses are not always the best judge of relative temperatures. Other factors influence our judgment as to which temperature is hotter.

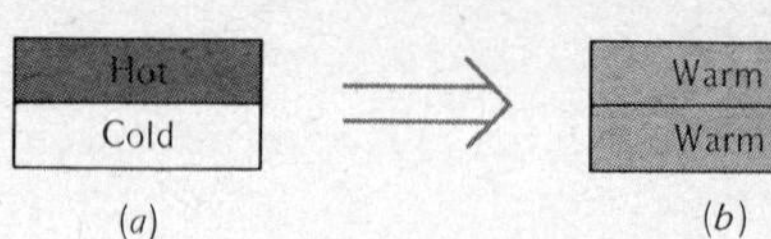

FIGURE 7.2 A hot brick and a cold brick placed together *(a)* will both end up warm *(b)*.

warmer and one cooler. Apparently, something has flowed from one brick to the other. This something is called *heat*. When heat is added to an object, its temperature rises. When heat is removed, its temperature falls. The experiment with the bricks points up an important characteristic of heat. It is never observed to flow from a cooler object to one which is hotter. Heat always flows in such a direction as to equalize the temperatures of two objects in contact.

Another aspect of heat can be shown by a simple argument. Suppose you boil a certain amount of water in a pot on the stove. Clearly this requires the addition of a certain amount of heat. Now suppose it is proposed to boil the Mississippi River. It is clear that this would require more heat than to boil a pot of water. Therefore the magnitude of the temperature changes produced by a given amount of heat is greater the smaller the mass of the object being heated.

TEMPERATURE SCALES

To make these ideas quantitative we need a numerical temperature scale. Then we can define a unit for heat. We will compare here three temperature scales that are in common use. These scales are based on the temperature at which water freezes and boils, since this measurement can be easily repeated in any laboratory.

On the Fahrenheit scale the temperature between the freezing point and the boiling point of water is divided arbitrarily into 180 degrees. Originally, 0° on this scale was chosen as the (then believed) lowest temperature obtainable using a mixture of ice and salt. The temperature of the human body was taken as 100°. This placed the freezing point of water at 32°F and the boiling point at 212°F. Today these latter two are the fixed points used to define the Fahrenheit scale.

The Celsius scale is a more logical one. There are 100 divisions, or degrees, between the freezing point and the boiling point of water. The freezing point of water is chosen to be 0°C (zero degrees Celsius). Therefore the boiling temperature of water is 100°C. The Celsius scale is sometimes also called the centigrade scale.

Both the Celsius and Fahrenheit scales have the defect that they are based on an arbitrary choice of the boiling and freezing points of water to specify the scale. In each, zero has no particular significance. The scale used predominantly in scientific work is called the Kelvin scale. Zero on the Kelvin scale is the true absolute zero of temperature. It can be proved that no lower temperature exists. The unit kelvin (K) is the same as the Celsius degree. Temperatures in kelvins then repre-

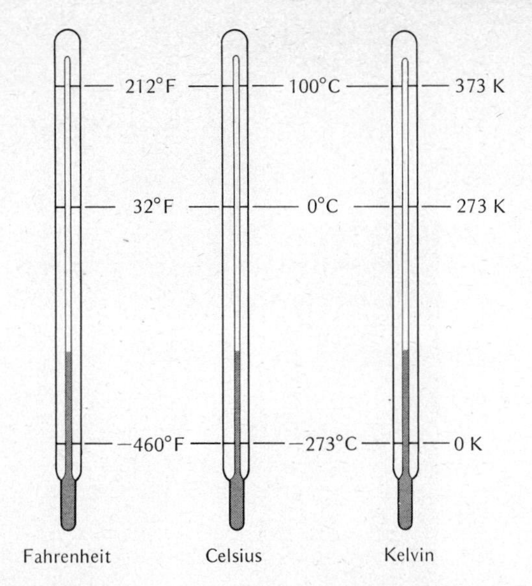

FIGURE 7.3 Comparison of Fahrenheit, Celsius, and Kelvin temperature scales.

sent degrees above absolute zero. Absolute zero is −273°C. Therefore 0 K is −273°C and 0°C is +273 K. 100°C is 373 K. The three temperature scales are shown for comparison in Fig. 7.3.

VERY HIGH AND VERY LOW TEMPERATURES

Scientists are very curious. They are always trying to gain new information. One of the things this means is that they do experiments under new kinds of experimental conditions. They have worked very hard to reach both very high and very low temperatures in the laboratory.

Very low temperatures have most usually been achieved using liquids that boil at low temperatures (Fig. 7.4). These liquids are better known to us as gases at ordinary temperatures—gases such as oxygen, nitrogen, hydrogen, and helium. Liquid nitrogen and liquid oxygen (a mixture of these is liquid air) boil at 77 K (−196°C) and 90 K (−183°C), respectively. Liquid hydrogen boils at 20 K (−253°C). The most difficult gas to liquefy, and the one which boils at the lowest temperature, is helium. It boils at 4.2 K and was first liquefied in Leiden, Holland, in 1908 by Heike Kamerlingh-Ohnes (1853–1926). For a number of years, Leiden was the only place in the world where such low temperatures could be reached.

It is impossible to describe here all the interesting behavior that occurs at very low temperatures. Two examples will be sufficient. Liquid helium at 4.2 K behaves pretty much like an ordinary liquid. But if its temperature is lowered still further, to 2.2 K, it begins to behave rather strangely. Below this temperature, its behavior is so unusual that it is called a *superfluid*. Among other properties of superfluid

FIGURE 7.4 Liquid nitrogen boils when poured on a block of dry ice. *(Courtesy Union Carbide.)*

helium it will flow *up* the wall of a container. A small container of superfluid helium will empty itself in a short time by this type of flow (see Fig. 7.5). Superfluid helium also will pass through holes too small for the liquid in its ordinary state to pass through.

Very shortly after the achievement of liquefying helium, Kamerlingh-Ohnes was studying the electrical properties of the metal mercury as he slowly lowered the temperature. He found that mercury's normal behavior was completely changed at low temperatures. Mercury became a *superconductor*. We will come back to this subject later in Chapter 10.

Interest in very high temperatures has a very practical basis. The temperature at the center of the sun is estimated to be about 20–40 millions of kelvins. The source of the sun's energy is thermonuclear fusion. If this process is ever to be used as a practical source of energy, we must learn to confine gases at such high temperatures. This field is called plasma physics. We will discuss this further in Chapter 21.

It is interesting to note that the practical use of thermonuclear

FIGURE 7.5 This photograph shows a glass flask with double walls, containing liquid helium. You can see the level of the helium about 3 centimeters from the bottom of the photograph. The small tube suspended in the flask also contains liquid helium that has climbed up the wall and is dripping off the bottom back into the flask. *(Courtesy Dr. Howard O. McMahon, A. P. Little, Inc.)*

fusion, which depends on very high temperatures, may depend on magnets made from superconductors, which are the result of research at very low temperatures.

7.2 UNITS OF HEAT

To define a unit of heat, we choose the amount of heat which will raise the temperature of a standard substance a certain amount. The substance chosen is water, because of its universal availability. The unit of heat is named the *calorie*, and is defined as follows. The calorie is the amount of heat necessary to raise one gram of water one degree Celsius. More precisely, it is the amount of heat needed to raise the temperature of 1 g of water from 15 to 16°C (from 288 to 289 K).

The amount of heat needed to change the temperature of water 1°C varies slightly as the temperature changes; so the precise definition of the calorie must include the temperature range 15 to 16°C. For our purposes we can neglect this small difference. The dieter's calorie is not the unit defined above, but is 1,000 calories, and is called the kilocalorie.

Another unit of heat that you may encounter in your newspaper is the British thermal unit (Btu). It is the amount of heat required to raise one pound of water one degree Fahrenheit. One Btu is about 250 calories. We mention it here because it is so commonly used in energy discussions in books and in the press.

Example 7.1

How much heat is needed to raise 100 grams of water 2°C?

From the definition given above, it takes 2 calories to raise each gram 2°C, and therefore 200 calories is required in all.

Example 7.2

Three calories of heat is added to 600 grams of water. How large a temperature change occurs?

These 3 calories must be shared among the 600 grams of water. Each gram of water then takes $\frac{3}{600} = \frac{1}{200}$ calorie. Therefore, the temperature rise is $\frac{1}{200}$ of a degree, or 0.005°C.

Now it is possible to clarify the distinction between heat and temperature. Temperature does not depend on the size or composition of an object. Heat flows from an object at a higher temperature to one at a lower temperature. The amount of heat that flows depends on both the size of the object and the material of which it is made. A large object requires more heat for a given temperature change than a small one of the same material. But if a large object and a small one are at the same temperature, no heat will flow between them when they are brought in contact with one another.

If it takes 1 calorie of heat to raise 1 gram of water 1°C, how much heat does it take to raise the temperature of other substances? For most materials it takes less than for water. The amount of heat required to raise the temperature of 1 gram of any material 1°C is called the *specific heat* or *heat capacity* of the material.

$$\text{Specific heat} = \frac{\text{heat absorbed per gram (calories)}}{\text{increase in temperature (°C)}} \qquad (7.1)$$

Table 7.1 lists the specific heats for a few common substances at room

TABLE 7.1 Values of specific heat for some common substances

Substance	Specific heat, calories/gram · °C
Water	1.0
Glycerin	0.6
Ice	0.5
Aluminum	0.22
Glass	0.16
Iron	0.10
Copper	0.09
Brass	0.09
Silver	0.06
Lead	0.03

temperature. We see from the table that the specific heat of copper is 0.09 calorie/gram · °C. This is read "calories per gram per degree Celsius." This means that 0.09 calorie will produce an increase of 1°C in the temperature of 1 gram of copper. Therefore 1 calorie of heat will raise the temperature of 1 gram of copper by

$$\frac{1.0 \text{ calorie/gram}}{0.09 \text{ calorie/gram} \cdot {}^{\circ}\text{C}} = 11^{\circ}\text{C}$$

rather than 1°C, as is the case for water.

One way of seeing the effect of the differences in specific heats of materials is to consider the flow of heat between two dissimilar materials. If 100 grams of copper at 100°C is placed in 100 grams of water at 0°C, we expect that the water will be warmed somewhat and the copper cooled. But by how much? The essential physics of the situation is that all the heat which goes to warm up the water must come from the copper (if care has been taken to eliminate heat from the outside environment). The temperature change in the water is due to the addition of a certain amount of heat; the temperature change in the copper is due to the loss of exactly the same amount of heat. Since water requires more heat for a 1°C temperature change than copper, the temperature of the water will change less than the temperature of the copper. The actual final temperature for the experiment is 8.3°C. The water is warmed 8.3°C; the copper is cooled 91.7°C (from 100 to 8.3°C). This results because water requires 11 times as much heat for a given temperature change as copper. The temperature change of the copper will be 11 times greater than the water, since we have used the same mass for each.

7.3 THE MECHANICAL EQUIVALENT OF HEAT

By the late 1700s a theory concerning the nature of heat existed which explained most of the observed facts and was accepted by most scientists of the time. The essential idea of this theory was that there exists a massless fluid, called caloric, or caloric fluid, and that the temperature of any object depends on the amount of caloric it contains. Caloric was presumed to exist in matter in two states, called sensible and latent. Only the sensible caloric showed itself as an increase in temperature. It was presumed possible to "squeeze out" the caloric from latent to sensible by mechanical pounding, or work. In this way the effects of friction, which always led to an increase in temperature, were accounted for.

Benjamin Thompson (1753–1814), later Count Rumford, made an observation which he was able to show contradicted the caloric theory, and paved the way for a consistent theory of heat. In 1798 Rumford was Minister of War in Bavaria and was involved in overseeing the boring of brass cannons. In the process he became aware of the truly prodigious amounts of heat generated in this process. The caloric theory of heat accounted for this fact, using the idea that latent caloric was squeezed out of the chips in the boring process and thus heated the cannon. Rumford disproved this idea by using a dull drill which produced very few chips but, if anything, more heat. Furthermore, heat was released as long as the horse who turned the apparatus could be persuaded to move. That is, the heat generated by the friction of the drill and the cannon was apparently inexhaustible. The caloric theory never postulated an infinite supply of caloric fluid, so this raised grave questions about the validity of that theory. Rumford's observation can be summarized briefly:

FIGURE 7.6 Benjamin Thompson, Count Rumford (1753–1814). *(Courtesy University of Pennsylvania Library.)*

> *Through friction, it is possible to create endless amounts of heat at the expense of mechanical work.*

Rumford did not turn his ideas into a quantitative or numerical theory. But he did relate them to an idea of Descartes, that heat is the motion of the atoms of a substance. The inference is that the mechanical work done does not show up as kinetic energy of the entire object, but rather as random kinetic energy of the motion of individual atoms. This is what we call heat. Rumford's work weakened the caloric theory and paved the way for its later demise. A fascinating biography of Rumford by Sanborn Brown is listed in the Selected Reading.

THE WORK OF JOULE

In 1843 an Englishman, James Joule (1818–1889), published the first results in a long series of experiments designed to measure the quantity of heat produced by a certain amount of mechanical work. This first experiment related the amount of mechanical work needed to generate an electric current to the amount of heat produced by the current.

FIGURE 7.7 James Joule (1818–1889). *(Courtesy British Information Services.)*

Later experiments were performed on the direct mechanical con-

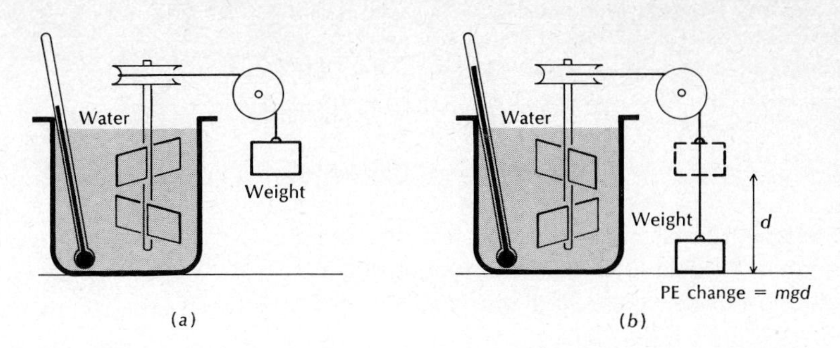

FIGURE 7.8 A schematic diagram of Joule's experiment. The weight falls a distance *d*, giving a potential-energy change of *mgd*. The potential energy lost is converted to the work of stirring the water. This work is converted to heat.

version of work into heat. One of the most famous of these involved the heating of a bucket of water with a paddle-wheel apparatus. Figure 7.8 is a schematic drawing of this device. The amount of work delivered to the paddle wheel is measured by the distance fallen by the weight. The potential-energy loss of the weight goes into work to turn the paddle wheel. Remember that work, potential energy, and kinetic energy are all convertible one into another. The increase in temperature of the water is a measure of the amount of heat generated by the paddles in the water. Joule's result was that 4.2 joules of work caused the same increase in the temperature of the water as the addition of 1 calorie of heat. An enormous number of different experiments in the years since then has led to the result that mechanical work or energy, when converted to heat, *always* gives

$$1 \text{ calorie} = 4.185 \text{ joules}$$

and

$$1 \text{ food calorie} = 1 \text{ kilocalorie} = 4{,}185 \text{ joules}$$

This relationship is called the mechanical equivalent of heat, or the joule equivalent. We will use the approximate value for this number of 1 calorie = 4.2 joules.

The result given above for the joule equivalent is found to be universal. Any time work, kinetic energy, or potential energy is dissipated as heat, the same conversion is found between energy and heat. Joule was sufficiently convinced of the universality of this idea that he took a thermometer to the Swiss Alps on his honeymoon. The idea was to measure the temperature difference between the top and the bottom of a waterfall. He reasoned that some of the kinetic energy of the falling water should be converted to heat as it struck the rocks at the bottom.

We can estimate the magnitude of the effect Joule expected to find. In a waterfall 100 meters high, each gram of water will lose an amount of potential energy given by PE = *mgd*.

$$\text{PE} = (10^{-3} \text{ kilogram})(9.8 \text{ meters/second}^2)(100 \text{ meters})$$

$$\text{PE} = 0.98 \text{ joule for each gram of water}$$

From the mechanical equivalent of heat, 0.98 joule is about 0.23 calorie.

If all the kinetic energy of falling is dissipated at the bottom of the waterfall as heat, a temperature rise of about 0.23°C is to be expected. Joule was able to verify the essential correctness of this proposal during his trip to Switzerland.

7.4 THE ENERGY PRINCIPLE

The work of Joule, and that of Rumford before him, suggests a very useful generalization. The idea was stated first by a German physician, Julius Mayer (1814–1878), and independently by Joule soon afterward. We have already seen how work, kinetic energy, and potential energy can be converted back and forth among one another. Now it seems that all these forms of energy can be converted to heat, and the amount of heat released always bears a constant relation to the amount of work or kinetic energy used. Heat, then, appears to be a form of energy. If we include heat as a form of energy, it seems that energy is never created or destroyed, but is merely changed from one form to another. As time has passed, this principle has withstood all experimental tests and now is accepted as a law of nature:

> *Energy is never created or destroyed. It is merely converted from one form to another.*

This principle is known as the *conservation-of-energy principle.*

It should be noted that, although Joule's work dealt with the conversion of kinetic or potential energy into heat, the reverse is also possible. Steam engines, gasoline engines, turbines, and rockets all convert heat energy into some form of useful work or kinetic energy. The same rate of exchange, 4.2 joules/calorie, is observed in every instance where heat is converted to work.

As the matter has developed, it turns out that energy appears in many forms: kinetic energy, potential energy, work, heat, chemical energy, electric energy, etc. But if *all* the energy in a system is accounted for, the conservation-of-energy principle is always found to be valid. The entire electric-power industry depends on generators in which the mechanical energy of water falling over a dam, or heat energy of steam, is converted into electric energy which is then converted back to useful work by machines at the other end of the transmission lines. Electric transmission lines thus transport energy from source to user.

PERPETUAL MOTION

If energy can never be created or destroyed, then it should be impossible to build a machine which will do useful work with no input of energy in any form. Such a machine is called a perpetual-motion machine. There

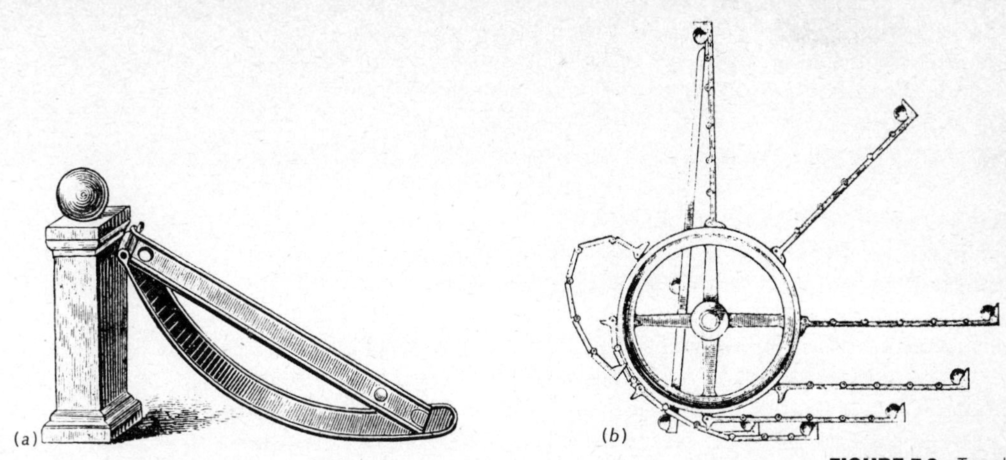

FIGURE 7.9 Two devices intended as perpetual-motion machines. In *(a)* the round lodestone at the top was expected to draw the steel ball up the straight ramp. It would then fall through the hole and go down the curved ramp. In *(b)* the hinged arms and rolling balls were expected to cause the wheel to turn clockwise. *(Courtesy Scientific American.)*

have been many attempts to make one during the years (including some famous hoaxes). All fail because of the energy principle.

7.5 HEAT ENGINES AND THE LAWS OF THERMODYNAMICS

In the discussion of energy conservation we pointed out that all forms of energy are fully convertible into all other forms, and that the rate of exchange is always 4.2 joules for 1 calorie. A large part of our industrialized society obtains energy from heat. We burn gasoline in our automobiles to be converted to kinetic energy. We burn coal, oil, and gas to produce steam to run electric generators in power plants, The conversion of heat energy to kinetic energy or other forms of useful work has been the subject of much study. A machine which performs this conversion is called a heat engine. The scientific discipline involved is called *thermodynamics*. It turns out that there are some rules here that are not at all obvious, and that have major implications for man's utilization of energy.

A heat engine, in schematic form, takes in heat from some source, such as the burning of fuel. A certain amount of useful work is obtained. As we will discuss further, not all the heat can be turned into work. Some of it must be rejected. In a car this is done by way of the radiator and cooling system. Figure 7.10 shows this general idea.

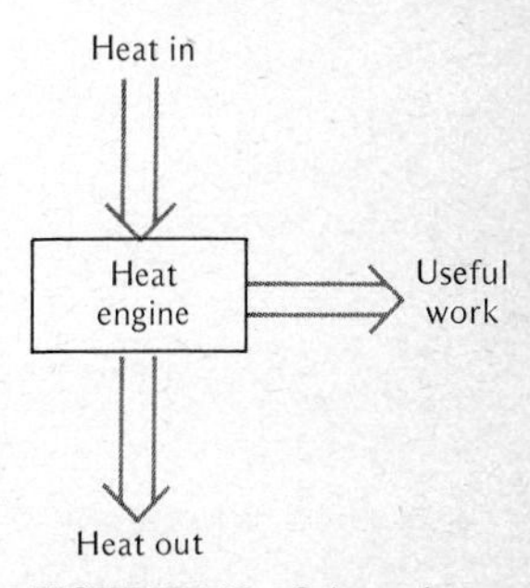

FIGURE 7.10 Schematic diagram of a heat engine.

The amount of work that can be obtained from such an engine is, of course, limited by conservation of energy. The maximum work can be no more than the difference between the heat in and the heat out. All practical engines achieve much less than this, because of both losses to friction and other inefficiencies in their operation. In the field of heat engines, the conservation-of-energy principle is called the *first law of thermodynamics*.

A principle called the *second law of thermodynamics* gives the greatest limitation to what can be achieved by heat engines. In this discussion we will talk about an ideal heat engine. The ideal engine achieves the theoretical maximum of useful work for a given set of conditions. Such an ideal engine is also called a Carnot engine, for the French engineer who first studied its properties. The second law of thermodynamics sets the limits on what can be achieved by the ideal engine. All practical engines fall short of this ideal.

A heat engine takes in energy in the form of heat, converts some to useful work, and then rejects the rest to the outside world. The fraction of this heat that can be converted to work is related to the temperature. Carnot showed, and all subsequent experiments have verified, that to do work an engine must take in heat at one temperature T_{hot} and reject heat at some lower temperature T_{cold}. The fraction of the heat input at the higher temperature that can be turned into useful work is called the efficiency, and is given by

$$\text{Efficiency} = \frac{T_{hot} - T_{cold}}{T_{hot}} \tag{7.2}$$

Here the temperatures *must* be in kelvins. Notice that this says that the only way the efficiency could be 1.0, or 100 percent, meaning that all the heat is converted to work, would be if the lower temperature T_{cold} were absolute zero. (This is one way of defining absolute zero.)

The second law of thermodynamics can be stated in several ways. For example,

> *For a heat engine operating between two temperatures, T_{hot} and T_{cold}, it is impossible to take in energy at T_{hot} and convert it to work without rejecting some heat at T_{cold}.*

In our experiment with the two bricks, we agreed that heat never runs uphill. That idea is a direct consequence of the second law. The cost of producing useful work by an engine is the cost of the fuel which produces the heat energy. The cost of fuel is pretty well determined by how much heat energy it can release. But if we use this heat energy in a heat engine, only a rather small fraction will be turned into useful work. All the rest will be rejected as heat that cannot be usefully used as mechanical work. This heat could be used to heat a building, but at present this is rarely done.

Example 7.3

Suppose we run a steam engine or turbine using steam at 120°C or 393 K, and dump the waste heat at some normal temperature such as 20°C or 293 K. What is the efficiency if the engine is ideal?

We have stated that efficiency can be calculated using Equation 7.2:

$$\text{Efficiency} = \frac{393\text{ K} - 293\text{ K}}{393\text{ K}} \cong 0.25 \text{ or } 25\%$$

Only 25 percent of the heat energy in the steam can be converted to useful work in an ideal engine operating between these two temperatures. Practical engines will seldom do better than 80 percent of this amount.

A partial solution to this dilemma is to raise the temperature at which the heat energy is put in. For steam engines and turbines this means going to high-pressure steam at 300 to 400°C or higher. Still only about 50 percent of the energy content of the fuel can be obtained as mechanical work.

It should be emphasized at this point that the difficulty we are discussing here is not a practical difficulty that can be overcome by sufficient engineering skill. The limitations on heat engine efficiency posed by the second law are limitations in principle. To defeat them would mean defeating a law of nature. Even if we were able to build an engine 100 percent perfect in terms of friction, and any other losses, the efficiency would still be limited by Equation 7.2.

One interesting proposal for generating electricity has been made based on these ideas. It is suggested that a heat engine be operated using heat from the waters of the Gulf Stream as the high temperature, and cold water 500 to 1,000 feet down as the low-temperature side. It might appear that this would be very inefficient, since the temperature difference T_{hot} minus T_{cold} would be very small. But in this case the energy at T_{hot} is free, so the only penalty is the large size of the apparatus needed. Serious consideration is being given to this proposal.

REFRIGERATORS AND HEAT PUMPS

A refrigerator is a heat engine of a somewhat different kind. Here, work is put into the engine in the form of electric energy to run the compressor. The refrigerator moves heat from a low temperature to a high temperature. That is, it moves heat from inside to outside the food storage space.

Another version of the principle is the *heat pump*. This device moves heat from a cold temperature to a warmer temperature. Its purpose is to heat a house at the higher temperature by pumping heat *in* from outdoors. It is interesting to calculate the efficiency of a heat pump. The efficiency here is given by

$$\text{Efficiency} = \frac{\text{heat moved}}{\text{work needed to move it}}$$

Thermodynamics tells us to calculate this as

$$\text{Efficiency} = \frac{T_{\text{hot}}}{T_{\text{hot}} - T_{\text{cold}}}$$

If T_{hot} is room temperature, about 300 K, and T_{cold} is 2°C (275 K), the efficiency is calculated as 1,200 percent! This means that the amount of heat energy that can be delivered from outside to inside is 12 times the electric energy needed to do it! Only the high cost of heat pump installation and the low cost of fossil fuel and other forms of energy have prevented heat pumps from being used more commonly for home heating. Practical heat pumps have efficiencies at present of only 200 to 400 percent.

Everything we have said about heat engines, and heat pumps is based on the laws of thermodynamics. A casual, but not altogether misleading statement of these laws can be made as follows:

First law: *You can't win.*
Second law: *You can't even break even.*

The third law, which relates to difficulties that occur as absolute zero is approached, can be stated in the same way:

Third law: *You can't get there from here.*

Frequently experimental scientists, frustrated by equipment difficulties, quote a "fourth law":

If something can go wrong, it will!

SUMMARY

Two objects have the same temperature if, when they are placed together, there is no change in the temperature of either, that is, if no heat flows. This is called *thermal equilibrium*.

The addition of heat will increase the temperature of an object. Heat is *never* observed to flow from a cold object to a warmer object.

The calorie is defined as the amount of heat necessary to raise the temperature of one gram of water one degree Celsius. The number of calories needed to raise the temperature of 1 gram of any substance 1°C is called its *specific heat.* When two objects come into thermal equilibrium by being placed in contact, the amount of heat gained by one is exactly the same as the amount of heat lost by the other.

Heat can always be created at the expense of mechanical work. The mechanical equivalent of heat, obtained by Joule, is

$$4.185 \text{ joules} = 1 \text{ calorie}$$

or, approximately

$$4.2 \text{ joules} = 1 \text{ calorie}$$

The law of conservation of energy states: Energy is never created or destroyed. It is merely converted from one form to another. The first law of thermodynamics is a statement of energy conservation.

The second law of thermodynamics places restrictions on the amount of work that can be obtained from an ideal heat engine. It is a generalization of the idea that "heat does not run uphill."

The efficiency of an ideal engine is calculated from

$$\text{Efficiency} = \frac{T_{\text{hot}} - T_{\text{cold}}}{T_{\text{hot}}}$$

SELECTED READING

Brown, Sanborn: "Count Rumford, Physicist Extraordinary," Science Study Series S28, Doubleday and Co., Garden City, New York, 1962. A biography of a complex personality who was far more than a scientist.

QUESTIONS

1 Can an unlimited amount of heat be removed from any object?
2 Does the amount of heat in an object depend on its size?
3 Would you expect the behavior of physical systems to be more or less complex at extremely low temperatures than at higher temperatures?
4 When heat is added to something its temperature increases and when heat is removed its temperature decreases. Therefore, are heat and temperature the same thing?
5 Imagine someone claimed to invent a material that had a negative specific heat. Describe how its temperature would change if you added heat to it.
6 If 1 g of water and 1 g of aluminum, both at the same high temperature, are placed in a refrigerator, which would you expect to cool faster?
7 At what temperature, in °C, does liquid helium boil?

8 Is there a highest temperature, just as there is a lowest temperature? Can you explain the answer?

9 Why is 0 K called "absolute zero?"

10 If the temperature of an object changes by 1°C, by how many °F does it change?

11 When two objects achieve the same temperature have they both had equal amounts of heat added to them?

12 Does heat ever flow in such a way as to cool an object?

13 Steam engines for trains were not very efficient in terms of energy usage. Can you explain why?

SELF-TEST

__________ temperature
__________ heat
__________ thermal equilibrium
__________ Celsius scale
__________ kelvins
__________ Fahrenheit scale
__________ superconductor
__________ superfluid
__________ calorie
__________ Btu
__________ specific heat
__________ caloric fluid
__________ Rumford
__________ Carnot
__________ Joule
__________ Mayer
__________ joule equivalent
__________ conservation of energy
__________ thermodynamics
__________ heat engine
__________ heat pump
__________ first law of thermodynamics
__________ second law of thermodynamics

1 Studied heat engines

2 Discovered unlimited amounts of heat can be obtained from mechanical work

3 A machine which takes in heat and produces mechanical work and gives off heat
4 Heat needed to raise 1 lb of water 1°F
5 A material that becomes a perfect electric conductor at low temperatures
6 The state between two objects where no heat flows back and forth
7 First stated the conservation-of-energy principle
8 1 calorie = 4.2 joules
9 Energy is never created or destroyed
10 Heat needed to raise 1 g of water 1°C
11 Temperature scale based on absolute zero of −273°C
12 Peculiar flow property of liquid helium at low temperatures
13 Measured the relationship between mechanical work and heat
14 Temperature scale based on water freezing at 32° and boiling at 212°
15 Amount of heat to raise 1 g of a substance 1°C
16 Another statement of energy conservation
17 The science of heat, temperature, and heat engines
18 Another form of energy
19 Sets limits on the usefulness of heat engines
20 Determines the flow of heat (independent of size or composition)
21 Scale based on water freezing at 0° and boiling at 100°
22 The so-called "fluid of heat"
23 A machine that moves heat at the expense of mechanical work.

PROBLEMS

1 How many calories are required to raise the temperature of 100 g of aluminum from 10 to 30°C?
2 How many calories are required to raise the temperature of 1,000 g of silver from 0 to 100°C?
3 1 kg of lead is cooled from 100 to 80°C. How much heat is removed?
4 120 calories of heat is added to 100 g of an unknown substance. The temperature of this substance is found to rise 4°C. What is its specific heat?
5 The addition of 2000 calories of heat to 100 grams of steam (water vapor) causes its temperature to rise from 100 to 140°C. What is the specific heat of steam?
6 A 100-kg object moving at 10 m/second strikes a stationary object of the same mass. If the collision is totally inelastic, calculate the energy lost as heat, in joules and calories.
7 An object slides down the incline from *A* to *B*. It loses 2.5 joules of heat. At the bottom its kinetic energy is 7.3 joules. What is its mass?

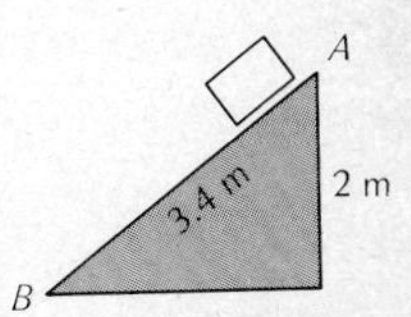

8 An object slides down the incline from *A* to *B*. Its mass is 1 g. If 9.0×10^{-3} joule is dissipated as heat, what is its kinetic energy at the bottom?

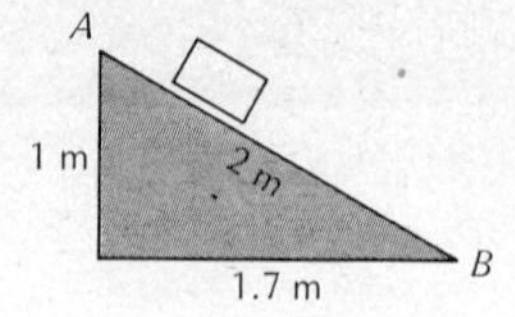

9 What is the efficiency of a heat engine operating between 27 and 7°C, as might occur in a system using the Gulf Stream?

10 What is the efficiency of a heat engine operating between 450 and 300 K? If 4×10^{8} joules of energy is put into this machine, how much useful work can be obtained? How much heat is rejected?

11 A 1-kg mass is pushed up the incline as shown. (*a*) If the incline is frictionless, how much work (in joules) is required? (*b*) The object is released at *B* and slides down to *A*. If 1 joule of energy is dissipated as heat during the sliding, what is the velocity of the object at *A*?

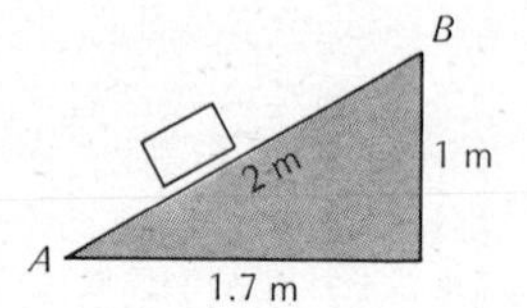

12 A steam turbine electric generator is running at 40% efficiency. If heat is rejected at 27°C (300 K), what must the input steam temperature be (neglecting all frictional and other energy losses)? (Express in °C and K.)

13 A force of 10 newtons is directed uphill as shown. (*a*) If 16 joules of heat is dissipated in moving the object from bottom to top, what is the potential energy of the object at the top? (Its velocity at the top is zero.) (*b*) What is the mass of the object?

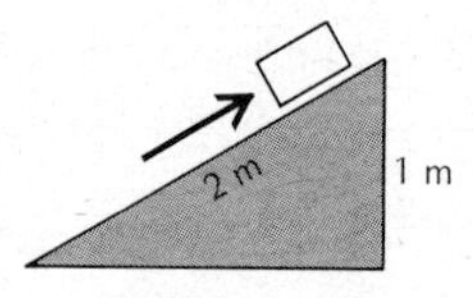

14 A 4,000-kg car strikes a 1,000-kg car in a completely inelastic collision. Initially, the 4,000-kg car was traveling 30 m/second, and the 1,000-kg car 10 m/second, both toward the north. (*a*) Calculate the final velocity of the wreckage. (*b*) Calculate the amount of kinetic energy lost. (*c*) Convert this to calories.

15 A stone of mass 2 kg is dropped 2 m into 10,000 cm^3 of water at 0°C. What is the final temperature of the water? (Ignore the heat capacity of the stone.)

16 A 3-kg rock drops 15 m into 3×10^{4} g of water. If the water is initially at 0°C, what is the final temperature? (Ignore the heat capacity of the rock.)

17 A 10-kg mass drops 10 cm into 100 g of water at 0°C. If all the energy goes into heating the water, what is its final temperature?

18 A 100-g rock falls 100 m into 2,000 g of water at 0°C. Calculate the temperature rise of the water if no heat is lost. (Neglect the heat capacity of the rock.)

19 A 1,000-g mass of lead, whose temperature is 50°C, is placed in 1,000 g of water at 0°C. What is the final temperature of the combination?

20 A 1,000-g mass of copper at 0°C is placed in 1,000 grams of water at 100°C. What is the final temperature of the combination?

In this chapter we will try to show how the principles developed in the previous two chapters apply to man's use of energy. It should be clear that the energy problem is not merely a technological problem. Many difficult political decisions must be made. If these decisions are to be made correctly, it requires that people be adequately informed. It is particularly necessary to be well informed, because so many different points of view are being presented. Many experts and pseudo-experts are proposing solutions and assigning blame for the energy problem. The temptation to seek easy solutions is very great. The only defense is as much factual material and understanding on the part of the electorate as possible. This chapter is only a small beginning.

THE ENERGY PROBLEM

8.1 THE SOURCES OF ENERGY

The United States, with 6 percent of the world's population, is responsible for about one-third of the present world consumption of energy. If every citizen of the world were to have our standard of living, as measured by energy consumption, energy use would increase by a factor of 16. Most of our sources of energy are finite, nonrenewable resources, so that vastly increased consumption cannot long be sustained.

In the United States today over 90 percent of our energy comes from what are called *fossil fuels*. They are coal, petroleum, and natural gas. Figure 8.1 shows the distribution among various sources of United States energy use. Coal, oil, and natural gas represent the end product of the decay in the absence of oxygen of organic plant matter. This has occurred over millions of years. We have learned that energy is never created nor destroyed. Where, then, did this energy in plant matter come from? The answer is the sun.

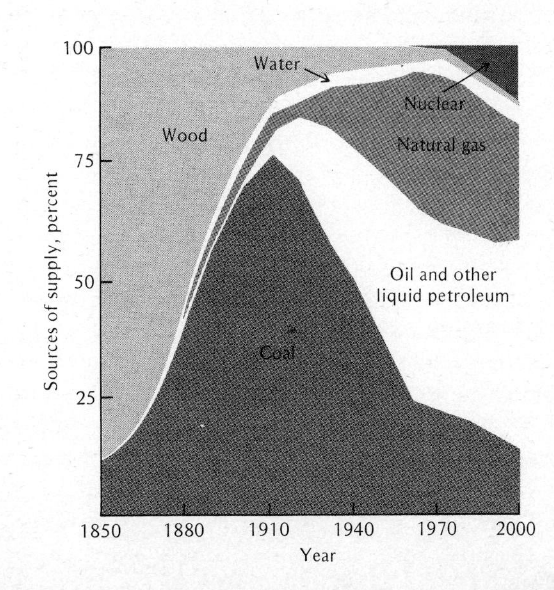

FIGURE 8.1 The importance of various sources of energy for the United States at different dates in our history. *(From A. J. Priest, "Problems of Our Physical Environment," Addison-Wesley, Reading, Mass., 1973. By permission of the publisher.)*

In the process of photosynthesis, energy from the sun is used in the creation of complex organic molecules. During this process, free oxygen is given off by the plant. If plant matter is burned, it is recombined with the oxygen, and energy is released. When we burn wood, this is what is occurring. But if the organic matter decays in the *absence* of oxygen, such as at the bottom of a sea, lake, or swamp, this energy is not released but is stored in the chemical molecules produced. So today, when we burn coal, oil, or gas, we release the energy originally received from the sun and stored millions of years ago. We are burning these fuels at a vastly greater rate than they are being replenished.

At present, less than 5 percent of United States energy comes from water or hydroelectric power. This energy, too, has its source in the sun. Heat from the sun evaporates water from the oceans. Heat from the sun drives the atmospheric circulation that brings this vapor high into the atmosphere and over land where it falls as rain. We obtain water power from the kinetic energy of this water as it moves to the sea in rivers. Great dams are built to store the water and channel it through the turbines. Most of the good hydroelectric power sources in the United States are already in use, so that little additional energy from this source can be expected. (Little in comparison to the total demand.) When we consider proposals for dams in scenic locations such as the Grand Canyon, we must weigh the loss of scenic and recreational values against the contribution of additional energy. No program of dam building will come close to "solving" the energy crunch. For example, the Glen Canyon dam on the Colorado River has a maximum generating capability of 1,035 megawatts (1.035×10^6 kilowatts), but its average output is more like 600 megawatts. (The limitation is the amount of water flowing in the river.) In 1970 the growth rate of electric usage in the United States was 15,500 megawatts per year. So in a period of approximately two weeks the capacity of this dam was completely used up by the growth in consumption of electric energy.

Today there are many proposals for the direct use of solar energy. Some of them seem practical, and others are almost in the realm of science fiction. The primary barrier is that most cost significantly more than the sources of energy we now use. The simplest scheme is to use solar heat to heat hot water, and to use this hot water to heat homes. Next more complex would be to use large collectors to focus sunlight, and to use the heat to run power plants. But 10 square kilometers of collector are needed for a 1,000-megawatt power plant. It is estimated that an area one-half the size of the state of New Mexico would supply the United States power needs through the year 2,000. The direct conversion of sunlight to electricity is presently used in the solar batteries employed in the space program. If these devices could be made cheaply enough, they could be used to generate significant amounts of electric energy. It has even been proposed that a large array of such solar cells

FIGURE 8.2 The Glen Canyon dam on the Colorado River. This dam can produce approximately 1,000 megawatts of electric power. Lake Powell, behind the dam, is hundreds of miles long and is heavily used for boating recreation. On the other hand, one of the most beautiful canyons on the river has been lost forever. Evaporation from the lake increases the salinity of the river and in as little as one or two centuries, silt will fill the lake and make the dam unusable for power generation. Is too high a price being paid for energy from such dams? *(Courtesy of Bureau of Reclamation.)*

be put in orbit above the earth, and the energy beamed to earth as microwaves (radar).

Our second major source of energy is of recent discovery: the energy available from atomic nuclei. Nuclear processes will be discussed in more detail in Chapter 21. For the moment let us merely say that this is energy which was stored in nuclei at some time early in the development of the universe. It can be released in two ways. The first, and the only one in practical use at this time, is called *fission*. In this process nuclei of heavy atoms such as uranium (element 92) are broken apart into nuclei of lighter atoms with the release of very significant amounts of energy. The controlled release of this energy occurs as heat in nuclear reactors, as compared to its uncontrolled release in bombs. This heat is used to make steam, which is used to generate electricity using steam turbines. Unless particular reactors, called breeders, can be developed as practical devices, the supply of usable nuclear fuel may soon be exhausted.

The second way energy from nuclei can be released is called *thermonuclear fusion*. In this process hydrogen, or other light nuclei, are fused together to form heavier nuclei, again with the release of energy. This process requires temperatures in excess of 10 million kelvins to make it work. It has only been achieved in bombs, where fission bombs are used to achieve the high temperatures necessary. Fusion occurs naturally in the sun. The controlled release of fusion energy still seems far away although, if it can be achieved, there exist essentially inexhaustible amounts of fuel.

There are only two other sources of energy available to man. Neither is being exploited in a significant way at this time. The first is the kinetic energy of the earth-moon-sun system. Using the energy in the tides would tap this source. It appears that there are only a few places on earth where the tides are sufficiently high for this scheme to be practical. The second of these sources is geothermal energy. Here one would take advantage of the hot interior of the earth to provide steam for electric generation or home heating. Limited use of this source now occurs in Iceland, California, and Italy.

All the many different kinds of energy technology that one hears discussed involve using energy from one of these four sources. The differences lie in the manner and efficiency of the utilization of energy by the various processes. Some of the technology we hear discussed involves energy storage or transportation, rather than new sources. The use of hydrogen as a universal fuel is an example. Hydrogen would be made from sea water, using electric energy. It would be transported elsewhere and then used as a fuel. It may well be a better fuel than gasoline, but it is not a source of energy, since we would have to expend as much energy to make it as we would obtain from it.

8.2 THE CURIOUS CASE OF THE EXPONENTIAL

At this point we make a short digression into mathematics. We begin with the old story of the Persian nobleman who wished to reward his slave, who had invented the game of chess. The man requested the rather modest-appearing reward of one grain of wheat for the first square on the chessboard, two grains for the second, four for the third, etc. When the totals were computed, the number of grains of wheat for the last square was found to be more than 9×10^{18}. This would have made a pile roughly 10 miles long, 10 miles wide, and 10 miles high!

You may have heard this story often enough that it comes as no surprise. What we are now interested in is how the numbers become so big so quickly. The answer is that we took the number of grains of wheat on the first square (one) and doubled it 63 times. Apparently, continuous doubling of even a relatively small number results in a very large number quite fast. If one starts with a large number and begins doubling, the numbers become startlingly big very quickly. For example, the present population of the earth is about 3 billion. It is observed to be doubling every 30 or 35 years. At this rate, one predicts a world population of 48 billion by the year 2100!

A common example of this kind of situation, although it may not appear to be similar, is compound interest paid by a savings bank. If you put your money in a bank at 5 percent compound interest, it will

FIGURE 8.3 A portion of the Geysers power plant in Sonoma County, California. Underground steam is used to provide energy to turbines which generate electricity. *(Courtesy of Pacific Gas and Electric Co.)*

grow at the rate of 5 percent per year. At the end of 15 years \$1.00 will have grown to \$2.08. At the end of another 15 years, it will have more than doubled again. So compound interest at the rate of 5 percent has a doubling time of between 14 and 15 years. This is true of any situation in which growth is expressed as a percentage of an existing amount. World population is growing at a rate of about $2\frac{1}{2}$ percent/year, which gives a doubling time of about 30 years. A simple way of calculating the approximately doubling time is the following:

$$\text{Doubling time (years)} = \frac{70}{\text{percentage rate}}$$

This gives an answer accurate enough for almost all purposes.

Growth at a fixed percentage, compounded, is called exponential growth. It can be represented in graphical form in two ways. One of these emphasizes the overwhelmingly fast, almost explosive, nature of exponential growth. The other allows simple extensions into the future, but somewhat conceals the magnitude of the growth rate. You should become familiar with both kinds of graphs, since both turn up frequently in discussions of energy. At times one is led to believe that the choice between graphs is based on whether one wants to frighten or soothe the reader.

The first type of graph is the one you are most familiar with. An example is shown in Fig. 8.4. Here the population of the United States

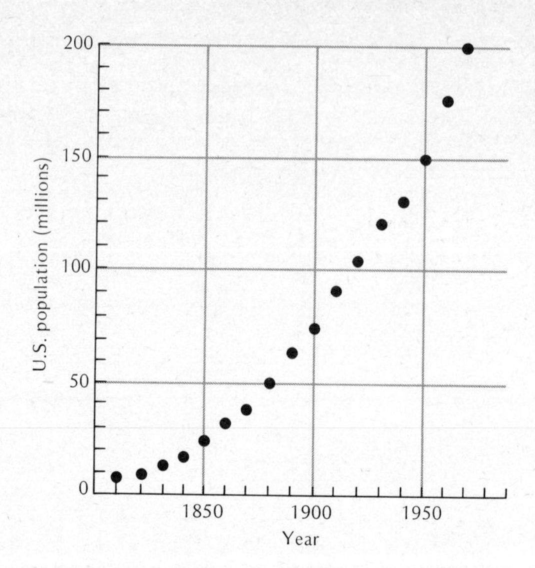

FIGURE 8.4 United States population versus time. *(From A. J. Priest, "Problems of Our Physical Environment," Addison-Wesley, Reading, Mass., 1973. By permission of the publisher.)*

is plotted on the vertical axis, and time on the horizontal axis. Notice that the curve grows ever more steep. This is characteristic of exponential growth. It "blows up" or "explodes" because of the compounding effect.

A second type of graph is called a semilogarithmic plot. In Fig. 8.5 we see such a plot for the population of the United States. Time is again plotted simply on the horizontal axis, but notice the vertical axis. Each major unit is a factor of 10 bigger than the one before. A straight line on a semilogarithmic graph represents exponential growth with a constant doubling time. Changing the percentage rate of growth changes the slope of the straight line. We see that the slope of United States population growth changed about 1900. Before that it had a doubling time of about 20 years, corresponding to a growth rate of about $3\frac{1}{2}$ percent. Since 1900 the doubling time has been about 50 years, representing a $1\frac{1}{2}$ percent growth rate. The dashed straight line in Fig. 8.5 represents an extension into the future of the rate that has prevailed since 1900. It predicts a population of roughly 500 million by the year 2050. This extension of a graph is called extrapolation.

8.3 THE USE OF ENERGY

Now that we have looked at exponential curves, we can begin to see how much energy is now being used, how much will be used in the future if we fail to change our ways, and if this much will be available. Figure 8.6 shows a semilogarithmic plot of United States electric energy production (in billions of kilowatthours) since 1930 (1 kilowatthour is shown in Chapter 6 to be equal to 3,600,000 joules). We see that, since 1930 (the author's lifetime), electric energy use in the United States has increased from 90 billion kilowatthours to nearly 2,000 billion kilowatt-

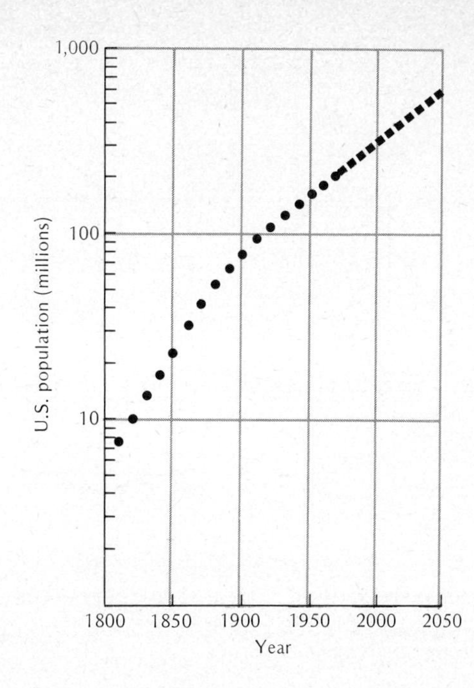

FIGURE 8.5 Semilogarithmic plot of United States population versus time. *(From A. J. Priest, "Problems of Our Physical Environment," Addison-Wesley, Reading, Mass., 1973. By permission of the publisher.)*

hours. This is more than a factor of 20, while the population has approximately doubled. The rate of electric energy usage is increasing 10 times as fast as the population. The doubling time is about 10 years—an average growth rate of more than 7 percent annually. Utah Power and Light has projected a growth rate in their area of 11.4 percent for the period 1974–1978, as an example.

A similar story of exponential growth of consumption can be told for all other forms of energy. But the sources of energy we now use are finite. The rate at which new sources of fossil fuel are being discovered is not exponential. Therefore, in the very near future, even if all the political problems of the Middle East are solved, the exponential growth in the use of energy will collide with the finiteness of our resources. In fact, the message is clear. Where a finite resource, such as fossil fuels, is concerned, exponential growth will sooner or later defeat any program of conservation, exploration, increase in efficiency, or what have you.

8.4 THE EFFECTS OF ENERGY GENERATION AND ITS USE

There are several physical principles we have already studied that apply to the question of energy generation and use. Let us first start at the back, so to speak. What happens to all the energy we use?

We have already learned that energy is neither created nor destroyed. It is merely shuffled around from one form to another in a sort of cosmic shell game. Does this mean we can use the same energy

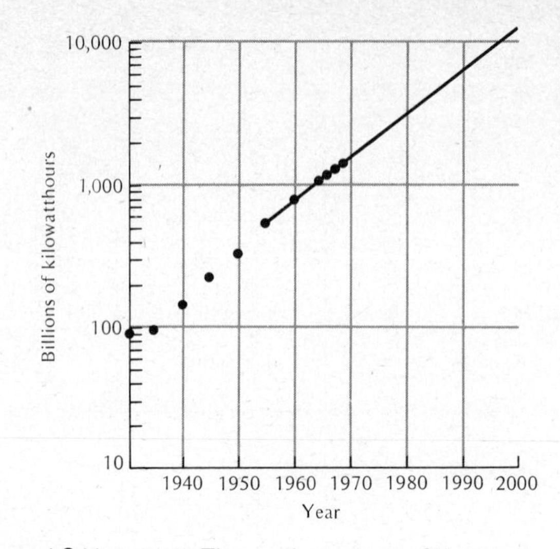

FIGURE 8.6 Semilogarithmic plot of United States annual energy production versus time. *(From A. J. Priest, "Problems of Our Physical Environment," Addison-Wesley, Reading, Mass., 1973. By permission of the publisher.)*

over and over to do useful work? Not at all. The end product of the overwhelming fraction of our energy use is heat. Over the entire earth the total amount of heat generated this way is less than 0.1 percent of the heat energy we receive from the sun. But, to use one local area as an example, in the Los Angeles Basin it is estimated that the heat released from energy use is about 5 percent of that received from the sun. No one knows for sure what will happen to the climate on earth if our global energy use climbs to 1 percent or more of that received from the sun. It is believed that a few degrees rise in the average temperature of the earth could cause melting of the ice caps of Greenland and Antarctica, causing the oceans of the world to rise several hundred feet. It is an interesting exercise to discover how many of the world's major cities lie within a few hundred feet of sea level. So one possible end result of exponential growth in energy use is a disruption of the climate of the earth due to excess heating.

There are two products of the burning of fossil fuels that may also produce changes in our climate. One is the carbon dioxide produced from the burning of these fuels. One century ago, the concentration of carbon dioxide in the atmosphere was 290 parts per million (ppm). Now it is 320 parts per million. Carbon dioxide traps heat and prevents it from radiating into space. Therefore, it produces a "greenhouse effect." Increases in the amount of carbon dioxide in the atmosphere are predicted to increase the average temperature of the earth. The concentration of carbon dioxide is presently increasing at a rate of 0.7 part per million per year.

Particulates, or very fine pieces of ash released principally in the burning of coal, produce an opposite effect. They cause more of the sun's energy to be reflected into space, and can act to cool the earth. No one is yet certain how these two effects will balance out. What is certain is

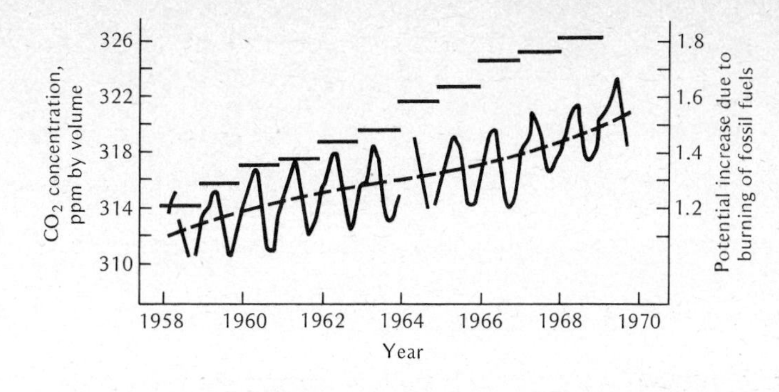

FIGURE 8.7 Long-term variation in the amount of carbon dioxide in the atmosphere. The horizontal lines show the amount that fossil fuel burning produces. Although some of the potential increase is being absorbed in the environment, clearly the long-term trend is steadily upward. *(From A. J. Priest, "Problems of Our Physical Environment," Addison-Wesley, Reading, Mass.,* 1973. *By permission of the publisher.)*

that at present we are using the entire earth as an experimental laboratory in which to find out.

We have already discussed the fact that heat engines must exhaust some heat to the environment. Therefore we cannot beat the heat problem by making more efficient electric generators, for example. Almost all the energy that goes into an electric generator turns up as heat, either as waste heat at the generator or as heat resulting from friction in the machines run by the electric energy.

Why don't we recycle all this heat back into useful work and use it over and over, since energy is never destroyed? Here we are defeated by the second law of thermodynamics. It says that a heat engine takes in heat at one temperature and exhausts heat at another temperature. The exhaust heat must be at a lower temperature than the heat put in, or no useful work can be done. Since the waste heat is emitted at a relatively cool temperature, an even colder sink must be provided if a second engine is to be run using this waste heat. There is clearly a limit to this process.

Another consequence of energy generation is pollution. The generation of electric power by nuclear fission or fusion generates radioactive isotopes that we will discuss in a later chapter. The burning of fossil fuels sends into the environment a wide variety of contaminants. Sulfur, in the form of sulfur dioxide, is one which has been much in the news. Many states have passed laws forbidding the burning of high-sulfur coal. The only catch is that there is not enough low-sulfur coal to go around. Some western coals that are very low in sulfur have significant amounts of radioactive impurities. Natural gas, or methane—the most perfect fuel of all—is in exceedingly short supply. For political reasons the price of this fuel has been kept low. As a result it has been used for many purposes where other fuels would have served as well.

There are some more subtle results of energy usage. High-voltage electric power lines require significant amounts of land. Presently the amount is about 0.4 percent of the United States land area. If there is an increase by another factor of 10, which is clearly possible if present trends continue, the amount of land devoted to this purpose will become a major burden. Burying the lines increases their cost by more than a factor of 10. Power plants require water for cooling which is in short supply in many places.

What are the limits to man's energy use? There are many possible answers to this question, but it appears that there is a limit; and that limit will be reached in the foreseeable future. The factor which will determine the limit is not clear. It may be that we will run out of fuel if fusion power does not work. It may be that the disposal of waste heat will be the limiting factor; we could quite easily reach a situation in which enough waste heat is generated to bring every river and stream in the United States to the boiling point if they are used for cooling. The amount of land area devoted to energy generation and transmission is becoming a significant fraction of our total area. The eventual constraint will be determined by which forms of energy we turn to, and how we handle the problems produced by each.

8.5 ANY MORE TRICKS FROM THIS BAG?

There is a common feeling in our society that science can solve any problem it sets its mind to. As one example, we point to the Manhattan Project in World War II in which the atomic bomb was developed at a cost of about $2 billion. As another, we remember that President Kennedy, in 1961, committed the United States to put a man on the moon during the 1960s. Neil Armstrong stepped out on the moon on July 20, 1969. Therefore it is very easy to conclude that, if proper financing and effort are provided, any problem can be solved. We hear many exhortations about "solving" the energy crisis. It should be kept carefully in mind that science and technology have accomplished many amazing feats, but never in any way has it been possible to violate the basic laws of nature. In this case the laws of interest are the laws of thermodynamics, which tell us what can and cannot be done with energy and heat.

The first law of thermodynamics says that energy is never created or destroyed. This means that we are limited to the energy that has been provided to us: fossil fuels, which are stored solar energy and will be exhausted soon; nuclear energy, which is available from a rather small number of fissionable isotopes that are not abundant; nuclear fusion energy, whose controlled release is not yet possible; solar energy; and heat energy from the interior of the earth. The question can be asked, are there any as yet unknown sources of energy still to be discovered? Negative statements about science are among the most risky of all predictions, but it appears to me that there are no more tricks in the bag comparable to the discovery of nuclear energy. There will undoubtedly be technological improvements in our present utilization of various forms of

FIGURE 8.8 Sending a man to the moon was a tremendous technological achievement, but no law of nature was broken in the process. *(Courtesy of NASA.)*

energy. But I anticipate that no major new sources of energy will be found. If this is true, and certainly until evidence is found to the contrary, we must live with what we have. This is clearly going to require major adjustments in our energy-oriented society. By the time this book appears in print, some of the necessary adjustments may be apparent. But over the next 10 years it is quite possible that more and more adjustments will be required, some of them in surprising and unexpected areas.

SUMMARY

There are four sources from which man presently does or possibly can obtain energy: (1) from the sun, either stored as fossil fuels, or collected and used directly; (2) from atomic nuclei, either through fission or fusion; (3) from the kinetic energy of the earth-sun-moon, by using the tides; or (4) from geothermal energy, heat sources within the earth.

The doubling time for exponential growth can be estimated by

$$\text{Doubling time} = \frac{70}{\text{percentage rate}}$$

Heat from the use of energy may change our climate. Excess carbon dioxide in the atmosphere may also change the climate, making it warmer as a result of the "greenhouse effect." Excess particulates in the atmosphere may cause cooling, because energy from the sun is reflected off into space.

SELECTED READING

Hammond, A, L., W. D. Metz, and T. H. Maugh, III: "Energy and The Future," American Association for the Advancement of Science, Washington, 1973. A good discussion of all energy sources, present and projected for the future. Pros and cons of all methods of generation, storage, and transmission are discussed.

Meadows, D. H., D. L. Meadows, J. Randers, and W. W. Behrens, III: "The Limits of Growth," Signet Books, New American Library, New York, 1972. Extends the ideas of exponential growth to a "world model" in which all interrelated effects are taken into account. This may be one of the most important books of the century. Unless its message is understood by both leaders and citizens all over the world, it will be difficult to avoid catastrophic mistakes.

QUESTIONS

1 Why does energy usage in the United States grow at a rate ten times that of population growth?

2 The weight of a child might double during the first eight months of its life. Assuming that this was true for you, extrapolate to how much you would weigh now had you continued to grow at the same rate?

3 What single feature characterizes all the common schemes put forward as possible solutions to the "energy crisis?"

4 From a world almanac, find what percentage of the electric energy produced in the world comes from hydroelectric power.

5 In 1880, what percentage of our energy requirements came from oil and other light petroleum products?

6 What are the most common objections to the use of nuclear fission reactors for generating electric power?

7 It has been proposed that the United States be independent of foreign sources of energy by 1980. Is this kind of "energy isolationism" reasonable in view of our needs and capabilities?

8 Can you name any politician (or other person in a position to influence matters) who has stressed the dangers of exponential growth?

SELF-TEST

__________ fossil fuel energy
__________ hydroelectric power
__________ nuclear fission
__________ solar energy
__________ thermonuclear fusion
__________ geothermal energy
__________ doubling time
__________ semilogarithmic graph
__________ extrapolation
__________ greenhouse effect
__________ particulates

1 Energy from the internal heat of the earth
2 Extension of a graph with a straight line to make predictions
3 Energy from the splitting of heavy nuclei
4 Energy from water stored behind dams
5 Characteristic of the rate of exponential growth
6 Atmospheric heating due to excess carbon dioxide
7 Energy from the sun
8 Energy stored in decayed organic matter
9 Energy from the merging of light nuclei
10 Graph on which exponential growth gives a straight line
11 Finely divided ash from fossil fuel burning

PROBLEMS

1 At 6 percent interest, compounded, how long does it take for your money to double?

2 The population of the world increased from 1 billion to 2 billion between about 1830 and 1940. What rate of growth does this imply? It will be 4 billion by about 1980. What is the more recent growth rate?

3 If a 10-km^2 solar-energy collector is needed to provide the heat for a 1,000-megawatt electrical plant, how much energy can you expect to obtain from a solar-energy collector on the roof of a typical house?

ELECTRICITY: THE FUNDAMENTAL EXPERIMENTAL LAWS

Electricity is an important part of the life of everyone. We depend on it for light, heat, and power to run machines. We use electricity in innumerable different ways. But, in addition to these practical aspects, electrical effects are fundamental to our understanding of the structure of matter. The forces that hold atoms together are electric forces. The atom itself is composed of electrified particles. To understand the structure of the atom, we must understand how electric charges interact with one another.

Therefore, to gain some understanding of the electrical aspects of the structure of matter, to understand something of the practical aspects of electricity, and to appreciate the intrinsic fascination of the phenomena themselves, this section is devoted to a study of electricity. In particular, we shall focus on the basic experimental laws of Coulomb, Oersted, Ampère, and Faraday, since these laws are sufficient to describe and understand almost all electrical phenomena.

Chapter 9 is an historical survey of the origins of electrical science and the development of Coulomb's law. Chapter 10 discusses electric current, electric potential, and the field concept. Chapter 11 shows the relationship between electrical and magnetic effects as demonstrated by the experiments of Oersted, Ampère, and Faraday.

ELECTRIC CHARGE AND COULOMB'S LAW

Effects we now consider electrical have been known for thousands of years. The spectacular nature of lightning must have terrified the earliest human. The ability of certain rocks, called lodestone, to attract iron nails led to the invention of the mariner's compass. But, until relatively recent times, these various phenomena were considered separate and distinct, and no connection was seen among them. One reason the development of electrical science came later is that the experiments in electricity and magnetism that we will discuss are not as close to our everyday experience as is mechanics. Many of these phenomena would not have been found unless specifically sought.

9.1 ELECTRIC CHARGE

A phenomenon known to the Greeks was called the *amber effect.* A piece of amber (petrified resin) rubbed with a piece of fur has the property of attracting small pieces of dust and chaff. From the Greek word for amber, *elektron*, comes our word electricity. The amber effect is the simplest of the electrical effects.

In 1600 William Gilbert, court physician to Queen Elizabeth I, published a book on magnetism, which included only a single chapter devoted to the amber effect. In this chapter he pointed out his discovery that many substances other than amber produce the same effect. His list included glass, diamond, sapphire, opal, amethyst, and a number of other materials. We could add many more materials, such as hard rubber, nylon, and other plastics, to such a list today.

The fact that the amber effect seemed to involve one object exerting a force on another some distance away was very troubling to seventeenth-century science. The idea of action at a distance was, by and large, considered unacceptable. This was particularly true in the time before Newton, when all forces were believed to be the result of direct mechanical interactions. The story of the attempts, many of them highly ingenious, to explain this paradox is fascinating reading. We shall move ahead, however, and deal with the matter on a somewhat different level. We shall accept the obvious experimental fact that such forces do exist, and build a description of these forces without inquiring about the underlying mechanism.

Gilbert overlooked in his experiments one of the most significant facts concerning the amber effect. This was discovered a few years after his death. If a piece of amber or hard rubber is rubbed with fur, it will attract a similar piece of amber or hard rubber which has not been rubbed. If, however, we rub two pieces of amber with fur, we find that they strongly repel one another. A force is observed which pushes them apart. There seem to be two kinds of electrical effects. In one, a rubbed object attracts other nearby objects. In the other, two identical rubbed objects repel one another. Suppose we now suspend a rubbed hard-rubber rod from a string and bring other rubbed objects near it.

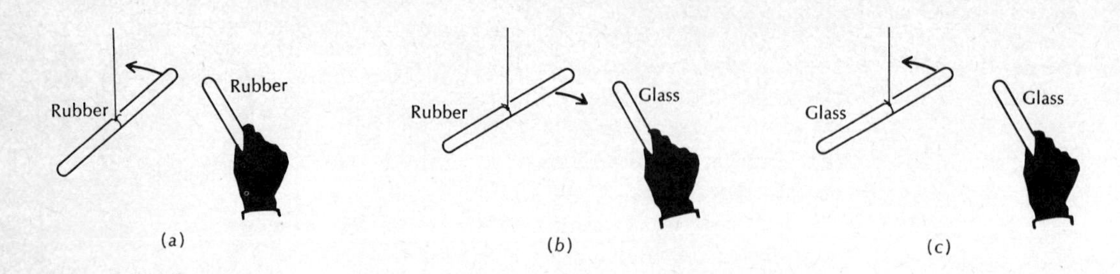

FIGURE 9.1 (*a*) Rubber rods which have been rubbed repel one another. (*b*) Rubber and glass which have been rubbed attract each other. (*c*) Glass rods which have been rubbed repel each other.

If we bring an identical rubber rod near the suspended one, we will find that there is a force pushing them apart, as before, But if we bring a rod of glass which has been rubbed with silk nearby, we will find that there is a force attracting the glass and rubber rods. If we suspend a glass rod which has been rubbed with silk, we will find that it is repelled by another glass rod which has been rubbed with silk (Fig. 9.1). These experiments lead to the conclusion that there are two different kinds of electricity or, as it is called, *electric charge*, and that the following rule holds for the forces between electric charges:

Like charges repel one another; unlike attract one another.

The two kinds of electricity are called *positive* and *negative*. Negative electricity is that acquired by amber or rubber rubbed with fur, and positive is that acquired by glass rubbed with silk. By the middle of the eighteenth century these experiments had been interpreted in terms of two kinds of electric fluids. It was believed that an electrified object had been given a quantity of one of these fluids by the process which electrified it. All the electrical effects could then be explained by the idea that like electric fluids repel one another and unlike electric fluids attract one another. That is, positive repels positive and attracts negative.

These effects can be easily demonstrated. Make two small balls of aluminum foil, and suspend each from several feet of thread so they are about an inch or two apart. Touch both with a charged rubber rod. They will fly apart. Now touch one with a charged glass rod (positive). It will be attracted to the negatively charged ball. If the two balls touch, they will share their charge equally and move apart (unless, quite by accident, they have exactly the same amount of charge of opposite sign).

During the middle and latter part of the eighteenth century, electrical phenomena became the subject of popular lectures and demonstrations. One of these lectures brought electrical effects to the attention of Benjamin Franklin. Independently of the work in Europe, Franklin developed a model which had only one fluid. An object which had an excess of this fluid, he said was positively charged, and an object which had a deficiency of this fluid was negatively charged. Franklin's

ELECTRO CHEMISTRY,

OR

PHILOSOPHICAL EXPERIMENTS WITH ELECTRICITY

MR. RICHARDS,

Respectfully informs the Ladies and Gentlemen of this vicinity, that he will deliver a Lecture on the science of

Electricity,

This evening, at

in which will be explained in the most familiar manner and illustrated by a costly and powerful Electrical Machine and apparatus, the Phenomena of *Electricity*, produced by *Friction*;

LEYDEN PHIAL,

Positive and Negative Electricity, Electrical Attraction and Repulsion, by which many objects will be put in motion, such as moving vertical wheels, ringing bells, &c. &c.

Electrical Orrery;

showing the motion of the earth and moon round the sun; imitation of celestial fire (or lightning); construction of lightning rods; the electric shock.

The operator will insulate a person in such a manner that brilliant sparks of fire may be drawn from his body. Electric Pistol, showing the cause of Thunder, Illumination of water, Dr. Franklin's Electrical Dancing Figures, Illuminated Jar, Miser's Plate, lighting of a candle, the *Explosion* of a *Powder Mill*; with many other pleasing experiments.

The phenomena of Electricity, and the most important laws of nature, are capable of being made familiar even to children, and in this enlightened age, an acquaintance with the first principles of physical science, is justly deemed an important part of education. It is, therefore, believed that an evening may be agreeably and profitably spent on this interesting subject, by all classes of people.

☞Admittance cents; children half price. Tickets may be had at the bar. Doors open at o'clock. ***Ladies and gentlemen are respectfully invited to attend precisely at the hour.

From the numerous testimonials received, the following only is selected.

MR. RICHARDS,

DEAR SIR—We were agreeably entertained, last evening, with your lecture on Electricity, and are confident that both the matter and manner gave general satisfaction.

We wish you success in your laudable effort to substitute Philosophical demonstration, (which improves the understanding and tends to secure the lives and property of our fellow citizens by teaching them how to control the volatile fluid.) for the illegal and trifling exhibitions frequently made in our country without any other object or end than to impose on the public and corrupt the morals of society.

We are your obedient servants.

JOHN S. PETERS, *Lieut. Governor.*
LYMAN STONG, *Minister of Hebron*

Hebron, Con. February 26, 1829

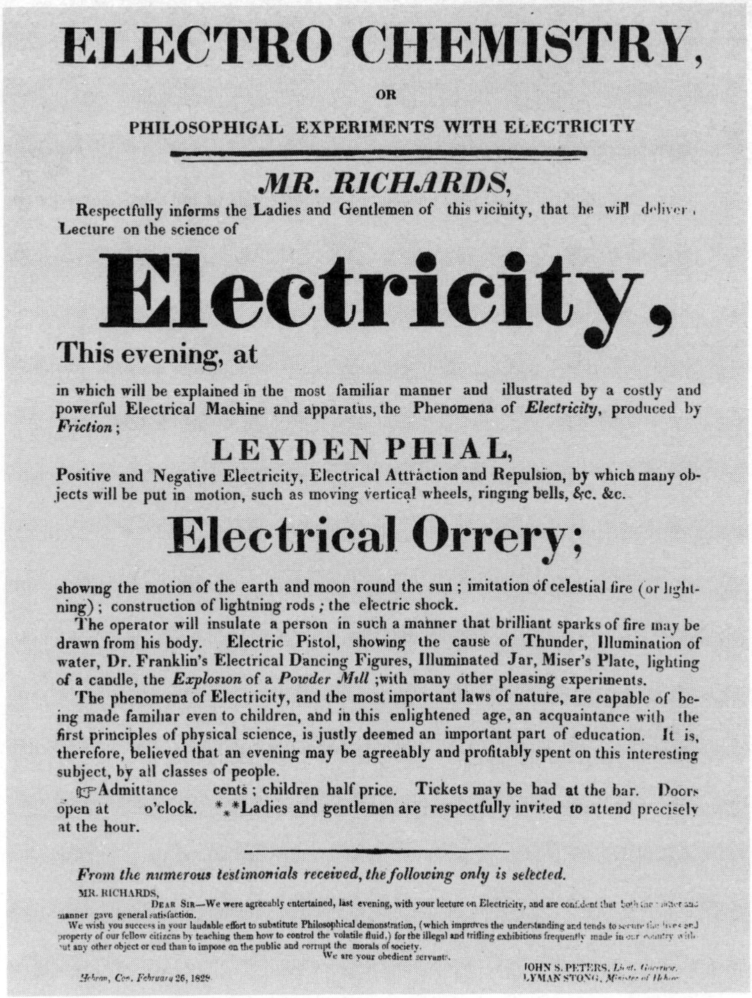

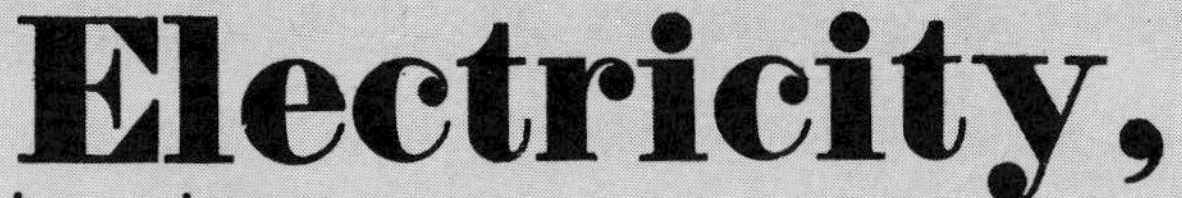

FIGURE 9.2 Poster of the early nineteenth-century advertising demonstration of various electrical effects by Mr. Willard Richards. (*From Claire Noall, "Intimate Disciple—A Portrait of Willard Richards," University of Utah Press, 1957.*)

choice of a glass rod which had been rubbed with silk as being positively charged established the choice of sign we use for electric charges today. This choice for the sign has some unfortunate consequences. When we discuss electric currents using Franklin's sign convention, a positive current is considered as flowing in the direction in which positive charge flows. In fact, when current flows in real metals, the only charge which moves is electrons, which are negatively charged. Therefore the sign convention tends to create confusion about what is actually going on. Both the one-fluid and two-fluid models of electricity can be used to describe all the electrical effects known in Franklin's time.

The explanation of electrical effects in terms of fluids is characteristic of the ideas of that time. Many other fluids had been proposed in science. Caloric fluid, which was used to explain the properties of heat, is one that we have already discussed. Phlogiston is the name given to another fluid that was supposed to be involved in the process of combustion.

LIGHTNING AND OTHER EFFECTS IN NATURE

Franklin was also the first to demonstrate that lightning is an electrical effect. In 1752 he flew a kite in a thunderstorm and collected electric charge in a Leyden jar, a device for storing electric charge that was then in common use. He found that charge from the thundercloud was identical with that produced in the laboratory. Franklin was very lucky. Shortly after he performed this experiment, a Russian scientist was killed attempting to repeat it.

Lightning occurs when large amounts of positive and negative charge are separated. This separation causes an unbalance of charge. The flash of lightning cancels out the charge unbalance. But how does this charge unbalance occur in the first place? Apparently, small drops of water become charged as they form. In general, the clouds containing the drops become positively charged with respect to the ground, which carries a negative charge. When the clouds become too highly charged, a stroke of lightning occurs. This carries charge between the ground and the cloud and removes the unbalance.

Another interesting example of the same kind of effect occurs when volcanoes erupt under the surface of the ocean or near the ocean shore. When hot lava is cooled in sea water, clouds of steam are formed. The drops of moisture in this steam are found to be strongly charged. Violent lightning displays are observed near such clouds. This was observed

FIGURE 9.3 Franklin's experiment with the kite in June 1752. (*Courtesy The Bettman Archive, Inc.*)

FIGURE 9.4 Lightning in the clouds above the volcanic island Surtsey, near Iceland, December 1, 1963. (*Courtesy Wide World Photos.*)

during the eruption of the volcano Surtsey near the south coast of Iceland in 1963 (Fig. 9.4).

Cars and planes moving through the air (particularly when it is dry) can acquire large amounts of charge. The air "rubs" the car, so to speak. Trucks frequently have drag chains to get rid of this charge. There are also special devices that are used to leak the charge off planes so that radio communication will not be interfered with. At toll booths on highways, there are devices to prevent the toll collector from receiving nasty shocks. These devices are sharply pointed because it has been observed that electric charge will leak off at sharp points.

FIGURE 9.5 Benjamin Franklin invented the lightning rod which is used to protect buildings such as the barn and silo shown here. (*Courtesy of Grant Heilman.*)

A *lightning rod* is another example of the use of this observation. Because the lightning rod is pointed, it will leak charge and prevent a large build-up of charge. If lightning does strike, it will strike the lightning rod because the rod provides the easiest path for lightning to travel. The lightning rod is connected by a heavy wire to the ground, and the building is saved (see Fig. 9.5).

ELECTROSTATIC PRECIPITATION

The electrical experiments we have been studying are from a branch of electricity called *electrostatics*. This means electric charge at rest.

There is one major practical use of electrostatics—in the area of controlling air pollution. Much of our air contains small particles of soot, ash, and other substances. These pollutants frequently injure the lungs if they are breathed for any length of time. It is important, then, to dispose of them.

If the particles of smoke or dust in a factory chimney can be given an electric charge, they can be attracted to a charged metal plate and trapped. This method of removing particles from smoke is called *electrostatic precipitation*. It can be made very efficient, but there are practical difficulties that prevent its use in all desirable situations (see Fig. 9.6).

Interestingly enough, the inventor of this process—Frederick Cottrell (1877–1948)—did not use his patents to become wealthy. He gave them to a company that he formed, called the Research Corporation, and the income from these and other patents has been used to support scientific research in many different fields.

9.2 ELECTRIC CONDUCTION

Another class of experiments involving electrical phenomena was first reported about 1731 by Stephen Gray in England. These experiments dealt with the communication of electrical effects over significant distances. Gray found that the effects of attraction and repulsion could be communicated from one object to another if they were connected by a piece of thread.[†] As Gray continued his experiments, it became clear that there are two classes of substances with respect to electrical effects. There are those substances which can be electrified by rubbing, which Gilbert had called *electrics*, and there are those which cannot be electrified by rubbing but which can conduct electrical effects from one place to another. These two classes are what we now call *nonconductors* and *conductors*. Conductors are primarily metals (although

[†]Actually, thread was a rather bad choice for communicating electrical effects; a metal wire is a much better conductor of electricity. If the experiments had not been carried out in the damp climate of England, the thread would not have performed nearly as well in communicating or conducting electrical effects.

FIGURE 9.6 These pictures show a smoke stack with its precipitator turned on (left) and off (right). (*Courtesy of Commonwealth Edison.*)

Gray used string), and nonconductors, or insulators, are substances like glass or hard rubber.

Another way of stating the difference between conductors and nonconductors in terms of electric fluids or, as we shall now call it, *electric charge*, is the following. Electric charge placed on a nonconductor remains where it is placed. If, however, electric charge is placed on a conductor, it flows freely to all parts of that conductor. In fact, most nonconductors allow electric charge to move, but very slowly. So we do not have an either/or situation, but rather a whole range of materials from very good conductors to very bad conductors.

The modern view of electric conductors, such as metals, is that there are negative electric charges (called *electrons*) which are free to move about in a metal. A piece of metal is viewed as consisting of a number of fixed positive charges, and negative electrons free to move. This "sea"

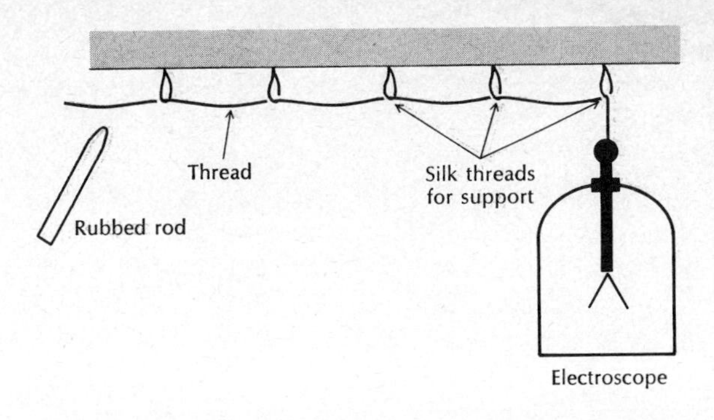

FIGURE 9.7 Schematic illustration of Gray's experiment. The thread carried the electrical effect to the electroscope, a sensitive detector of electrical effects discussed below.

of negatively charged particles corresponds to the electric fluid. We shall use this picture to describe what happens in electrical experiments.

THE ELECTROSCOPE

We can understand the operation of a device frequently used to detect electrical effects by applying these ideas. This device is called an *electroscope*. Shown in the drawing are the essential parts, a metal ball connected with a metal rod to two pieces of gold leaf. Gold is used because it is a conductor and can be made into foils which are extremely light and thin. Usually, the device is enclosed in glass, as shown in Fig. 9.8*a*, to protect the frail gold leaves and keep them from blowing in the breeze.

If we put some electric charge on this apparatus, what will happen? Since it is composed entirely of electric conductors (metals), the charge will spread over all the metal parts of the apparatus, including the gold leaf. This occurs because like charges repel one another, and therefore the charge placed on the apparatus tries to spread out as much as possible. Because the gold foils are light and easily moved, the repulsion between the charges on the leaves will cause the leaves to spread apart, as in Fig. 9.8*b*, in which the electroscope is shown negatively charged. Since the negative charge is mobile, it is easy to see how it spreads evenly over the apparatus. In the opposite case, when a positively charged rod is touched to the ball, some negative charge is drawn off

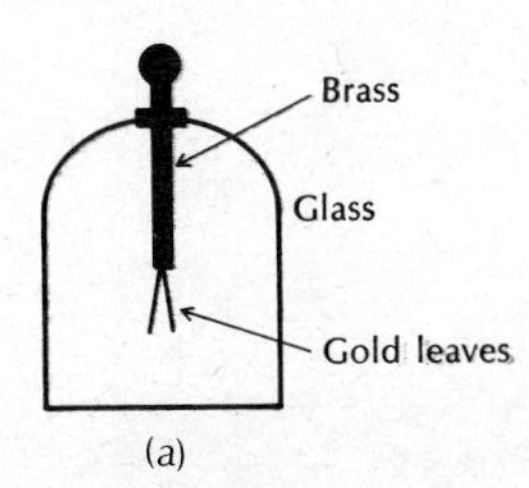

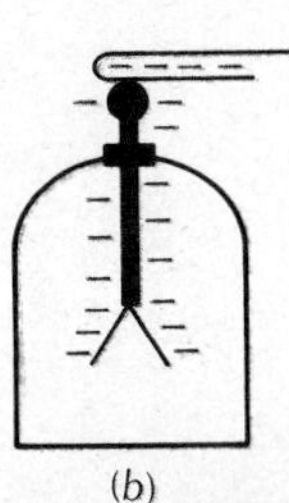

FIGURE 9.8 (*a*) An uncharged electroscope. (*b*) A charged electroscope.

the electroscope into the rod, leaving the electroscope with a net positive charge. The mobile negative charges which remain readjust themselves for the most uniform distribution of charge, just as they did in Fig. 9.8*b*. Therefore the electroscope behaves in essentially the same manner when either positively or negatively charged.

Now suppose we bring up to a ball on an uncharged electroscope, *without touching it*, a charged object such as a hard-rubber rod which has been rubbed with fur. This is shown in Fig. 9.9. If we perform this experiment, we observe that the leaves of the electroscope move apart. As we remove the rod from the vicinity, the leaves come together again. This experiment verifies the idea that there are charges which are free to move in the metal. The negatively charged rod in Fig. 9.9 repels some of the mobile negative charge in the ball. This puts an excess of negative charge on the leaves of the electroscope, which respond by moving apart. If the negative rod is moved away, the charges redistribute themselves and the leaves come together.

FIGURE 9.9 A negatively charged rod held near an uncharged electroscope drives negative charge from the ball. Therefore, the leaves spread because they have acquired an excess of negative charge.

THE SIGN OF ELECTRIC CHARGE

We can use the electroscope to tell quickly the sign of the electric charge on any object. Suppose we place a positive charge on the electroscope, using a rubbed glass rod. This will cause the leaves to separate, as in Fig. 9.10*a*. Now if we bring a negatively charged rod up near the ball, negative charge is driven from the ball to the leaves, and they will move together, as in Fig. 9.10*b*. If a positively charged rod is brought near a positively charged electroscope, the leaves will attempt to move further apart, as in Fig. 9.10*c*. If the electroscope had been negatively charged in the beginning, the effects in each experiment would have been exactly the opposite.

9.3 COULOMB'S LAW

All the experiments we have described so far have been qualitative; that is, they describe electrical phenomena but do not allow for any quantitative or numerical calculations of the magnitude of the effects. A quantitative description of the forces between electric charges was presented by Charles Augustin de Coulomb (1736–1806). The results

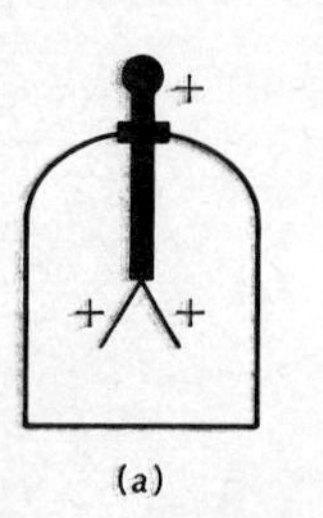

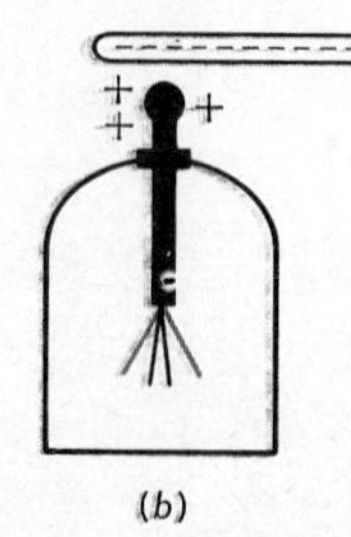

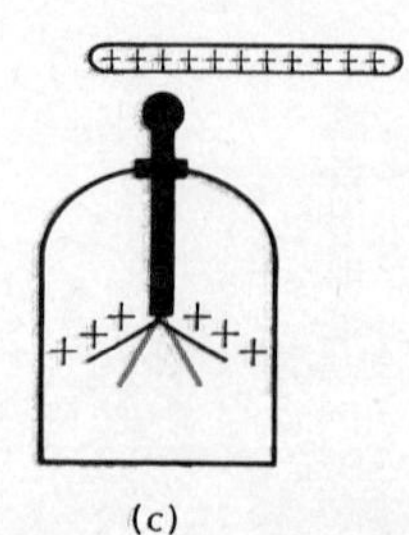

FIGURE 9.10 (*a*) A positively charge electroscope. (*b*) A negatively charged rod brought nearby causes the leaves to move closer together. (*c*) A positively charged rod brought nearby causes the leaves to move farther apart.

he achieved had been proposed or implied by several others before him, but Coulomb is given the honor of the discovery, and the law is named for him because he made a direct, mechanical measurement of the forces between electric charges in terms of the distance between them and the magnitude of the charge.

FIGURE 9.11 Charles Augustin de Coulomb (1736–1806). (*Courtesy of Burndy Library.*)

Coulomb used a device called a *torsion balance* that allowed him to measure forces as small as 10^{-8} pound. The balance takes advantage of the fact that the force required to twist a small fiber is proportional to the amount of twist, and that these forces are very small if the fiber is very fine. Fibers finer than human hair can be made from melted quartz. Therefore, very small forces will cause significant deflections of a measuring apparatus using such a fiber. Figure 9.12 shows schematically how this works. The amount of twist is a direct measure of the amount of force on the ball. These experiments of Coulomb came before those of Cavendish on gravity that were described in Chapter 4. Cavendish used Coulomb's torsion balance idea.

The first aspect of Coulomb's experiments was a determination of how electric force varies with the distance between two charged objects. He charged one of the suspended balls and then brought another charged ball nearby, as shown in Fig. 9.14. He found the following result. If he measured a force F when two objects were 1 centimeter apart and then moved them to 2 centimeters apart, *without changing the amount of electric charge on either object*, the force was reduced to $\frac{1}{4}F$. Tripling the distance reduced the force to one-ninth of its previous value. Since we know that $1^2 = 1$, $2^2 = 4$, and $3^2 = 9$, we see that the force varies inversely as the square of the distance between the two objects. In mathematical language, the force F between charges is proportional to $1/r^2$, where r is the distance between the objects. That is,

$$F \propto \frac{1}{r^2} \qquad (9.1)$$

Note that this result is similar to that for gravity.

Coulomb's careful experiments also showed how the force depends on the magnitude of the charge. He used a clever technique to obtain known variations in the amount of charge on the balls with which he was experimenting. When he had a charged ball and touched it with an

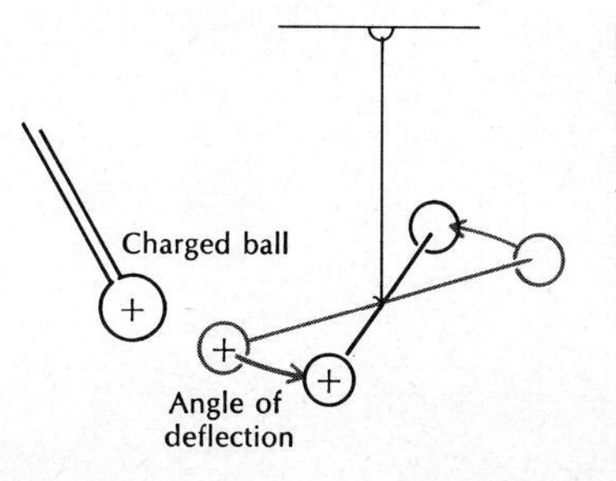

FIGURE 9.12 Schematic drawing of the torsion balance apparatus. The electric repulsion between the like charges causes the suspended ball to twist away. The amount of twist is proportional to the force.

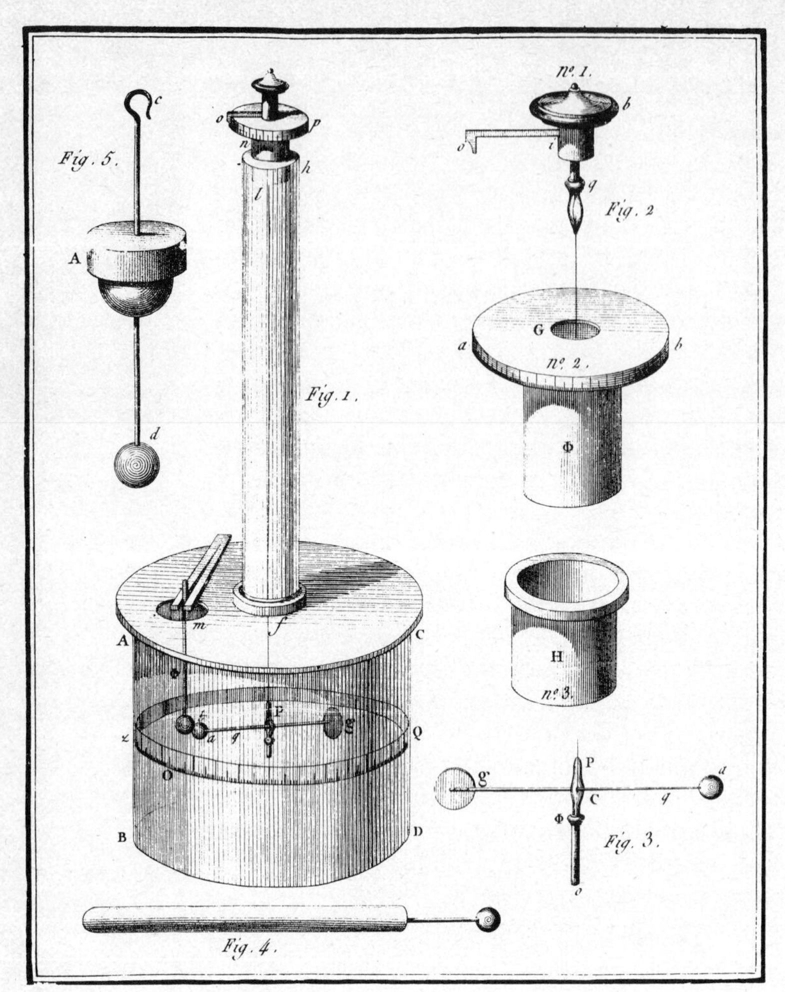

FIGURE 9.13 Drawings of Coulomb's torsion balance apparatus.

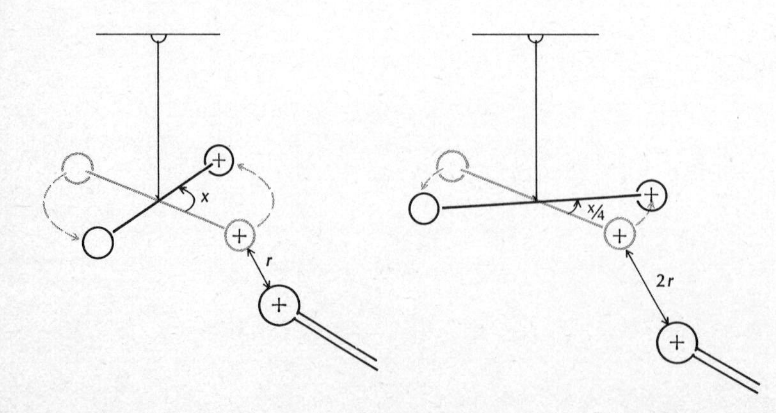

FIGURE 9.14 Doubling the distance between two charged balls reduces the force to one-fourth its previous value.

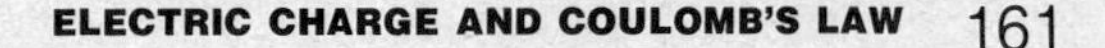

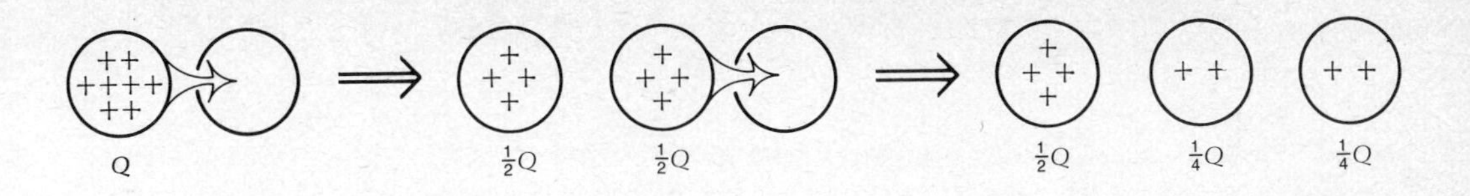

FIGURE 9.15 Coulomb's scheme for making known variations in the amount of electric charge on the balls. A ball with charge Q is touched by an identical uncharged ball. The result is two balls with charge $\frac{1}{2}Q$. If one of these shares its charge with a third ball, its charge is cut in half again to $\frac{1}{4}Q$.

identical uncharged ball, the charge was equally shared between the two balls, and thus the first ball now had exactly one-half its original charge, as in Fig. 9.15. This provided a precise method for varying the charge on either or both balls by a known amount. He simply brought up a third, uncharged, ball. Now the charge on one of the original two was reduced by one-half. If he wished, he could repeat this procedure again and again with either ball, and thus do a series of experiments with known variation in the charge on the balls. These experiments led to the conclusion that the force between charged objects is proportional to the magnitude of the charge on *each* object. Expressed mathematically,

$$F \propto Q_1 Q_2 \qquad (9.2)$$

where Q_1 is the charge on object 1, and Q_2 is the charge on object 2. If we cut Q_1 in half, the force is cut in half. If we cut both charges in half, the force is reduced to one-fourth its former value. Note that the sign of the charge is included in this result. If both Q_1 and Q_2 have the same sign, the force will be positive. If they are of opposite sign, the force will be negative. We define a positive *force* as a force pushing the two charged objects apart, since charges of like sign repel one another. A *negative force* pulls them together.

If we combine Equations 9.1 and 9.2, we have

$$\text{Force is proportional to } \frac{\text{charge on object 1} \times \text{charge on object 2}}{(\text{distance between objects 1 and 2})^2}$$

$$F \propto \frac{Q_1 Q_2}{r^2} \quad \text{or} \quad F = K\frac{Q_1 Q_2}{r^2} \qquad (9.3)$$

where the numerical constant K is chosen so as to make the equality hold in whatever system of units is chosen for charge, distance, and force. The choice of K defines the system of units for electric charge. Equation 9.3 is the mathematical statement of *Coulomb's law* for the force between electric charges. As yet we have no system of units for charge, which we need to give K a numerical value. K plays a role in Coulomb's law similar to that of G in the law of gravity.

UNITS FOR ELECTRIC CHARGE

We will only consider one system of electric units. The unit for charge we use leads to the units with which you are probably familiar—volts

(V) and amperes (A)—for other quantities. This is the reason for what may appear to be a rather arbitrary definition.

The unit of charge we shall use is called the *coulomb* (C). The coulomb is defined in such a way that two objects one meter apart, on each of which is one coulomb of charge, will have a force between them of 9×10^9 newtons. If we put this in Equation 9.3, we have

$$9 \times 10^9 \text{ newtons} = K \frac{(1 \text{ coulomb})(1 \text{ coulomb})}{1 \text{ meter}^2} \tag{9.4}$$

Therefore $K = 9 \times 10^9$. A very logical question arises; Why not choose the size of the coulomb so that two objects, each carrying 1 coulomb of charge and 1 meter apart, experience a force of 1 newton? Then K would numerically be equal to 1, and things would be much simpler. The unit defined this way turns out to be extremely small for most uses. This will be clearer in Chapter 10. Therefore we accept the arbitrary definition, and write Coulomb's law as

$$F = 9 \times 10^9 \frac{Q_1 Q_2}{r^2} \tag{9.5}$$

Here Q is in coulombs, r is in meters, and F is in newtons. Equation 9.5 *defines* what we mean by 1 coulomb of charge.

Just to use this once, let us do a simple calculation.

Example 9.1

Suppose that a small force of 4×10^{-3} newton is measured between two identically charged objects 0.1 meter apart. How much charge does each have?

We write Coulomb's law:

$$F = 9 \times 10^9 \frac{Q_1 Q_2}{r^2}$$

$$4 \times 10^{-3} \text{ newton} = 9 \times 10^9 \frac{Q_1 Q_2}{(0.1)(0.1)}$$

$$\frac{(4 \times 10^{-3})(0.1)(0.1)}{9 \times 10^9} = Q_1 Q_2$$

$$\tfrac{4}{9} \times 10^{-3-1-1-9} = Q_1 Q_2 \qquad \text{since } 0.1 = 10^{-1}$$

$$\tfrac{4}{9} \times 10^{-14} = Q_1 Q_2$$

But the statement of the problem was that $Q_1 = Q_2$. So we have

$$Q^2 = \tfrac{4}{9} \times 10^{-14}$$

$$Q = \tfrac{2}{3} \times 10^{-7} \text{ coulomb on each object}$$

It may be useful at this point to refer back to the material in Section 4.8 and the Appendix on power-of-ten numbers.

9.4 THE ELECTRON

We have mentioned that in metals electric charge flows because an object called an electron moves. So far as is now known, the charge on an electron is the smallest amount of electric charge found in nature. The magnitude of this charge was measured in an elegant and famous experiment in 1911 by an American, Robert A. Milliken. The details of this experiment will be discussed in Chapter 14. The magnitude of the charge on an electron is 1.6×10^{-19} coulomb. Therefore 10^{19} electrons have a charge of 1.6 coulombs. The sign of the electron's charge is negative.

The distance between electrons in matter is small, of the order of atomic dimensions, which is about 10^{-10} meter. Let us calculate the magnitude of the forces between electrons.

$$
\begin{aligned}
F &= 9 \times 10^{9} \frac{Q_1 Q_2}{r^2} \\
&= 9 \times 10^{9} \frac{(-1.6 \times 10^{-19})(-1.6 \times 10^{-19})}{(1 \times 10^{-10})(1 \times 10^{-10})} \\
&= 9(-1.6)(-1.6) \times 10^{+9-19-19+10+10} \\
&= 23 \times 10^{-9} \text{ newton} \\
F &= 2.3 \times 10^{-8} \text{ newton}
\end{aligned}
$$

SUMMARY

The amber effect is the effect produced by rubbing amber, rubber, or plastics with fur, or glass with silk. The objects rubbed become electrically charged.

Like electric charges repel each other; unlike charges attract each other.

Rubber rubbed with wool or fur is negatively charged; glass rubbed with silk is positively charged.

Electrostatic precipitation is a technique for removing particulate matter from industrial emissions.

Electric conductors, such as metals, can conduct electrical effects over large distances. Conductors are not charged by rubbing. In a metal, it is the motion of negative charges, called electrons, that allows charge to move.

Coulomb's law says that electric forces vary as the inverse square of the distance between charged objects, and directly as the amount of charge in each object. This can be written

$$\text{Force is proportional to } \frac{\text{charge on object 1} \times \text{charge on object 2}}{(\text{distance between objects 1 and 2})^2}$$

$$F = K\frac{Q_1Q_2}{r^2}$$

If Q is in coulombs, F is in newtons, and r is in meters, then K is, by definition, 9×10^9.

The electron carries the smallest amount of electric charge yet measured, -1.6×10^{-19} coulomb.

SELECTED READING

Blanchard, D. C.: "From Raindrops to Volcanoes," Science Study Series S50, Doubleday and Company, Garden City, N.Y., 1966. A story of raindrops and atmospheric electricity, told in nontechnical terms, ending with the discovery that volcanic lava produces positively charged clouds of steam when it flows into the sea.

Moore, A. D.: "Electrostatics," Science Study Series S57, Doubleday and Company, Garden City, N.Y., 1968. A collection of experiments on electrostatics, including instructions for building several kinds of electrostatic generators.

QUESTIONS

1 How can a rubbed piece of amber attract an unrubbed object?

2 Is the force between two charges ever zero?

3 If electrical effects are so strong compared to gravitational effects, why are gravitational effects dominant in the motion of the planets?

4 Does the behavior of an electroscope when a charged rod is brought near it demonstrate that it is the negative charges that are free to move?

5 How can the electrons in a metal move if they are strongly attracted by the stationary positive charges?

6 A silk cloth is used to rub a glass rod. What will happen when the silk cloth is brought near an amber piece which has been rubbed with fur?

7 How could early experimenters have distinguished electrical effects from gravitational effects?

8 When you comb your hair and the comb become charged, is the charge on the comb positive or negative? If you do not know, could you devise an experiment which would tell you?

9 Is it possible for there to be no net electric force on a charge in the presence of other charges?

SELF-TEST

__________ amber effect
__________ Gilbert
__________ Franklin
__________ electrostatic precipitator
__________ Coulomb
__________ Gray
__________ electroscope
__________ conductor
__________ nonconductor
__________ coulomb

1. Developed the law for forces between electric charges
2. Introduced the present use of positive and negative charges
3. Discovered electric conductivity
4. Wrote on the amber effect
5. Uses electric charge to fight pollution
6. A material in which electric charge moves freely
7. The unit of electric charge
8. Device for studying charge; can be used to determine the sign of charge
9. Material in which charges remain where they are placed
10. Produced by rubbing rubber with fur

PROBLEMS

1. A force of 10^{-3} newton is observed between two charges of $+10^{-5}$ coulomb each. How far apart are the two charges?
2. Calculate the force between two charges of +15 coulombs that are 1,000 m apart.
3. The force between two equal charges 2×10^{-2} m apart is measured to be -10^{-7} newton. What are the magnitude and sign of these charges?
4. The force between two equal charges 10 m apart is measured to be $+9 \times 10^{-3}$ newton. (*a*) What is the magnitude of these charges? (*b*) What can one say about their sign?
5. Calculate the force between two charges of $+10^{-6}$ coulomb 1 centimeter apart.
6. Calculate the force between two objects if one has a charge of $+1.5 \times 10^{-7}$ coulomb and the other a charge of -2.0×10^{-6} coulomb and they are 0.1 m apart.

7 Two charges, separated by 1 m, exert a 1-newton force on each other. How large are the charges?

8 An object of mass 2 kg is 3 m from a second, immovable object. Each has a charge of $+3 \times 10^{-6}$ coulomb. Calculate the acceleration of the 2-kg object.

9 An electron has a mass of 9×10^{-31} kg and a charge of -1.6×10^{-19} coulomb. Calculate its acceleration toward a charge of $+4.8 \times 10^{-19}$ coulomb that is 10^{-10} m away.

10 Calculate the force between two electrons (charge $= -1.6 \times 10^{-19}$ coulomb) that are 10^{-8} m apart.

11 (*a*) Calculate the force between an electron ($Q = -1.6 \times 10^{-19}$ coulomb) and a proton ($Q = +1.6 \times 10^{-19}$ coulomb) that are 10^{-10} m apart. (*b*) If the mass of the electron is 9×10^{-31} kg and the proton is assumed fixed, calculate the acceleration of the electron.

12 Two electrons are 1 m apart. (*a*) Calculate the electric force one exerts on the other. (*b*) If they were free to move, would they move together or apart? (*c*) How would your answer to part (*a*) change if one of the electrons were a proton? (*d*) How would your answer to part (*a*) change if there were two protons instead of two electrons?

13 The mass of a proton is approximately 1.6×10^{-27} kg, and its charge is $+1.6 \times 10^{-19}$ coulomb. Calculate the electric and a gravitational force (Chapter 4) between two protons 10^{-10} m apart. Compare the magnitude of these two forces.

VOLTS, AMPERES, AND SUCH

Our major interest in electricity probably stems from the fact that we are able to use it to perform useful work. Our technological society has become totally dependent on electricity as a means of transporting energy from one place to another. It turns out that the electrostatic effects discussed in the previous chapter have some, but not many, practical applications. In this chapter and the next, we will discuss some other aspects of electricity which give it practical value.

10.1 ELECTRIC CURRENT

We have previously referred to Gray's experiments on the conduction of the amber effect for some distance along pieces of string or thread. In this section we shall elaborate on the conduction of electricity from point to point. The amount of electric charge obtained by rubbing rods or even by more sophisticated apparatus based on the same principle is very small, and therefore the electric currents are very small. The study of electric current really began with the invention of the chemical battery by Alessandro Volta in 1800. A battery delivers vastly greater amounts of charge than can be achieved with electrostatic devices.

Electric current is the motion of electric charge, usually through a conductor. It is usually given the symbol I. The unit of electric current is the *ampere*, which is defined as one coulomb of charge passing a given point every second. Therefore

$$I = \frac{Q}{t} = \frac{\text{charge}}{\text{time}} = \frac{\text{coulombs}}{\text{second}} = \text{amperes} \tag{10.1}$$

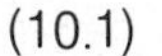

By convention, the direction of the current is the direction of motion of positive charge. The source of current may be a battery, or any of a variety of electric generators which we have not yet studied. Let us focus on the battery.

A common flashlight battery or dry cell has two terminals labeled positive (+) and negative (−). When a wire or any other conductor is connected between the two terminals, electric charge flows through the conductor. In metals, the actual motion is that of electrons, which are negatively charged. The current I, as defined above, flows in a direction opposite the electron flow. This bit of confusion arises from Franklin's unlucky choice for the sign of electric charge. This particular choice is so deeply embedded in texts and custom that it will not be changed easily. Figure 10.2 shows the conventional electric current and the electron flow for a simple case. The point to remember is that the actual effects of the current are the same, however we define it.

FIGURE 10.1 Alessandro Volta (1745–1827). *(Courtesy University of Pennsylvania Library.)*

It is necessary to distinguish carefully electric current effects from the electrostatic effects discussed in Chapter 9. When electric charge flows in a wire, the wire acquires virtually no net charge. Charge enters and leaves the wire at the same rate. The wire will not attract bits of paper or dust, as will a rubbed rod. We can compare a copper wire carrying a current to a hose carrying water. The hose can deliver large quantities of water, quantities much larger than the amount of water in the hose. In fact, each drop of water inserted into one end of the hose causes a drop to come out the other end. Similarly, a charge is "inserted" into the wire at one terminal of a battery, and another charge is forced out the other end, if a complete connection is made. Very little charge is stored in the wire.

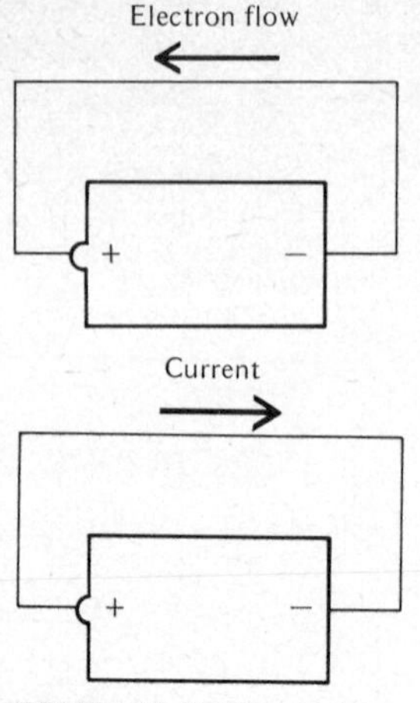

FIGURE 10.2 The electron flow and the conventional current in a wire attached to a battery.

THE CHEMICAL BATTERY

Most chemical processes occur with an exchange of electrons from one atom to another. The chemical battery is an arrangement whereby the electrons that are exchanged are carried through wires external to the battery. As an example, zinc metal can lose electrons easily to form zinc ions:

$$Zn \rightarrow Zn^{2+} + 2e$$

Copper ions can gain two electrons to become copper metal:

$$Cu^{2+} + 2e \rightarrow Cu$$

These two reactions can form the basis of a simple battery, or cell, as shown in Fig. 10.3. In this cell electrons flow in the external wire from the zinc electrode to the copper electrode. In a practical battery some means must be provided to prevent copper ions in the solution, or electrolyte, from moving to the zinc electrode and obtaining their electrons directly.

In a battery the total amount of charge available depends on the number of atoms of reacting material. The electric potential (see Section 10.2) depends on the choice of electrode material.

10.2 ELECTRIC POTENTIAL

A common flashlight battery is labeled $1\frac{1}{2}$ volts. An automobile battery is labeled 12 volts. Does this mean that the automotive battery is eight times better? Does it do eight times as much work? Could it be replaced by eight flashlight batteries?

Voltage is a measure of the capacity of an electrical system to do useful work. But it is not the whole story. The other part of the story is how much charge is available to do this work. The *volt* is the measure of how much work is available for each coulomb of charge the system can deliver. A battery is labeled 1 volt if each coulomb of charge it delivers does 1 joule of work (or produces 1 joule of heat energy).

One volt is then defined as the capacity to do one joule of useful work for each coulomb of charge.

An ordinary D battery can deliver about 5,000 coulombs of charge, and therefore about 7,500 joules (1.5 volts $\times$ 5,000 coulombs) of work.

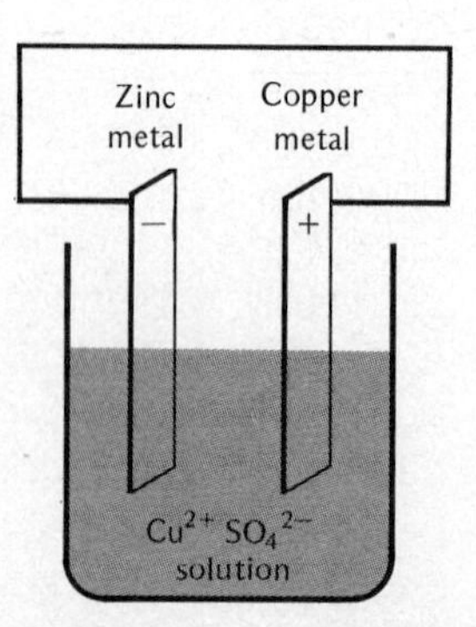

FIGURE 10.3 A battery whose electrodes are copper and zinc.

A 12-volt automobile battery can deliver about 300,000 coulombs of charge, and therefore 3,600,000 joules of useful work. As you know, an automotive battery is very heavy. The greatest practical problem in the creation of a battery-powered automobile is to find a battery that delivers at least 10 times as much energy for the same weight of battery. Otherwise too much work is expended just lugging the batteries around. A battery using liquid sodium metal and liquid sulfur is being intensively studied as one possible alternative to the presently available types of batteries.

One question always arises in the discussion of electricity, What is dangerous: voltage or current? The answer is current. High voltages are not dangerous, so long as the amount of current is small. Even very low voltages, like 110 volts, can be lethal if they drive enough current through the body. Ordinary household circuits are quite capable of providing dangerously high currents. Electrostatic generators commonly used in lecture demonstrations can create potentials of 500,000 volts, but the current in such a device is very small. Although the device can startle, it is not harmful.

The human body creates small electric potentials in its operation. Both the heart and the brain can be studied by measuring these potentials. Electrocardiography is the study of the electric potentials produced

FIGURE 10.4 Photograph of "high-voltage" student using an electrostatic generator. His hair stands completely on end because of the repulsion of like charges. (*Photograph by Alexis Kelner.*)

by the heart, and electroencephalography is the study of the potentials produced by the brain. In both cases the potentials are much less than a volt. The reason the heart is sensitive to electric shock is that its action is triggered by small electric potentials. Devices called pacemakers are inserted into the chests of patients whose heartbeats have lost their rhythm. The pacemaker delivers small and regularly timed electric potentials to the heart muscle to stimulate its action.

POWER

Suppose a piece of copper wire is connected between the terminals of a 1-volt battery. The length and size of the wire are chosen so that exactly 1 ampere flows in the wire. Each coulomb of charge that flows does 1 joule of work, or gives up 1 joule of heat in the wire, since it is driven by a 1-volt potential, as shown in Fig. 10.5. Since 1 ampere is 1 coulomb/second, a 1-volt battery driving a 1-ampere current is doing work at the rate of 1 joule/second. The rate at which work is done is called *power*, and the unit of power is the watt, as previously defined in Chapter 6. One watt is defined as one joule/second. In the circuit we are considering, the 1-volt battery is said to deliver 1 watt of power to the external circuit. Since no mechanical work is done in this particular example, all this is converted to heat in the wire. The amount of heat generated by a current in a wire was first measured by Joule in his study of the mechanical equivalent of heat, and is called the joule heat. In an ordinary 100-watt light bulb, 100 joules/second of energy is released, part as light and part as heat.

It is now possible to show the relationship between the power delivered to a circuit, the voltage, and the current. *Power* is defined as work divided by time, so that

$$\text{Power} = \frac{\text{work}}{\text{time}} = \frac{\text{joules}}{\text{second}}$$

But electric work is given by the product of charge Q and the potential through which it is moved (from the definition of potential).

$$\text{Power} = \frac{\text{charge} \times \text{potential (volts)}}{\text{time}}$$

And charge divided by time is current, so that

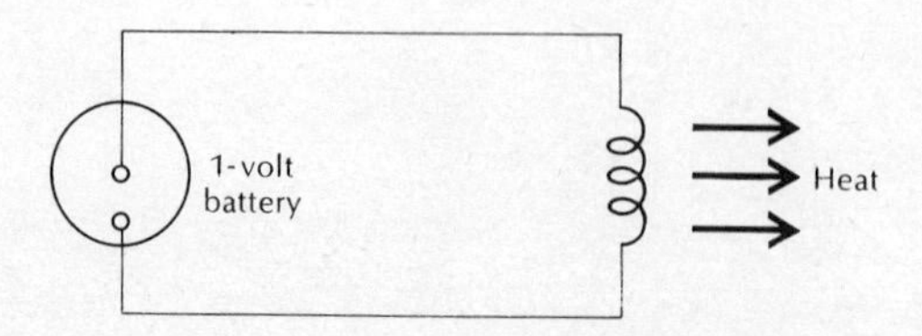

FIGURE 10.5 In this example the electric work done by the current turns up as heat.

$$\text{Power} = \text{current} \times \text{potential} \tag{10.2}$$

$$P = IV$$

We can check the units to see if this is correct.

$$\text{Watts} = \text{volts} \times \text{amperes}$$

$$= \frac{\text{joules}}{\text{coulomb}} \times \frac{\text{coulombs}}{\text{second}}$$

$$= \frac{\text{joules}}{\text{second}}$$

And since the watt is defined as 1 joule/second, everything is consistent.

10.3 OHM'S LAW AND RESISTANCE

The behavior of metal wires when an electric current is passed through them was investigated by Georg Ohm (1789–1854), and resulted eventually in a law that bears his name. This is an experimental law which is obeyed by most, but by no means all, electric conductors. Wires of all metals at all but the very lowest temperatures show a simple relationship between the current and the driving voltage. The current is found to be proportional to the voltage:

$$I \propto V$$

A plot of current versus voltage for three different lengths of wire showing this relationship is given in Fig. 10.6. Another way of stating this result is to say that voltage divided by current is a constant that depends on the specific conductor. This constant is called the electric resistance of the conductor and is usually designated R. Its name is meaningful in that a conductor with a high resistance "resists" the flow of current. It takes a larger voltage to drive a given current through a high resistance than through a low resistance. The unit of resistance is called the *ohm*, if the current is in amperes and the potential in volts.

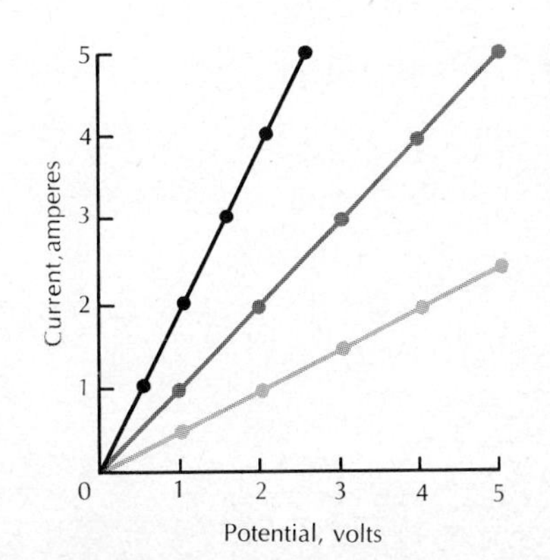

FIGURE 10.6 Current plotted as a function of voltage for three specimens of wire. The straight-line results show that current is proportional to voltage. The lightest line represents the wire with the most resistance, because the least current flows at any value of the voltage. The black line represents the wire with the smallest resistance of the three.

If a one-volt battery drives one ampere of current through a wire, the resistance of the wire is defined as one ohm. Ohm's law can thus be stated

$$\frac{\text{Voltage}}{\text{Current}} = \text{resistance}$$

or

$$\text{Voltage} = \text{current} \times \text{resistance} \qquad (10.3)$$
$$V = IR$$

The flow of water in a hose provides a useful analogy for electric current. The amount of water flowing corresponds to the current, and the water pressure at the tap to the voltage. A larger hose will carry more water per second for a given pressure, and therefore its "resistance" is lower.

If we take Equation 10.2 for the power delivered by a current I driven by a voltage V,

$$P = VI \qquad (10.2)$$

and combine it with Ohm's law, we can obtain two other relationships for the power used or dissipated in a particular circuit. These are

$$\text{Power} = I^2R \qquad (10.4)$$

and

$$\text{Power} = \frac{V^2}{R} \qquad (10.5)$$

It appears from these equations that, if the resistance is increased, the power can be greater or less, depending on which equation one uses. Clearly this is nonsense. Remember that an increase in resistance leads to a decrease in current. So both current and resistance in Equation 10.4 change when the resistance is changed. The correct conclusion is that power is decreased when resistance is increased, if voltage is kept constant. Which equation to use of the three presented depends on what data are available.

Example 10.1

Calculate the current needed for an electric dryer that uses 2,000 watts (2 kilowatts) at a voltage of 220 volts.

$$P = IV$$
$$2{,}000 \text{ watts} = I \times 220 \text{ volts}$$
$$I = \frac{2{,}000}{220} \text{ amperes}$$
$$\cong 9 \text{ amperes}$$

Example 10.2

Calculate the resistance in the heater of the dryer in Example 10.1. This can be calculated either from Ohm's law, or from the power equations:

$$P = \frac{V^2}{R}$$

$$R = \frac{V^2}{P} = \frac{(220 \text{ volts})^2}{2{,}000 \text{ watts}}$$

$$\cong 24 \text{ ohms}$$

or

$$P = I^2R$$

$$R = \frac{P}{I^2}$$

$$= \frac{2{,}000 \text{ watts}}{(9 \text{ amperes})^2}$$

$$\cong 24 \text{ ohms}$$

or

$$V = IR$$

$$R = \frac{V}{I}$$

$$= \frac{220 \text{ volts}}{9 \text{ amperes}}$$

$$\cong 24 \text{ ohms}$$

SUPERCONDUCTIVITY

Although it applies to most conductors at room temperature, Ohm's law is not obeyed for a number of metals at very low temperatures. When cooled to very low temperatures, metals such as lead, mercury, and aluminum abruptly lose all electric resistance. They are called *superconductors*. The resistance is as close to zero as any available

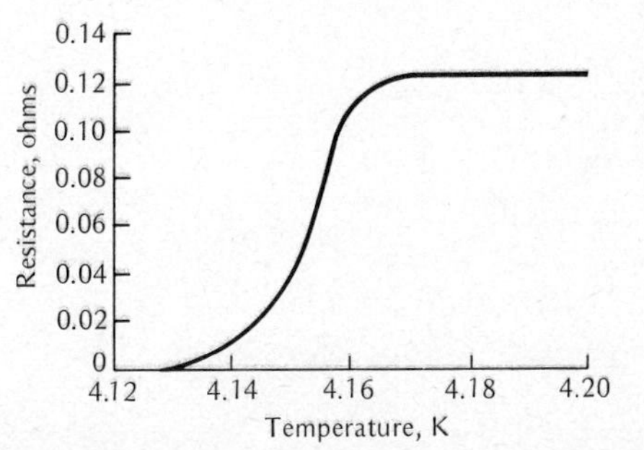

FIGURE 10.7 This graph shows how the electric resistance of mercury vanishes at about 4.15 K.

instrument can measure. The temperature where this transition occurs varies with the material, and is 7.22 K for lead, 4.15 K for mercury, and 1.14 K for aluminum. Table 10.1 lists some transition temperatures.

The fact that all metals at normal temperatures have some resistance means that electric energy cannot be transported over long distances without some loss of energy to heat in the power lines. Electric transmission takes place at high voltages and low currents, so as to minimize these losses. But in a 500-mile line a significant amount of the total energy is lost in transmission. This has led to proposals to use buried lines made of superconducting materials. The drawback is that the highest temperature at which any known material is superconducting is about 22 K. The possibility of achieving greatly higher transition temperatures is considered not to be very great. Technology has not advanced to the point where this appears to be a practical alternative to conventional power lines, except perhaps within metropolitan areas where overhead lines are not practical, and buried conventional lines are very expensive.

10.4 THE ELECTRIC FIELD

In Chapter 9 we discussed briefly the difficulties with the action-at-a-distance concept. One way to avoid these difficulties is to propose that there is some physical entity present at the point in space where the force is applied. That is, if we consider the electric force between two charges A and B, instead of saying that A and B exert a force on each other acting through the distance that A and B are separated, we can say that A somehow disturbs the space at B and that this disturbance creates the force on B. This disturbance is called the *electric field*. The field concept was introduced by Michael Faraday (1791–1867).

The electric field is a quantity like velocity, in the sense that it has

TABLE 10.1 Temperatures at which some materials become superconducting

Material	Temperature, K
Aluminum	1.20
Zinc	0.91
Tin	3.73
Mercury	4.15
Lead	7.22
Thorium	1.39
Lead-bismuth alloy	7.3–8.8
Niobium-tin alloy	18
Vanadium-gallium alloy	15
Aluminum-germanium-silicon alloy	22

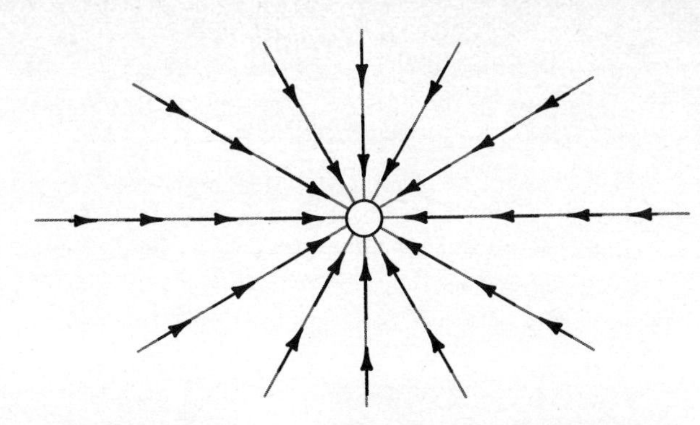

FIGURE 10.8 The size of the arrows indicates the magnitude of the electric field. The grey lines indicate the lines of force.

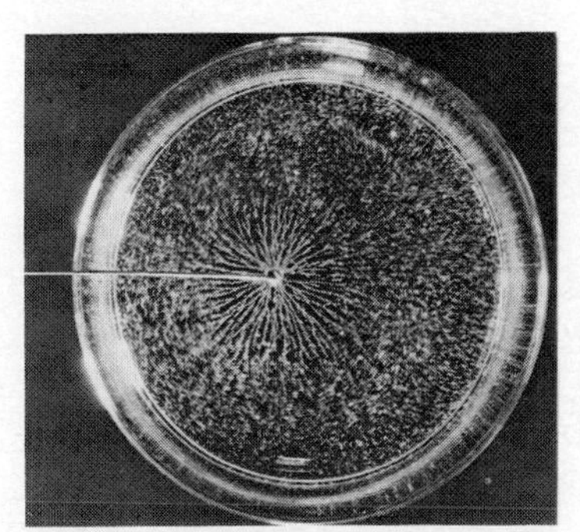

(a)

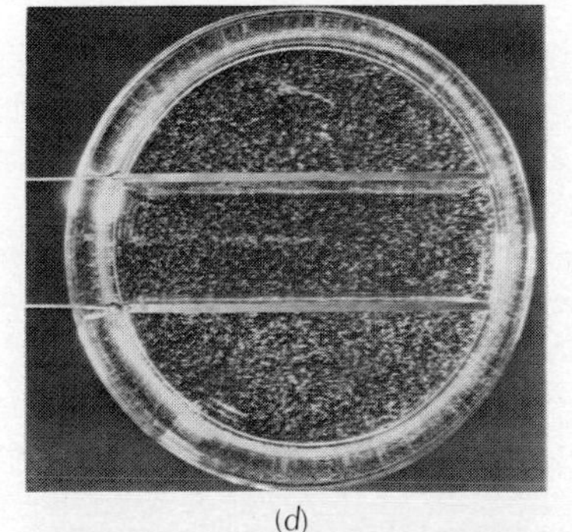

(d)

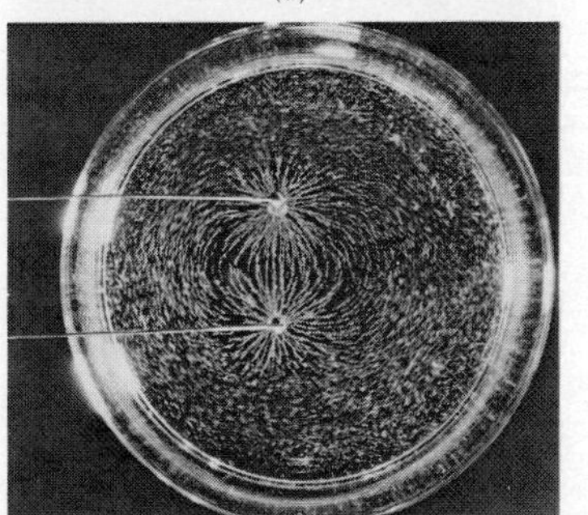

(b)

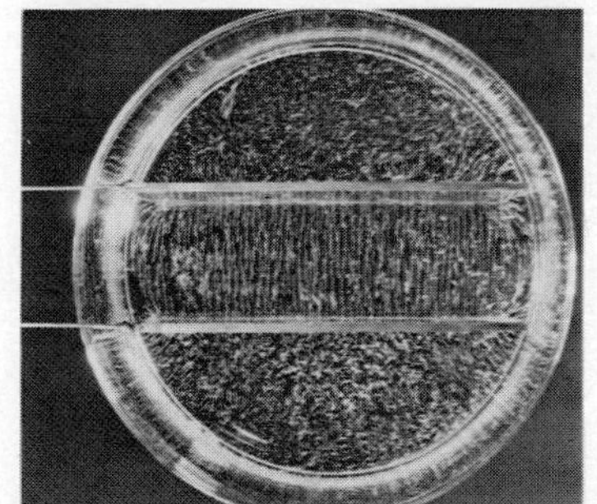

(e)

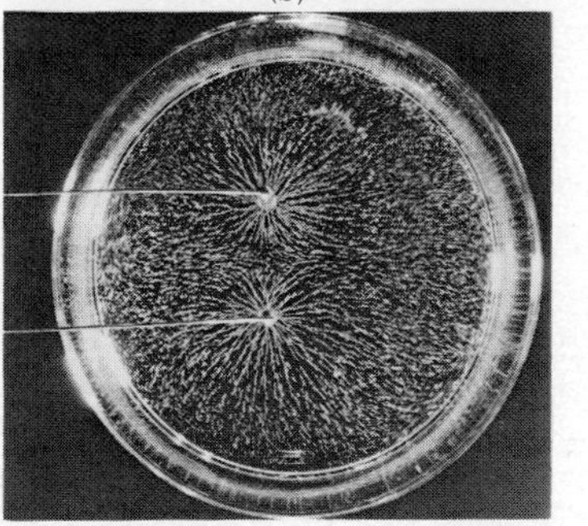

(c)

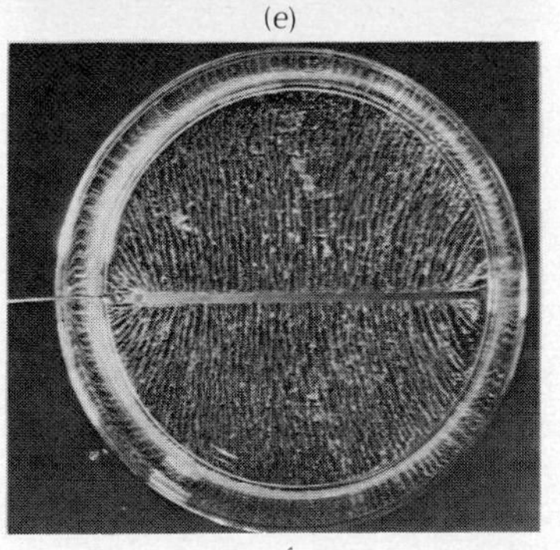

(f)

FIGURE 10.9 Grass seeds floating in an insulating liquid are found experimentally to line up parallel to an electric field. These six photographs show the electric field resulting from several distributions of electric charge. (*a*) A single charged rod. (*b*) Two rods with equal and opposite charges. (*c*) Two rods with the same charge. (*d*) Two parallel plates, no electric field. (*e*) Two parallel plates with opposite charges. (*f*) A single charged metal plate. (*Courtesy PSCC, "Physics," D. C. Heath and Company, 1965.*)

both a magnitude and a direction. Figure 10.8 shows the electric field near an electric charge confined to a small region of space. The direction of the arrows shows the direction of the field, and their length is an approximate indication of the magnitude of the field. The direction of the arrow indicates the direction of the force on a positive charge. Faraday introduced the concept of lines of force, indicated in grey. The line of force shows the direction a charge will move when subjected to the electric force in question. Note that the closer together the lines of force, the greater the magnitude of the electric field. Lines of force begin or end on an electric charge.

It is worthwhile to indicate the utility of the electric field concept. After all, why not use Coulomb's law to calculate the force on any charge anywhere? But if there is an array of electric charges in a shape which is not mathematically simple, this will be difficult. The electric field is the measure of the total effect of all charges anywhere on a charge at a point. We can state and measure the electric field without describing the distribution of charge which causes it.

For many years after the introduction of the field concept by Faraday, there existed a difference of opinion as to whether the field is a physical reality or simply a convenient mathematical device. The modern view is that the electric field is a physical reality. In Chapter 18 we shall see that light consists of particles called *photons*. It is now believed that the electric force between two charged objects is caused by the exchange of a type of photon. These photons are passed back and forth between charged objects, and this results in the electric force. The electric field at a point, then, is a measure of the presence of photons at that point. It is not possible here to develop these ideas further, because of their mathematical complexity.

The precise numerical definition of electric field can be seen using Fig. 10.10. If we consider the point *A* in Fig. 10.10, it is clear that any charge placed at *A* will experience a force due to the charge at *B*. We define the electric field at point *A* as the force on 1 coulomb of charge if it were to be placed at *A*. From the definition above, the units of electric field are newtons per coulomb. The electric field gives the force on each coulomb of charge placed at *A*. We can describe experimentally the way to find the electric field at *A*. Take a small amount of electric charge (small enough so that the field we are trying to measure is not disturbed) and place it at *A*. Measure the force on that charge. The electric field is then the magnitude of the measured force divided by the value of the small test charge. The force on any charge placed at *A* is given by the magnitude of that charge times the electric field *E* at point *A*. [In an exactly analogous sense, the gravitational field at a point is given by *g*. Its units are newtons per kilogram. To calculate the total force (weight) we multiply *g* by the mass.]

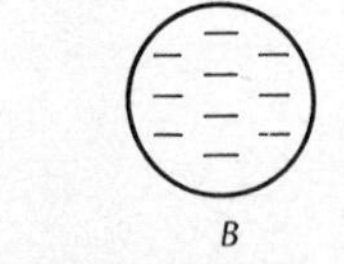

FIGURE 10.10 The electric field at *A* is given by the force on a unit charge placed at *A*.

For example, the electric field due to a single charge can be calculated from Coulomb's law. Coulomb's law gives the force between two charges Q_1 and Q_2 as

$$F = 9 \times 10^9 \frac{Q_1 Q_2}{r^2}$$

The electric field E at the position of charge 1 is given by

$$E = \frac{F}{Q_1} = 9 \times 10^9 \frac{Q_2}{r^2} \qquad (10.6)$$

Equation 10.6 gives the electric field due to the charge Q_2 at the position of Q_1.

The force on a charge due to an electric field is always given by

$$\text{Force} = \text{charge} \times \text{electric field}$$

$$F = QE \qquad (10.7)$$

THE RELATIONSHIP BETWEEN ELECTRIC POTENTIAL AND ELECTRIC FIELD

From the definition of electric potential, the work required to move a charge from A to B, if there is a potential of V volts between A and B, is

$$W = QV$$

$$W = \text{coulombs} \times \frac{\text{joules}}{\text{coulomb}} = \text{joules}$$

But the force on a charge Q can also be calculated as QE, where E is the electric field. Therefore the work done on a charge Q, if it is moved a distance d, is

$$W = F \times d = EQd$$

The work must be the same no matter how we calculate it; so the two expressions for work can be set equal to each other. The result is

$$EQd = QV$$

$$Ed = V$$

or

$$E = \frac{V}{d} = \frac{\text{volts}}{\text{meter}} \qquad (10.8)$$

Therefore another way of expressing the electric field in mks units is volts per meter. One volt/meter is the same as one newton/coulomb.

This discussion of electric potential has left ambiguous the choice of sign. A point B has a positive potential with respect to A if it takes an outside force to move a positive charge from A to B, that is, if the electric force on a positive charge is directed from B to A.

Example 10.3

Two flat plates 2 centimeters apart have a voltage of 100 volts applied between them. What is the electric field between them?

$$E = \frac{V}{d}$$

$$= \frac{100 \text{ volts}}{0.02 \text{ meter}}$$

$$= 5{,}000 \text{ volts/meter} = 5{,}000 \text{ newtons/coulomb}$$

Note that the separation must be expressed in meters.

10.5 ELECTRICITY IN THE ATMOSPHERE

The ideas of electric field and electric potential can be used to discuss some interesting features of atmospheric electricity. On a normal day there is an experimentally observed electric field near the ground of about 130 volts/meter. It is directed vertically downward (a positive charge will move downward). The magnitude of the field decreases at higher altitudes, so that the total potential between ground level and 10 kilometers is, roughly, 300,000 volts. Because the air is a very poor

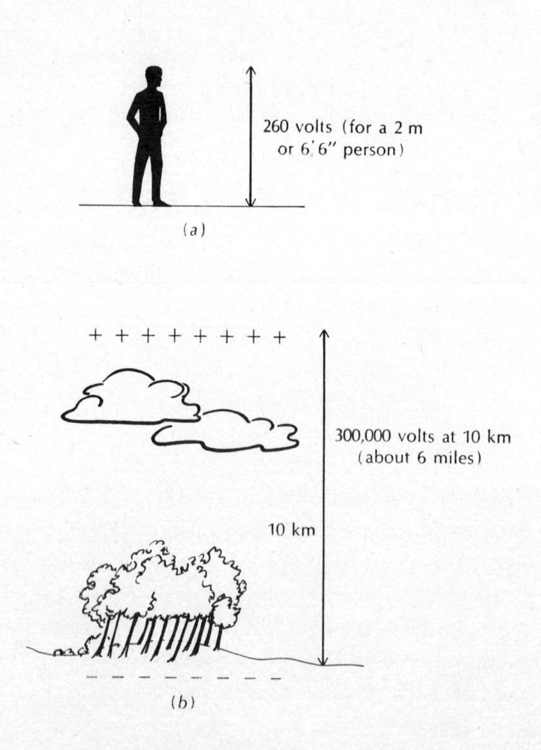

FIGURE 10.11 (*a*) The electric field near the surface of the earth is normally about 130 volts/meter. (*b*) The total potential to a height of 10 kilometers is normally about 300,000 volts.

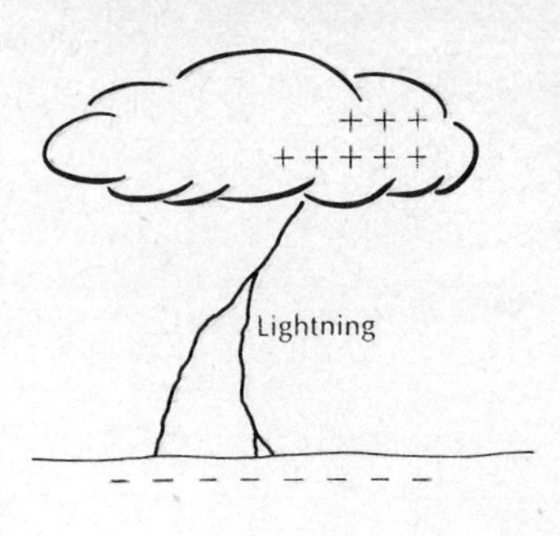

FIGURE 10.12 Lightning is the flow of large amounts of current to restore the electric balance which has been upset in the thundercloud.

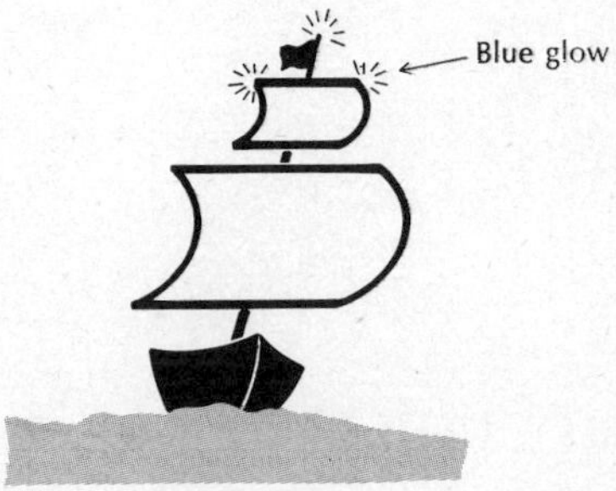

FIGURE 10.13 St. Elmo's fire was observed by sailors on stormy nights. It occurs because the ship is a better conductor of electricity than air. The blue glow is evidence that substantial currents are being carried through the ship and discharged into the air at sharp points on the rigging.

conductor, the currents associated with this electric field are very small —too small for any but the most sensitive instruments. It is believed that thunderstorms all over the earth are the mechanism which maintains the potential, but the details are not well understood.

During an electric storm the electric field at ground level rises and then returns to normal with each stroke of lightning. Some process in the storm clouds separates positive and negative charge. This creates large electric fields, which then produce the explosive surge of electric current we call lightning. Currents of over 100,000 amperes have been observed in a stroke of lightning. The total charge transferred ranges between 10 to 100 coulombs.

Sailors in sailing vessels frequently observed the phenomenon of St. Elmo's fire on stormy nights. This is a blue glow which appears on the pointed ends of spars and rigging. It is caused by the flow of electricity due to the large electric fields which exist during storms. The glowing discharge from a pointed object is called a *corona discharge*, and can be created in the laboratory by placing a large amount of electric charge on any object with a sharp point. All generators of high voltages are made with smoothly curved surfaces to prevent charge from leaking off through such discharges.

High-voltage electric power lines will also show corona discharges, particularly in stormy weather when the electric resistance of air is low. Since this is a source of energy loss, power companies try to minimize

FIGURE 10.14 Corona discharge on a 345,000-volt electric power line in a light rain. (*Photograph by Thomas C. Brown.*)

the effect, but it cannot be entirely eliminated. Figure 10.14 shows a corona discharge on a 340,000-volt power line in Utah.

SUMMARY

Electric current is the flow of electric charge from point to point. Its unit is the ampere, defined as one coulomb of charge passing a point per second.

The chemical battery permitted the generation of large electric currents, compared to electrostatic generators.

One volt is the capacity to do one joule of useful work for each coulomb of charge delivered.

Power is defined as the rate of doing work. The unit of power, the watt, is one joule per second. In an electric circuit power is given by

$$\text{Power (watts)} = I \text{ (amperes)} \times V \text{ (volts)}$$

Ohm's law states that current and voltage are related by the resistance of a wire:

$$R \text{ (ohms)} = \frac{V \text{ (volts)}}{I \text{ (amperes)}}$$

Two other relationships for power combine Ohm's law with $P = VI$. They are

$$P = I^2R \qquad \text{and} \qquad P = \frac{V^2}{R}$$

Superconductors are materials whose electric resistance vanishes at some low temperature, called the transition temperature.

The electric field at a point A is defined in terms of the force on a (hypothetical) test charge placed at A. The units of electric field are newtons per coulomb, or also volts per meter. The force on a charge at A, if the field at A is E_A, is

$$F = QE_A$$

If the potential difference between two points A and B is known, the electric field can be calculated as

$$E = \frac{V_{AB}}{d_{AB}}$$

(Strictly this is true only for flat plates a distance d apart.)

QUESTIONS

1 If a 1-ampere current is flowing through a wire, how many electrons are passing a given point every second?

2 Why does a chemical battery eventually discharge?

3 One coulomb of charge moves from one terminal of a 12-volt battery to another. How many joules of electric energy are delivered by the battery in this process?

4 Name a few devices in your home which depend on the heat generated in a conductor by an electric current to operate.

5 How much more current flows through a 100-watt bulb than a 50-watt bulb?

6 Which has a greater resistance: a 50-watt bulb or a 100-watt bulb?

7 Will a superconducting wire carrying a current become warm?

8 Why is it more efficient to transport electric power at low current and high voltage rather than the other way around?

9 Does a negative charge placed at a point in space at which an electric field exists experience a different electric field than that experienced by a positive charge at the same point? Does the negative charge experience a different electric force?

10 If there is an electric field of 130 volts/m at the earth's surface, does this mean that currents are continually flowing from your head to your toes?

11 Do the electrons in the wire connecting the terminal of a battery flow away from or toward the negative terminal?

12 A 10-watt bulb is powered by a D battery. How long will the bulb burn?

13 Why do you think automobile batteries work poorly in very cold weather?

14 Why should a line of force begin and end on a charge?

15 In a typical lightning stroke about 25 coulombs of charge move. Before the discharge the electric potential difference between earth and sky is about 30 million volts. How much energy is stored in the system of separated charges?

SELF-TEST

__________ Volta
__________ Ohm
__________ Faraday
__________ electric field
__________ electric current
__________ chemical battery
__________ electric potential
__________ power
__________ watt
__________ volt
__________ ohm
__________ ampere
__________ superconductor
__________ corona discharge

1 Developed the concept of electric resistance
2 Invented the chemical battery
3 Developed the field concept
4 The unit of power; equal to 1 joule/second
5 The unit of resistance
6 A device to generate electric current through chemical processes
7 A metal whose electric resistance vanishes at low temperatures
8 Emission of electric charge with an accompanying glow of light
9 The flow of electric charge; measured in coulombs per second
10 The unit of electric potential; equal to 1 joule/coulomb
11 The rate of doing work
12 The capability of doing electric work; measured in volts
13 The force on a charge placed at a point can be calculated if this quantity is known

14 The unit of electric current; given by coulombs per second

PROBLEMS

1 Calculate the electric field 10^{-3} m from an electron ($Q = -1.6 \times 10^{-19}$ coulomb).

2 Calculate the electric field 2 m from a charge of 10^{-6} coulomb.

3 Calculate the electric field 10^{-10} m from an electron ($Q = -1.6 \times 10^{-19}$ coulomb).

4 Calculate the electric field 6 m from a charge of 10^{-9} coulomb.

5 A charge of 10^{-6} coulomb is moved 0.01 m parallel to an electric field of 10 newtons/coulomb. (*a*) How much work is done? (*b*) What is the potential difference between the first and last point of this motion?

6 A charge of 3×10^{-9} coulomb is moved 2 cm parallel to an electric field of 10^5 volts/m. How much work is done?

7 A current of 3 amperes flows through a potential difference of 1,000 volts. Calculate the power.

8 (*a*) What is the electric field 10 m from a charge of 10^{-10} coulomb? (*b*) If a second charge of 10^{-16} coulomb and mass 10^{-20} kg is placed at this point and released, what is its initial acceleration?

9 The electric field at a point 1 m from a charge is 10^{-5} newton/coulomb. (*a*) What is the charge? (*b*) At what distance from the charge is the field twice the value given above? (*c*) What is the direction of the field at the point found in part (*b*), away from or toward the charge?

10 (*a*) What is the electric field $\frac{1}{2}$ m from a charge of 10^{-5} coulomb? (*b*) What force acts on a charge of 10^{-12} coulomb placed at this point? (*c*) What is the direction of this force, toward or away from the charge found in part (*a*)?

11 The electric field in the "electron gun" of a TV picture tube is found to be 10^5 newtons/coulomb. (*a*) What is the electric force acting on an electron in this field? (*b*) What is the gravitational force acting on the electron? (That is, what is the weight of the electron?) (*c*) From your answers to parts (*a*) and (*b*), do you think it is important to consider the gravitational force acting on an electron in such a situation?

12 An electron is released at rest and accelerates through a voltage of 10,000 volts. (*a*) What was the original electric potential energy of the electron? (*b*) What was the final kinetic energy of the electron? (*c*) If the mass of the electron is 9.1×10^{-31} kg, what is the final velocity?

13 A 100-watt light bulb is driven by a potential of 220 volts. Calculate the current flowing in the bulb.

14 A 500-watt iron is driven by a potential of 110 volts. Calculate the current.

15 A current of 10 amperes flows through a potential difference of 10 volts. (*a*) What is the power? (*b*) How much work is done in 1 minute?

16 An electric stove has a self-cleaning oven. When the oven is being cleaned, a current of 35 amperes is drawn from a 220-volt line. It takes 0.5 hour to clean the oven. (*a*) What power is used in cleaning the oven? (*b*) What is the resistance of the oven-cleaning element? (*c*) How much energy in joules and kilowatthours is required to clean the oven? (*d*) If power costs 2.5¢/kilowatthour, how much does it cost to clean the oven?

17 An air conditioner operating off a 110-volt line draws 15 amperes current when it operates at maximum cooling power. (*a*) How much electric power does this air conditioner require? (*b*) How much energy is used by this device in 1 hour? (*c*) If the power company charges 2.5¢/kilowatthour for electricity, how much does it cost to run this device for 1 hour?

18 A 12-volt battery is used to power a 100-watt light bulb designed to be used with the battery. (*a*) What current does the bulb draw? (*b*) What is the resistance of the bulb? (*c*) If a 120-volt potential difference is used, what current flows through the bulb?

19 An "instant on" TV set uses 15 watts of power from a 120-volt power line when it is just plugged in but not turned on. (*a*) What current does this set draw under these conditions? (*b*) How much electric energy does this set use in 1 day of just sitting idle? (1 day $= 8.6 \times 10^4$ seconds) (*c*) If power costs 2.5¢/kilowatthour, how much does it cost to have this set just plugged in for 1 day?

20 A student studies for 4 hours with the aid of a 200-watt light bulb. The line voltage is found to be 110 volts. (*a*) What electric current does the bulb carry? (*b*) What is the electric resistance of this bulb? (*c*) Electric energy costs about 2.5¢/kilowatthour. How much does it cost the student to burn his bulb for 4 hours? (*d*) How many joules of electric energy are expended by the bulb during this 4-hour period?

21 An electron starts from rest in an electric field of 100 volts/m. Calculate its energy after it has moved 0.01 m.

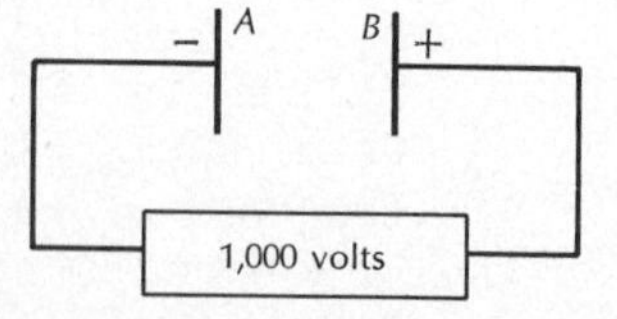

22 Electrons with a charge of -1.6×10^{-19} coulomb are emitted at *A*. The mass of the electron is 9×10^{-31} kg. Calculate the energy the electrons have when they arrive at *B*.

23 An electron ($Q = 1.6 \times 10^{-19}$ coulomb) is moved 0.1 m parallel to an electric field of 5×10^4 newtons/coulomb. How much work is done?

24 An electric field *E* is set up as shown. A charge of 1 coulomb is moved from *A* to *B*, a distance of 10 cm. How much work is done?

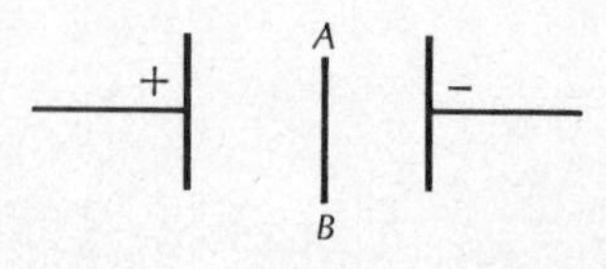

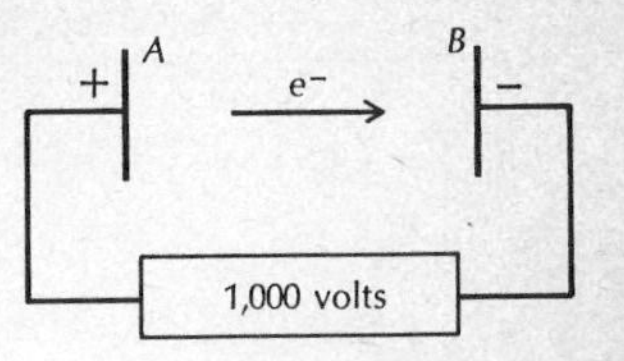

25 An electron is emitted from A in the direction shown with a kinetic energy of 2.0×10^{-16} joule. What is its energy when it arrives at B? The charge of the electron is -1.6×10^{-19} coulomb.

26 A kilogram of electrons contains 1.1×10^{30} electrons. (*a*) If they are all gathered at one place, what is the electric field 1 m from this bundle of electrons? (*b*) If a proton (charge 1.6×10^{-19} coulomb) were placed at this point, what force would act on it? (*c*) If the proton were released, would it move away from or toward the electrons?

27 The picture tube of a color TV set accelerates the electrons which aid in forming the image through a voltage of the order of 25×10^3 volts. The mass of the electron is 9.1×10^{-31} kg; its charge is 1.6×10^{-19} coulomb. (*a*) What is the work done on the electrons? (*b*) What is the kinetic energy of the electrons after they have been accelerated? Assume they start from rest. (*c*) What is the final velocity of the electrons?

28 The electric potential difference between two plates is 8×10^3 volts. The plates are separated by a distance of 2 cm. (*a*) What is the electric field between the plates? (*b*) What force acts on a charge of 10^{-15} coulomb when it is placed between the plates? (*c*) How much work is done on the charge of 10^{-15} coulomb as it moves from the positive to the negative plate? (*d*) Assuming the charge starts from rest, what is its velocity just before it strikes the negative plate? Assume the charge has a mass of 10^{-21} kg.

29 An electron is accelerated through an electric potential difference of 400 volts. The charge of the electron is -1.6×10^{-19} coulomb. The mass of the electron is 9.1×10^{-31} kg. (*a*) What is the work done on the electron? (*b*) What is the kinetic energy of the electron? Assume that it starts from rest. (*c*) What is the velocity of the electron? (*d*) If the electron is accelerated over a distance of 10^{-3} m, what is the electric field in the region in which the electron moves?

30 The electric field between two charged parallel plates, separated by a distance of 10^{-3} m, is found to be uniform and equal to 1,000 volts/m. (*a*) What is the electric potential difference between these plates? (*b*) A positive charge of 10^{-15} coulomb is released from rest at the positively charged plate. How much work is done on this charge as it moves across the gap separating these plates to the negatively charged plate? (*c*) What is the kinetic energy of this charge when it reaches the negatively charged plate? (*d*) If the charge has a mass of 10^{-20} kg, what is its velocity when it strikes the negatively charged plate?

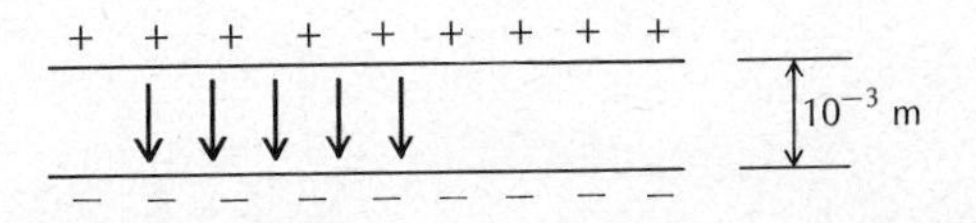

31 It is found that near the surface of the earth an electric field exists which is directed downward and has a magnitude of 130 volts/m. (*a*) A small charge of 10^{-16} coulomb is placed in this field. What electric force acts on this charge? (*b*) How much work is done on this charge if it moves a distance of 10 m under the influence of this force? (*c*) What is the electric potential difference over this 10-m distance?

32 The nucleus of a helium atom contains two protons and two neutrons. The protons are separated by a distance of about 1.3×10^{-15} m. (*a*) What is the force of electric repulsion between the protons? (*b*) What is the electric field due to these protons at a distance of 5×10^{-11} m from the nucleus? Assume that the two protons act as a single charge of 3.2×10^{-19} coulomb. (*c*) What is the electric force acting on an electron at a distance of 5×10^{-11} m from the nucleus?

33 Two charges, both of charge 10^{-10} coulomb, are placed at points *A* and *B* on the line drawn to the right. Point 0 is midway between *A* and *B*. (*a*) What is the electric field at 0 due to the charge at *A*? (*b*) What is the electric field at 0 due to the charge at *B*? (*c*) If a charge of charge 10^{-5} coulomb were placed at 0, in what direction would its initial acceleration be? (*d*) What is the net electric field at 0?

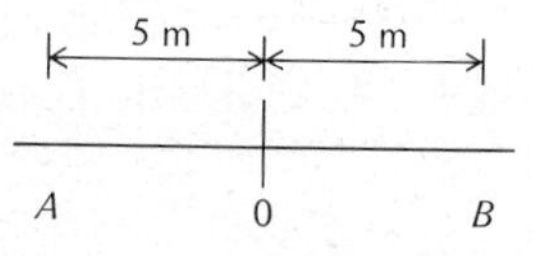

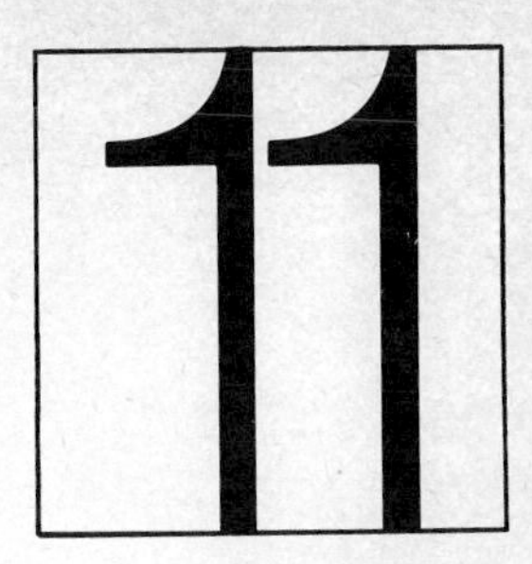

MAGNETS AND THE MAGNETIC EFFECTS OF A CURRENT

William Gilbert's book *De Magnete* (1600) summarized the knowledge on magnetism accumulated from ancient times up to that date. At some early stage in history the discovery was made that certain rocks had the property of attracting to them small pieces of iron. These rocks, called lodestones, are particularly prevalent in the region once known as Magnesia, in Asia Minor. From this comes the term magnetism, used to describe the phenomenon. The Chinese apparently first made the discovery that a specimen of lodestone orients itself in a north-south direction if suspended on string. Thus was invented the magnetic compass.

If a piece of iron or steel is rubbed in one direction by a lodestone, it acquires the properties of the lodestone. It will attract small bits of iron and, if suspended, will act as a compass. It is possible to demonstrate this rather easily by taking a steel needle (not a pin, which is brass) and stroking it in one direction (eye to point or point to eye) with a lodestone or any other magnet, as in Fig. 11.1. (This process involves organizing, or lining up, atom-sized magnets. Rubbing in both directions would disorder them.) Now float the needle on a bowl of water. (The needle will float if lowered gently onto the surface, with the fingers or a fork, as shown in Fig. 11.2.) The floating, magnetized needle becomes a compass. One end will point north, and the other south, if other magnets or iron objects are kept away. We shall use the compass and its properties to describe some of the properties of magnets.

11.1 THE MAGNETIC FIELD

The needle we have stroked with the lodestone is said to be magnetized. It is an example of what is called a bar magnet—a rod of iron or steel which has been magnetized. It is possible to magnetize iron bars in other ways, but for the moment we shall simply study their properties.

One end of any bar magnet will point to the north. Let us call this end the north-seeking pole, or simply the north pole of the magnet. The other end is called the south pole. If we bring an ordinary compass near a strong bar magnet, we shall find that the compass, instead of pointing north, will point toward the *south* pole of the bar magnet. That is, the north-seeking end of the compass is attracted to the south-seeking end of any other magnet.

Gilbert was led by this type of observation to conclude that the earth

FIGURE 11.1 A needle can be magnetized by stroking with a magnet in one direction.

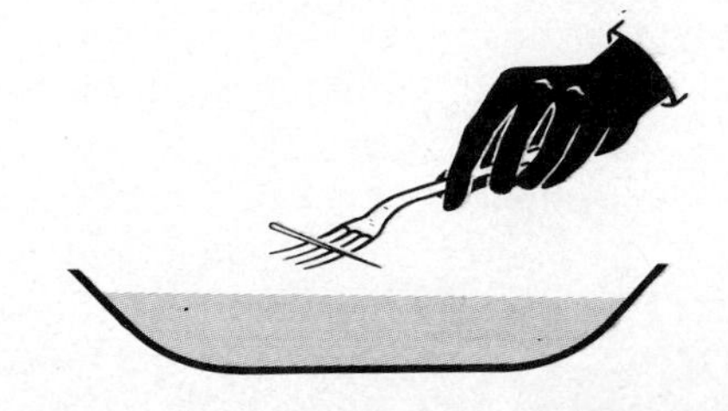

FIGURE 11.2 Floating a needle on the surface of a dish of water.

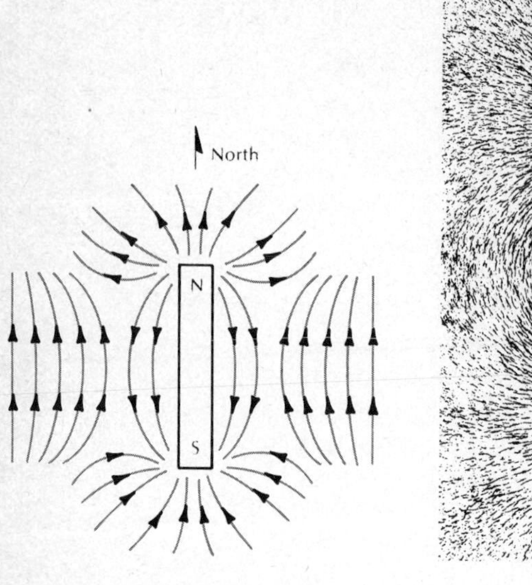

(b)

FIGURE 11.3 (*a*) Near a bar magnet a compass (small arrows) points toward the south pole of the bar magnet. Far from the magnet the compass points north in response to the earth's magnetic field. The lines are magnetic field lines. (*b*) Iron filings in the immediate vicinity of a magnet. The iron filings act like small compasses. (*Courtesy PSSC, "Physics," D. C. Heath and Company, 1965.*)

behaves like a huge magnet and has a *south* pole near the earth's geographic north pole. The exact location of the magnetic pole is, in fact, just north of Hudson Bay. There is also a magnetic pole near the earth's geographic south pole. If we use the system of naming described above, it is a *north* pole.

Suppose we now place a bar magnet on a table in a north-south direction and explore the tabletop in its vicinity with a compass. Figure 11.3*a* shows the results. Far away from the bar magnet the compass points toward the north. Nearer the magnet we see that the compass points more and more toward the magnet, and not toward the north. The compass then tells us something about what is going on in the vicinity of the bar magnet. The iron filings in Fig. 11.3*b* also demonstrate this. Figure 11.4 shows a similar pattern of iron filings around a lodestone. Recalling Fig. 10.8, which showed an electric field, we can use the compass result to define what we mean by a magnetic field. The *direction* of the magnetic field is shown by the direction of the compass arrow. The lines are drawn parallel to the direction in which the compass points, and are the magnetic field lines.

The compass experiment has indicated that the north and south poles of a magnet attract one another. What happens if we bring two south poles together? Experiment shows a strong force driving them apart. Therefore there is a similarity between electrical and magnetic effects. Like electric charges repel one another, and unlike charges attract. Like magnetic poles repel, and unlike poles attract. But there is a striking difference in that we never find a single magnetic pole, as

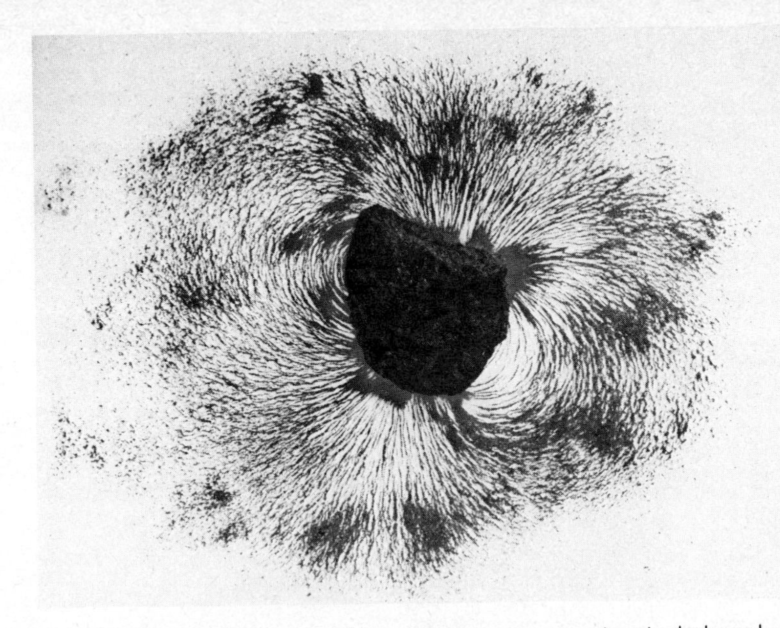

FIGURE 11.4 Iron filings have been sprinkled around this piece of lodestone. The filings line up parallel to the direction in which a compass would point. The pattern formed by the filings shows the magnetic field of the lodestone. (*Photograph by Alexis Kelner.*)

we do a single electric charge. All magnets ever constructed show both a north and a south pole.

11.2 THE MAGNETIC EFFECT OF A CURRENT

The similarity between electrical and magnetic effects, in that both attraction and repulsion exist, is striking. There are other bits and pieces of evidence that indicate, perhaps, that the two kinds of effects are related. Both show forces which act at a distance from their source. Bolts of lightning passing through iron rods are observed to leave the rods magnetized. Although the idea had been around for some time, the invention of the battery in 1800 made possible the experimental discovery of the connection between electricity and magnetism, because it made possible the generation of reasonably large electric currents.

FIGURE 11.5 Hans Christian Oersted (1777–1851). (*Courtesy University of Pennsylvania Library.*)

Hans Christian Oersted, a Danish professor of natural philosophy, set out to find the effect of electricity on a compass. In 1820 he reported the results of a series of experiments. He discovered that a steady electric current generates a magnetic field in the region surrounding the wire, as demonstrated by its effect on a compass. This result is indicated schematically in Fig. 11.6*a*. The center dot represents the wire, seen end on, which is carrying the current; the arrows indicate the direction a compass points in the vicinity of the wire; the lines represent the magnetic field lines, which show the direction the compass would point at any point in space. The result of a steady current is a magnetic field which goes in circles around the wire. If we investigate its strength, we will find that the field decreases as the compass is placed farther and farther from the wire. We will also find that the strength of the field de-

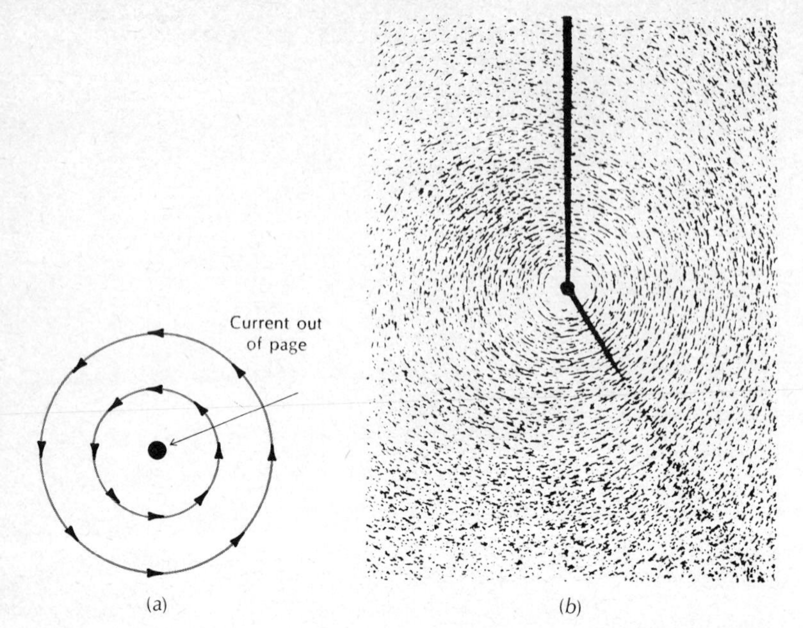

FIGURE 11.6 (*a*) Magnetic field around a current-carrying wire. (*b*) Photograph of iron filings near a current-carrying wire. (*Courtesy PSSC, "Physics," D. C. Heath and Company, 1965.*)

pends on the current in the wire. The larger the current, the stronger the magnetic field. Figure 11.6*b* is a photograph of iron filings near a wire that clearly shows this circular pattern.

The conclusion to be drawn from this experiment is that an electric current generates a magnetic field. If the experiment is done in various geometrical arrangements, the direction of the magnetic field is always found to be at *right* angles to the current. There is a simple rule for determining the direction of a magnetic field due to a current.

> *If the wire is grasped with the right hand, with the thumb in the direction of the flow of positive current, the fingers will point in the direction of the magnetic field.*

If magnetic fields are caused by currents, what is the explanation for the magnetism of an ordinary iron magnet? Clearly, there are no wires attached to the magnet. The modern view is that tiny current loops,

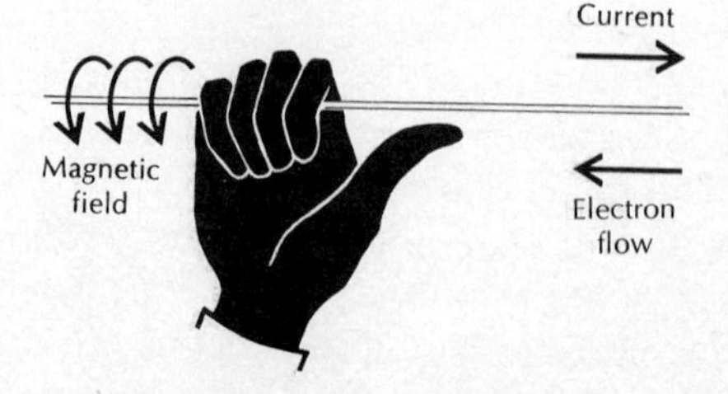

FIGURE 11.7 The right-hand rule for determining the direction of the magnetic field in a current-carrying wire. Electron flow is shown, to remind you that the definition of the direction of positive current is opposite that of electron flow.

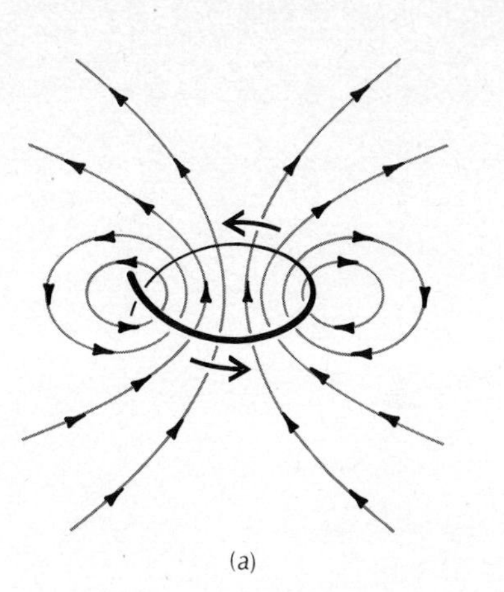

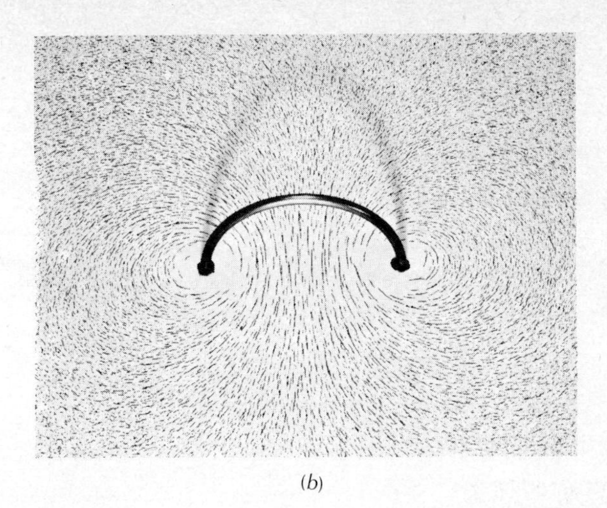

(a) (b)

FIGURE 11.8 (*a*) Magnetic field near a loop of wire. (*b*) Photograph of iron filings near a loop of current-carrying wire. (*Courtesy PSCC, "Physics," D. C. and Company, 1965.*)

the size of atoms, are the source of the magnetism of iron and other materials. Figure 11.8 shows the magnetic field near a loop of wire carrying a current. The shape of this field is remarkably similar to the shape of the field near a bar magnet, as was shown in Fig. 11.3.

Oersted's experiment shows that a force is exerted on a compass by the magnetic field of a wire. But Newton's third law says that there should be an equal and opposite force on the wire. Oersted and Michael Faraday sought and found such an effect by using a heavy magnet and a light wire. These two experiments are shown schematically in Fig. 11.9. In Fig. 11.9*a* the light compass needle (which is a magnet, as we have shown) moves in the magnetic field of the wire. In Fig. 11.9*b* a force acts so as to move the light wire vertically. The force on a current-carrying wire in a magnetic field is perpendicular to *both* the magnetic field and the direction of the current. The direction of the force depends on the direction of both the current and the magnetic field. If either is reversed, the force is reversed in direction. The use of the right hand again allows a determination of the direction of the force, as shown in

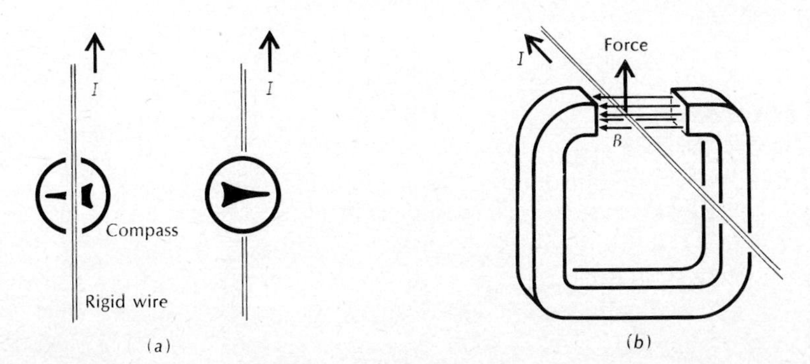

FIGURE 11.9 (*a*) A force is exerted on a compass by the magnetic field of a current-carrying wire. (*b*) A force is exerted on a current-carrying wire by a magnetic field.

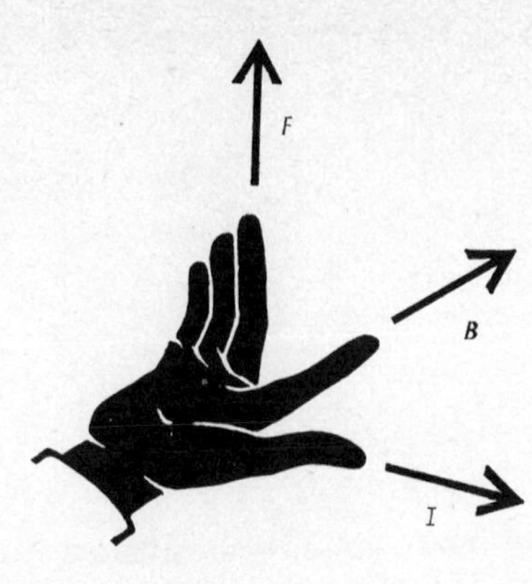

FIGURE 11.10 The right-hand rule for determining the direction of the force on a wire in a magnetic field.

Fig. 11.10. If the thumb and first two fingers of the right hand are extended at right angles to one another, with the thumb parallel to the current and the first finger parallel to the magnetic field, the direction of the middle finger shows the direction of the force.

Almost all devices for the measurement of electric current use the force on a wire in a magnetic field to indicate current. One of the simplest and most sensitive is called a *galvanometer*. It is indicated schematically in Fig. 11.11. A loop of wire is suspended in a magnetic field so that the loop is free to turn. When current is passed through the wire, there is a force downward at x and upward at y, and the loop turns. The amount of turning is a measure of the current.

All electric machinery, as well, depends on the force on a wire in a magnetic field. To make an electric motor requires only a simple modification of the galvanometer of Fig. 11.11. The force at point y is upward, even as the loop turns. Therefore, if a large current is passed through the loop, it will turn until it is vertical and then stop. If the direction of the current in the loop were reversed just as it passed this stopping point, the force on the wire would be in the proper direction to keep the loop moving. So, to make an electric motor, the arrangement must be similar to that in a galvanometer, except that the current direction is reversed each half-revolution. A motor, of course, is arranged with many turns of wire instead of a single loop, so that large forces can be achieved.

11.3 AMPÈRE'S EXPERIMENT: THE FORCES BETWEEN WIRES

Since electric currents produce magnetic fields which exert forces on magnets, and magnetic fields exert forces on currents, it is natural to suppose that two wires might exert a force on each other.

André Marie Ampère (1775–1836) made an experimental test of this question almost immediately after receiving word of Oersted's

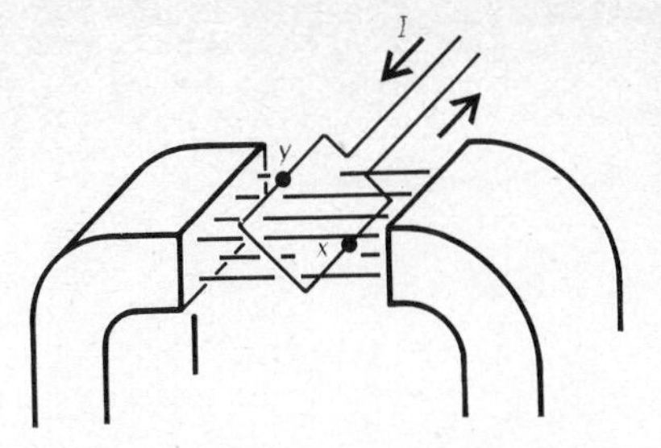

FIGURE 11.11 The force on the current-carrying loop due to the magnetic field causes the loop to turn. The force is down at x and up at y, if the magnetic field direction is to the right.

results. The results of his experiment are shown in Fig. 11.13. When the current is in the same direction in both wires, as in Fig. 11.13*a*, the wires are attracted toward one another. When the current is in opposite directions, as in Fig. 11.13*b*, the wires are repelled by one another. We can see from the magnetic fields generated by the wires and our two right-hand rules that the forces shown in Fig. 11.13 are in the direction to be expected. This is shown in Fig. 11.14. The right-hand rule applied to wire 1 in Fig. 11.13*a* shows that the magnetic field is directed *into* the paper. The second right-hand rule shows that such a field should produce a force upward on wire 2.

For practical reasons it is easier to measure accurately the force between wires carrying currents than the force between charged balls, as in Coulomb's experiments. Therefore the force between current-carrying wires is used to define the ampere of current. Since the ampere is one coulomb/second, the definition of the ampere also defines the coulomb. The ampere is defined as that current which will produce a force of exactly 2×10^{-7} newton on each meter of two wires exactly 1 meter apart.

FIGURE 11.12 André Marie Ampère (1775–1836). *(Courtesy University of Pennsylvania Library.)*

11.4 THE UNITS OF MAGNETIC FIELD

We can use the force on a wire in a magnetic field to show something about the strength of a magnetic field. Experimentally, it is found that the force on a current-carrying wire increases if the current increases, if the magnetic field increases, or if the length of wire in the field in-

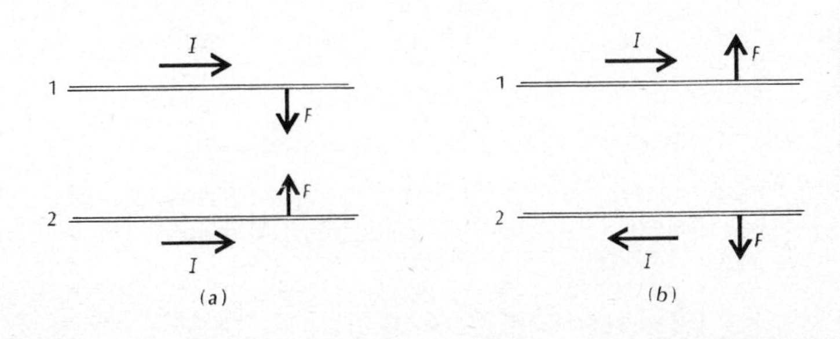

FIGURE 11.13 The force observed between current-carrying wires. (*a*) Currents in the same direction. (*b*) Currents in opposite directions.

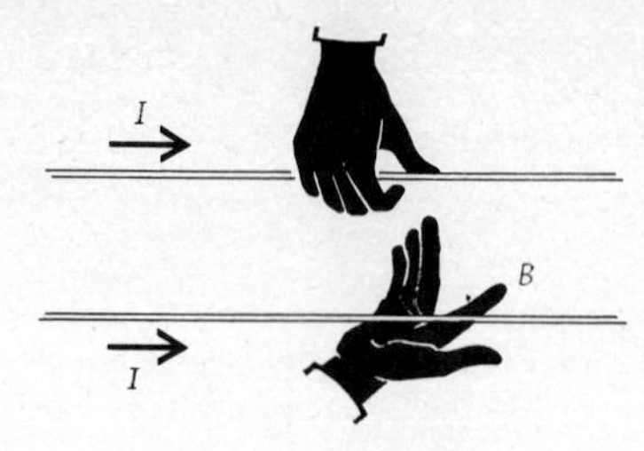

FIGURE 11.14 The two right-hand rules applied to show that the force between current-carrying wires is attractive.

creases. The force on a wire in a magnetic field is thus proportional to the product of these three quantities:

$$\begin{aligned}\text{Forces} &\propto \text{current} \times \text{length} \times \text{magnetic field} \\ F &\propto I\ell B\end{aligned} \tag{11.1}$$

Here B indicates the magnetic field, and ℓ the length of the wire. The force is found to be at right angles to both the magnetic field and the current. From relationship 11.1 we can write

$$B \propto \frac{F}{I\ell} \tag{11.2}$$

This relationship is used to define a unit of magnetic field. Our unit field will be that field which produces a force of one newton on a wire one meter long, carrying a current of one ampere. In these units the force on a wire of length ℓ carrying a current I and perpendicular to a magnetic field B is

$$F = I\ell B \tag{11.3}$$

The unit of magnetic field is then 1 newton/ampere-meter. The magnetic field of the earth (which varies from place to place) is between 10^{-5} and 10^{-4} newton/ampere-meter.

Another unit of magnetic field frequently encountered is the gauss (G). The relationship between gauss and newtons per ampere-meter is

$$1 \text{ gauss} = 10^{-4} \text{ newton/ampere-meter}$$

The gauss is thus the smaller unit of magnetic field. The earth's field is about 1 gauss.

Example 11.1

Calculate the force on a 2-meter-long wire carrying a current of 3 amperes perpendicular to a magnetic field of 10^{-4} newton/ampere-meter (1 gauss).

$$F = BI\ell$$

$F = (10^{-4}$ newton/ampere-meter)(3 amperes)(2 meters)

$F = 6 \times 10^{-4}$ newton

FIGURE 11.15 Michael Faraday (1791–1867). (*Courtesy the Granger Collection.*)

11.5 ELECTROMAGNETIC INDUCTION

Michael Faraday (1791–1867) was one of the most brilliant experimental scientists of the nineteenth century. Trained in chemistry under the tutelage of Humphry Davy, he became interested in electrical phenomena when he heard of Oersted's and Ampère's discoveries. The fact that an electric current generates a magnetic field led Faraday to speculate on and specifically seek the converse effect, which is the generation of electric currents by a magnetic field.

We have already shown, in the discussion of Fig. 11.8, that a loop of wire generates a magnetic field. A coil of many turns of wire generates a stronger field, because of the effect of each added turn of wire. Faraday used such coils, or solenoids, to generate magnetic fields, as shown in Fig. 11.16. He investigated the possibility that such a field might generate electric currents in nearby wires. In order to have as much nearby wire as possible, he constructed interleaved, or nesting, solenoids. One, circuit *A*, was used to generate the magnetic field. The other, curcuit *B*, was used to detect the expected electrical effects. A simplified view of this type of apparatus is shown in Fig. 11.17.

Faraday was unable at first to observe any evidence of a current generated by the magnetic field, until he observed a slight effect just as the magnetic field was turned on or off. Exploring further, he discovered the phenomenon of electromagnetic induction, which he first described in 1831. The key aspect of the phenomenon is that only when the magnetic field is changing is there any current generated in the surrounding wire or coil. In a device like that in Fig. 11.17, current is observed in circuit *B* only when the switch set at *A* is opened or closed, and at no other time. One way of stating this result is:

A changing magnetic field produces an electric field. This electric field will cause current to flow in any closed electric conductor in the vicinity.

When the magnetic field in coil *A* is increasing, the current in coil *B* is observed to flow in one direction. When the magnetic field is decreasing, it is observed to flow in the other. A very simple rule to determine the direction of the induced current was formulated in 1834 by H. Lenz (1804–1865), and is known as Lenz' law. This law can be stated:

The current induced by a changing magnetic field is always in

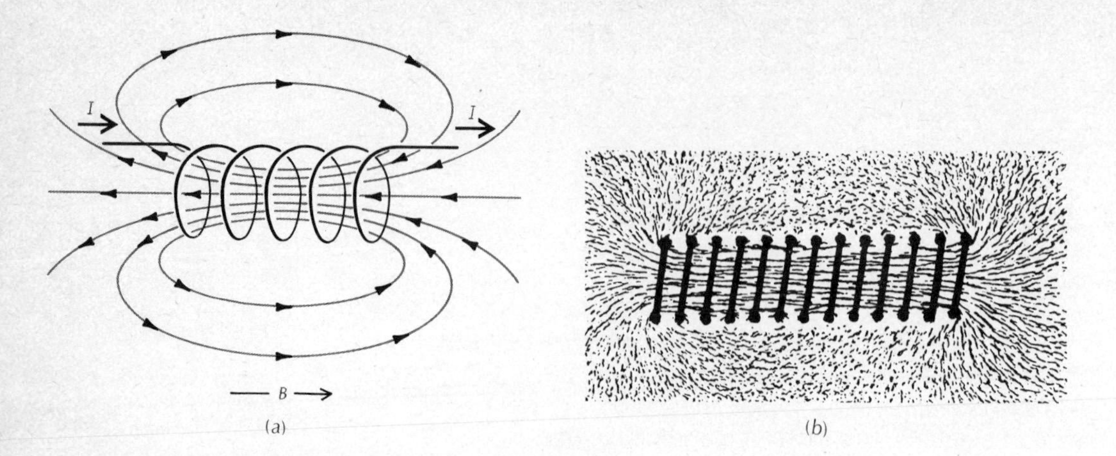

FIGURE 11.16 (*a*) Magnetic field near a solenoid (several loops of wire carrying a current). (*b*) Photograph of iron filings near a solenoid carrying current. (*Courtesy PSSC, "Physics," D. C. Heath and Company, 1965.*)

such a direction that the magnetic field due to the induced current opposes the change in the magnetic field.

We see an example of this in Fig. 11.18*a*. The switch has just been thrown, and current is flowing in *A* so as to generate the magnetic field indicated by the black arrow. The induced current in *B* will be in a direction such as to oppose the increase in the field of *A*. This is shown by the grey arrow. If the switch in *A* is opened, the current in *B* will be in a direction so as to maintain the magnetic field as shown in Fig. 11.18*b*.

The experimental laws of electricity covered in the last three chapters include, with one exception, all the electrical and magnetic phenomena known to occur. Any other electrical or magnetic effect can be understood in terms of the experiments described here. James Clerk Maxwell (1831–1879), during the period 1865 to 1873, developed a complete mathematical theory encompassing all these effects and including the

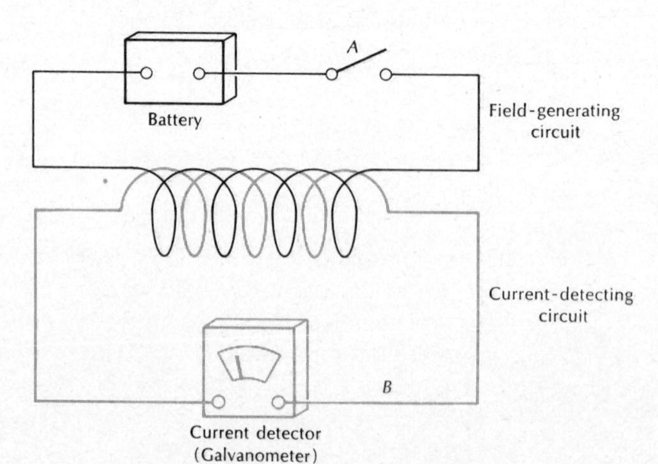

FIGURE 11.17 Schematic version of Faraday's apparatus for detecting electromagnetic induction. The galvanometer in circuit *B* detects current only at the time the switch at *A* is opened or closed. That is, it detects current in circuit *B* only when the magnetic field due to the current in circuit *A* is changing.

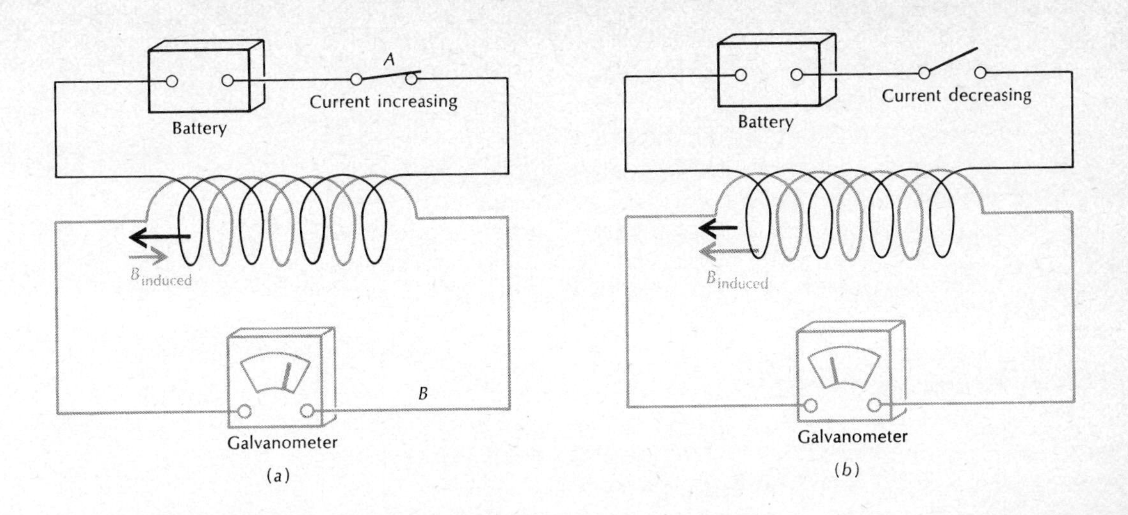

FIGURE 11.18 (*a*) When the current and magnetic field in circuit *A* are increasing, the magnetic field due to the induced current in circuit *B* opposes the increase in field. (*b*) When the current and field in circuit *A* are decreasing, the current and field in circuit *B* act to oppose the decrease in field.

exception mentioned above. He predicted the existence of electromagnetic waves, or light, using this theory. We shall take this up further in Part 5.

ELECTRIC GENERATORS

We can now understand the principle of electric generators. Figure 11.19 looks very much like Fig. 11.11. There is a loop of wire in a magnetic field. But here we turn the loop of wire. As we turn the loop, the magnetic field through the loop changes. It reverses its direction (with respect to the loop) each half-revolution. This changing magnetic field produces a current in the loop, just as though we were moving a magnet back and forth. The current in the loop can be brought to the outside and used for whatever useful purpose desired. The amount of electric energy produced is limited to the amount of mechanical work expended

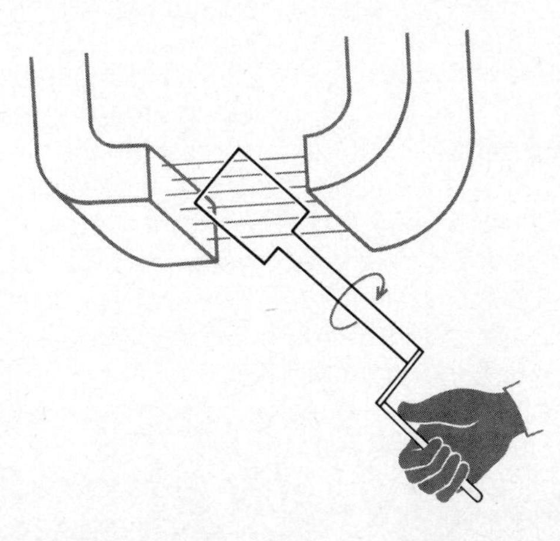

FIGURE 11.19 As the loop turns in the magnetic field, current is generated in the loop. The direction of the current alternates, because the field direction alternates.

in turning the loop. Large electric generators use steam turbines, or turbines run by the flow of water from great dams, to produce the mechanical work for electric power generation.

It is useful to point out that the current generated in the device of Fig. 11.19 is *alternating current*. Alternating current is current that sloshes back and forth many times per second. (The frequency is determined by the rotational frequency of the generator.) Direct, or one-way, current can be obtained from such a generator by adding a switching device called a *commutator*.

11.6 ALTERNATING CURRENT

Historically, the first electric power transmission for home and commercial use was *direct current*. In direct current the current flow is always in the same direction. A small part of New York City still has direct current, however, the overwhelming majority of the electric power in this country is alternating current. Alternating current reverses direction 60 times per second (50 times per second in Europe).

The major advantage of alternating current is that energy can be transported for long distances on transmission lines with less loss. To see how this comes about, we must remember the power lost in resistance, from Chapter 10. The power lost is given by

$$P = I^2R$$

The resistance of a power line is determined by its length and by the size of the wire used. There are practical limits to how big the wire can be, so the only way the power loss can be reduced is to reduce the current. Our other equation for power shows how this can be done:

$$P = IV$$

Power can be transmitted either at high voltage and low current, or at high current and low voltage. Clearly high-voltage transmission gives the lowest current, and therefore the least power loss.

Alternating current is converted to high voltages for transmission, and then back to low voltages for use by employing *transformers*. A transformer uses the principle of electromagnetic induction. The basic idea is shown in Fig. 11.20. The shaded region is made of iron and is called the *core*. On the left side the alternating current in the wires wound on the iron causes an alternating magnetic field. The iron keeps all of the magnetic field trapped. The alternating field in the iron induces a voltage and a current in the output winding. In the example shown the output voltage is five times the input voltage, because there are five times as many turns of wire on the output side. The current on the output side is one-fifth that on the input side. If this were not true, there would be a violation of the energy conservation law because the power in

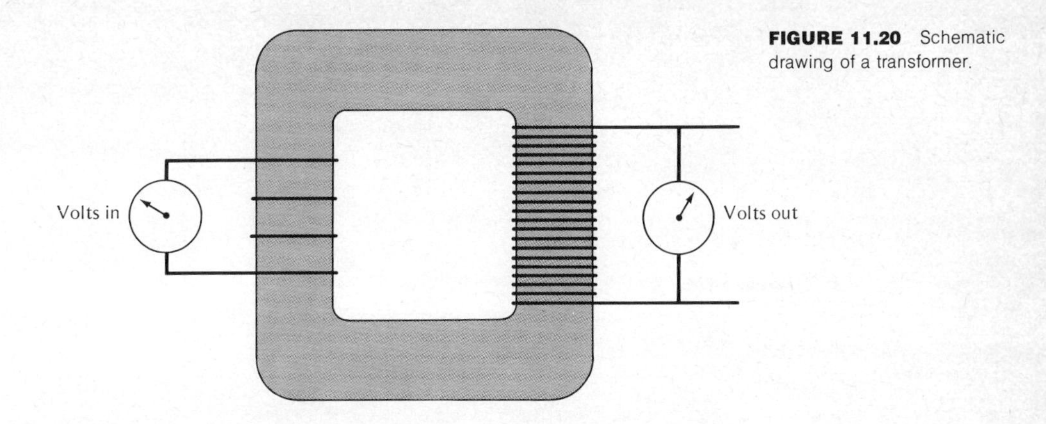

FIGURE 11.20 Schematic drawing of a transformer.

must equal the power out. (Practical transformers are about 99 percent efficient and lose only 1 percent of the power put through them.) To reduce a high voltage the transformer is operated in reverse: The input side is the one with the larger number of turns.

Recent technology has made possible high-voltage dc transmission line systems. Such a system has existed on the West Coast since the early 1960s, and others are being built. It is claimed that these systems have less of an environmental impact than ac systems. They need narrower right-of-ways, and they emit less electromagnetic energy. At the end of a dc line the power is converted to alternating current for distribution. The main holdup on dc transmission has been that the devices required to convert low voltages to high voltages are more expensive and less efficient than equivalent devices (transformers) in ac systems.

11.7 THE FORCE ON A MOVING CHARGE

We have already shown that an electric current experiences a force in a magnetic field. But current is electric charge in motion in a wire. Therefore it seems that a moving charge not in a wire should also experience a force. Experimentally, this is found to be the case. The most common example of this is in a TV set. In the picture tube a beam of electrons is generated. The path of these electrons is bent in magnetic fields, so that the beam will strike the screen at various points. The screen is coated with a phosphor that glows when struck by the beam of electrons.

Figure 11.21 shows an electron moving in a magnetic field perpendicular to the paper, and the direction of the force on it. The inward force causes the electron to move in a circular path. The force is at right angles to both the direction of the electron's motion and the direction of the magnetic field, providing a centripetal force, as discussed in Section 4.3.

FIGURE 11.21 Curved path of an electron as a result of the force from a magnetic field directed up out of the paper.

11.8 PARTICLE ACCELERATORS

It is now possible to understand the basic principle of operation of one of the most impressive pieces of equipment of physics today, the particle accelerator. In particular we shall discuss circular accelerators. The first circular accelerator, the *cyclotron*, was invented and constructed by E. O. Lawrence (1901–1958). Since that time these machines have grown from the rather modest size shown in Fig. 11.22 to rings about 2 kilometers in diameter. Even larger machines have been considered, but may never be built because of their cost.

The simplest of these machines to understand is the cyclotron, which uses a magnetic field to keep particles moving in a circle. The cyclotron takes advantage of the fact that the time required for a charged particle moving in a magnetic field to complete a circle does not depend on the size of the circle for a given magnetic field. The larger the circle, the faster the particle moves; it moves just fast enough to travel around the circle in the same amount of time required to travel around a smaller circle. The object in all particle accelerators is to achieve high energies, which means high velocities. The largest circles have the highest velocity and energy.

The idea behind a cyclotron is something like "running the gauntlet," but in a circle. The particle to be accelerated, usually a positively charged proton, is acted on by an electric field each time it comes around the circle, so that on each revolution it gains energy, and the circle becomes larger. (The proton is the nucleus of the hydrogen atom, and will be discussed in more detail in Chapters 19 and 20.)

Figure 11.24 shows the essential elements of a cyclotron. Two

FIGURE 11.22 The late E. O. Lawrence (right) the inventor of the cyclotron, and M. Stanley Livingston, standing in front of an early cyclotron, whose diameter was 27 inches. This photograph was taken in 1934. The cyclotron itself is between the poles of the magnet at approximately waist height. (*Courtesy University of California Radiation Laboratory.*)

FIGURE 11.23 Aerial view of the Fermi National Accelerator Laboratory, near Batavia, Illinois. The main accelerator, the largest component of the system, is 1.24 miles in diameter. The outline of the main ring can be seen at the top of the picture. (*Courtesy National Accelerator Laboratory.*)

hollow metal cans, called *dees*, are separated by a small space. The magnetic field is perpendicular to the dees, as shown by the arrows. The whole region is reduced to as good a vacuum as is possible. If an electric field is set up between the dees as shown, a positively charged particle will be accelerated when going to the left, and slowed down when going to the right. But if the direction of the electric field is switched in exactly the amount of time it takes the particles to make one-half revolution, they will be accelerated on *each* pass between the dees. The magnetic field keeps the particles in a curved path and coming back for more. Because of the synchronism between the period of the particles and the frequency that the dees are switched, the particles are accelerated until their orbit is as large as the dees, and they

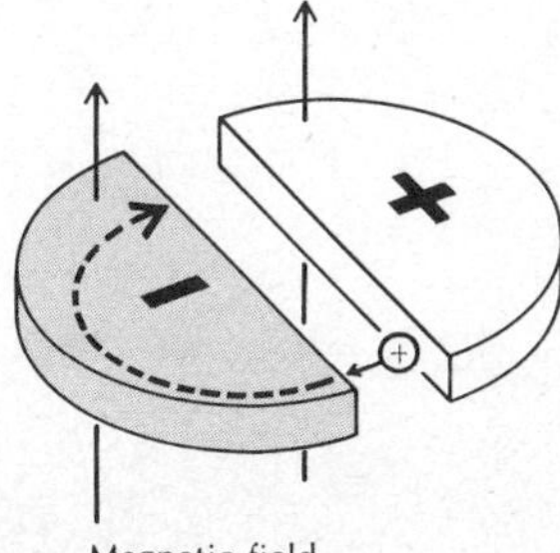

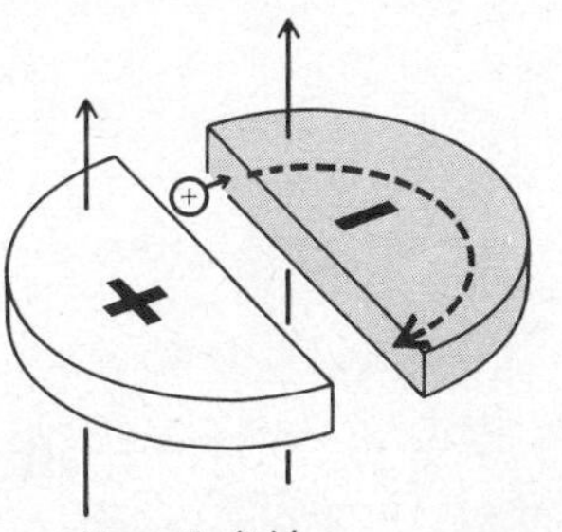

FIGURE 11.24 Schematic drawing of the operation of a cyclotron. The magnetic field keeps the particles moving in circular paths. The direction of the electric field created by the dees is reversed each time the particle comes by, so that it is given an additional kick twice in each round trip.

then strike a target. The particles are started near the center with a small velocity.

There are certain difficulties associated with the cyclotron that limit the maximum energy which can be obtained through its use. Most of these difficulties have been overcome by assorted sophisticated schemes in the more advanced accelerators. The discussion of these machines must be left to other reading.

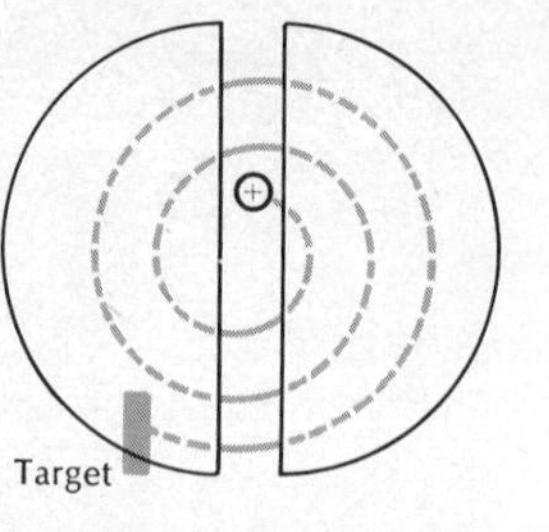

FIGURE 11.25 Path of a positively charged particle in a cyclotron as seen from above.

SUMMARY

There exists a magnetic field in the vicinity of any magnet. The direction of the magnetic field is the direction in which the north-seeking end of a compass points.

Oersted showed that a steady electric current generates a magnetic field at right angles to the current at any point.

A magnetic field exerts a force on a wire carrying current. If the wire and the magnetic field are at right angles, the force is at right angles to both and is given by

$$F = BI\ell$$

Ampère showed that two parallel wires carrying current exert a force on each other. Currents in the same direction produce an attractive force. Opposite currents cause repulsion between the two wires.

The units of magnetic field are newtons per ampere-meter. A magnetic field of one newton per ampere-meter produces a force of one newton on a length of wire 1 meter long carrying a current of one ampere if the current and magnetic field are at right angles to each other.

Electromagnetic induction: A changing magnetic field produces an electric field. This field will cause current to flow in any nearby conductor.

Lenz' law states: The current induced by a changing magnetic field is always in such a direction that the magnetic field due to the induced current opposes the change in the magnetic field.

A loop of wire rotating in a magnetic field will have an alternating current generated in it. This is the principle of an ac generator.

A transformer uses the principle of electromagnetic induction to change the voltage and current in an ac transmission line.

An electric charge moving in a magnetic field experiences a force. This force is at right angles to its velocity and at right angles to the magnetic field.

A cyclotron is a particle accelerator in which charged particles move in a circle because of a magnetic field, and are given ever-increasing amounts of energy.

SELECTED READING

MacDonald, D. K. C.: "Faraday, Maxwell, and Kelvin," Science Study Series S-33, Doubleday and Company, Garden City, N.Y., 1964. Short biographies of three of the giants of nineteenth-century physics.

QUESTIONS

1 What evidence can you cite that the magnetic field around a magnet varies with the distance from the magnet?

2 How do we know that stroking a piece of iron with a lodestone does not electrify the iron as rubber is electrified when rubbed with silk?

3 If a magnetic compass needle is placed perpendicular to a current-carrying wire, will any magnetic force act on it?

4 If a magnetic field exerts a force on a moving charge in a current-carrying wire, why do we see the wire as a whole responding to this force?

5 If you cut a bar magnet in the center, are you left with separate north and south poles?

6 Describe some possible sources of energy loss in a transformer. Why do some transformers hum?

7 Why is the aurora borealis (northern lights) more usually associated with the pole than with the equator?

SELF-TEST

__________ Faraday
__________ Maxwell
__________ Ampère
__________ Oersted
__________ galvanometer
__________ compass
__________ lodestone
__________ magnet
__________ north pole
__________ south pole
__________ transformer
__________ gauss
__________ electromagnetic induction

__________ Lenz' law
__________ induced current
__________ alternating current
__________ cyclotron

1 A bar of iron that creates a magnetic field
2 A sensitive device for the measurement of electric current
3 Discovered the magnetic field created by a current
4 A unit of magnetic field; equal to 10^{-4} newton/ampere-meter
5 A device used to vary the voltage and current in ac systems
6 The induced current creates a field to oppose the change in the magnetic field
7 A high-energy particle accelerator
8 Current which reverses its direction 60 times per second.
9 Created a set of equations that summarize all the experimental observations on electricity and magnetism
10 Discovered electromagnetic induction
11 A light magnet suspended so it can turn in the earth's field
12 Measured the force between current-carrying wires
13 Rock which is magnetized
14 The end of a magnet that points south
15 The end of a magnet that points north
16 The creation of an electric field by a changing magnetic field
17 The current created by a changing magnetic field

PROBLEMS

1 Calculate the force on each meter of a wire carrying a current of 3 amperes in a magnetic field of 0.1 newton/ampere-m at right angles to the wire.

2 Calculate the force on a wire 6 m long which carries a current of 10 amperes. There is a magnetic field of 0.02 newton/ampere-m at right angles to the wire.

3 A straight wire 3 m long experiences a force of 0.12 newton in a magnetic field at right angles to it. The current in the wire is 2 amperes. Find the value of the magnetic field.

4 The earth's magnetic field is about $\frac{1}{2}$ gauss. (*a*) Calculate the force per meter on a wire carrying a current of 100 amperes at right angles to the earth's field. (*b*) Express the force in pounds.

5 In the diagram shown, is the force on the wire directed into or out of the paper?

$I \longrightarrow$

ATOMS, MATTER, AND ELECTRONS

The structure of matter is an important part of the realm of physics. Atoms are the basic building blocks of matter. They are arranged in various ways that we call solid, liquid, and gas. And when one investigates further, one finds that the atom is made from smaller objects yet. One of these is the electron.

Part 4 consists of three chapters. The first is a survey of the ideas concerning the atom up to about 1870, including the construction of the periodic table. Chapter 13 is a discussion of the various states or arrangements in which atoms are found: solids, liquids, and gases. Chapter 14 is the story of the electron—its discovery and some of its properties.

THE ATOMIC THEORY

The idea that matter consists of atoms has been around for over 2,000 years. By an *atom* we mean a discrete, countable object, which is not changed, subdivided, or destroyed in the course of ordinary chemical processes. The evidence for their existence revolves mainly about the point that atoms are things which can be counted, in the sense that marbles are counted, rather than measured, in the sense that water is measured. This is true even though there are more atoms in a drop of water, for instance, than one would in fact care to count. The nature of atoms is such that they *can* be counted, not that one *must* count them.

Whole books have been written on the fascinating history of atomism from the Greek philosophers to modern times. One of these is listed in the Selected Reading at the end of this chapter. It covers far more than the discussion here. We shall only survey some of the high points in the story, in an attempt to see how science has handled such a problem. It should be clear, by the end of the story, that the ideas of science do not come ready-made and neatly packaged, as textbooks sometimes lead one to think, but rather that there is a long dynamic process of development before an idea reaches the point where it is accepted as obvious—a word too often used to describe particularly difficult concepts after a problem has been solved.

The atomic theory is the meeting ground between physics and chemistry. Here the two sciences overlap. Chemistry goes on to consider the joining of atoms into molecules, and their properties. Physics goes on to study the detailed structure of the atom, and the nature of its constituent parts.

12.1 THE GREEKS HAD A WORD FOR IT

The genesis of the atom concept is found in the work of the ancient Greek philosophers, primarily in the period 600–300 B.C. To understand their contribution we must realize that they approached the matter from a completely different point of view than that of a modern physical scientist. The modern viewpoint can be paraphrased as: What is the ultimate nature of matter? What are the building blocks of which it is made? The Greek point of view was more like: We see that matter exists. In a fundamental way, what does it mean to say that something exists? What characteristics distinguish something that exists from something that does not exist?

The earliest speculations on the nature of matter that we are aware of are those of Thales of Miletus (about 600 B.C.). He attempted to find

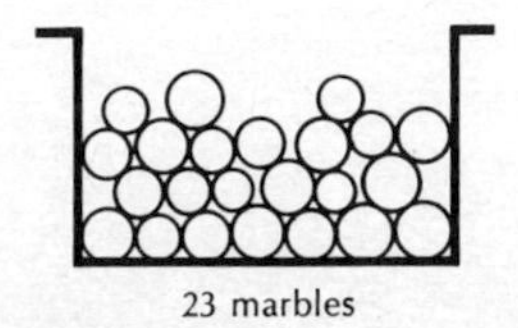

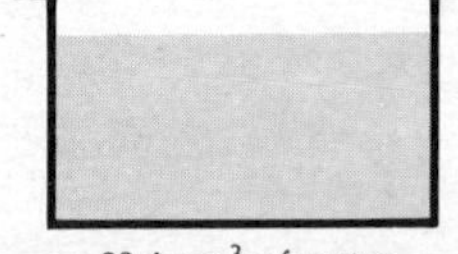

FIGURE 12.1 Atoms may be counted, like marbles, rather than measured, like water.

a unity in nature by proposing that all matter came forth from a single kind of primary matter. Thales proposed that water was this primary matter. Other speculations of this type were those of Anaximenes (about 585–525 B.C.) who proposed air as the primary matter of all things, and Heraclitus (480 B.C.) who suggested fire. Heraclitus' reason for choosing fire was that, to him, the primary characteristic of all things is their constant changing. Fire represented to him the principle of change.

In direct opposition to Heraclitus was Parmenides (500 B.C.). He came to the conclusion that change was impossible. He believed that the only change which could occur would be from a state of being to a state of nonbeing, or a state of existence to one of nonexistence. But nonbeing is *not*, and therefore becomes nothing. Thus change is an illusion, and change cannot occur. Whether one follows the logic or not, Parmenides' ideas posed an enormous dilemma. We observe that change occurs all around us, but if change is impossible, what is going on? The attempts to solve this dilemma led to the first atomic theories.

DEMOCRITUS

Democritus (about 460 B.C.) attempted to allow for the obvious fact of change without doing more violence than necessary to the ideas of Parmenides. He proposed that change is possible because of the motion and rearrangement of a large number of objects which are in themselves unchangeable and indivisible. He used the word *atomos*, which means indivisible, to describe these objects. So Democritus proposed that atoms are unchanged, preserving the idea of Parmenides, but that change is possible through the motion and rearrangement of atoms. His atoms had the following characteristics:

1. Atoms are identical, except for size and shape. All sizes and shapes are allowed.
2. Atoms are always in motion. This motion can be changed by what is done to them. This idea is central to the modern view of how atoms behave when they are heated.

One other aspect of Democritus' theory is important. He believed that everything is composed of atoms, including the soul. His soul atoms were smooth and round, so that they could slip easily through the body. This materialistic point of view is of importance, because it determined the attitude of the early Christian church toward Democritus' ideas.

Two contemporaries of Democritus also have atomic theories to their credit, Empedocles and Anaxagoras (about 450 B.C.). Empedocles taught that all matter is composed of combinations of the four elements, earth, air, fire, and water. It is tempting to be amused at the naiveté of this notion. Then one realizes that in the world around us we see matter

FIGURE 12.2 An old print showing the supposed relationships among the elements: earth (terra), air (aer), fire (ignis), and water (aqua).

in only four forms: solids (earth), gases (air), liquids (water), and the state we call fire, or flames. To the Greeks fire was the element of change, since fire was observed to work some change on the burning substance. Anaxagoras believed that all matter is composed of what he called *seeds*, but that these seeds can be subdivided without changing their essential character. Here he differed from Democritus.

ARISTOTLE

None of the original writings of Democritus have survived to the present day. Most of what we know of his work we find in the writings of the philosopher Aristotle (384–322 B.C.). Aristotle rejected Democritus' atomic theory and accepted earth, air, fire, and water as the four basic elements. Aristotle was certainly the leading scientist among the early philosophers, and there is scarcely any aspect of the science of his day on which he did not have an opinion. Because of his enormous reputation, his statements generally were not challenged. The result was that the things he said that were wrong lasted well beyond their time. Another reason that Aristotle's influence lasted so long was a political one, the influence of the Church.

FIGURE 12.3 Aristotle. (*Courtesy Burndy Library.*)

Another question considered by Aristotle is that of smallest parts. We have already seen that Democritus' smallest parts were the indestructible atoms. Aristotle's point of view can be seen in terms of a simple thought experiment. Suppose I take a bucket of water and divide it into two parts. Both parts are still water, by every test I can perform. Suppose I repeat the division—I still have water. But suppose I divide the bucket of water in half a very large number of times. Does a point ever arrive at which the two halves are no longer water? Have I, by the process of infinite division, been able to take water and make it something which is not water? Can I go one step further, and take something which exists, water, and by infinite subdivision make it not exist? Aristotle's answer to this question is the idea that there are natural minima, or smallest parts, into which any substance can be divided. This idea is not a complete atomic theory, because it does not view these natural minima as any sort of building block for matter. The modern atomic theory views atoms as the building blocks of matter. (The modern atom can also be subdivided into smaller parts, as we shall see.)

Two of the principal champions of Democritus' view in the period after Aristotle were the Greek Epicurus (347–271 B.C.) and the Roman poet T. Lucretius Carus (96–55 B.C.). Both believed in the atomic theory of Democritus. Following its materialistic ideas, they also believed in a materialistic, or pleasure-seeking, form of life. Our word epicurean comes from this philosophy. The early Christian church was somewhat puritanical, and disagreed violently with the epicurean philosophy and the materialism of Democritus. Since Aristotle was the principal foe of the ideas of Democritus, it was natural that his views were favored by the Church.

12.2 THE MIDDLE AGES

All Greek science and philosophy were lost to the Western world in the general convulsions of the early Christian era. The work of the Greeks was preserved by the Moslem world of the Arabs, who held Aristotle, in particular, in high esteem. There are many commentaries on Aristotle's work during this period that in fact amplify and extend it.

During the twelfth century the works of the Greeks were translated from Arabic back into Latin and again became available to the West. St. Thomas Aquinas (thirteenth century) used Aristotle's logic in his proofs for the existence of God. From this point on the Church stood even more strongly behind Aristotle, since to deny Aristotle would be, in effect, to deny the existence of God (or at least Aquinas' proof for his existence). Because of the support of the Church, Aristotle's science outlived its usefulness, so to speak. Aristotle made enormous contributions to the science of his day. He is somewhat maligned today, because of the difficulty in overcoming those of his ideas which were wrong. This was not the fault of Aristotle, but of those who followed him. They accepted every word he wrote as the truth without question, partly because his views coincided with those of the Church. The backing of the Church essentially fossilized the development of physical science until the seventeenth century.

About the sixteenth and seventeenth centuries, with the rise of true experimental science, the notion of atoms as building blocks began to be revived. This is not to say that there were not followers of Democritus in the 2,000 years following Aristotle. It is just that they were few and far between and had little or no influence on others around them. There were also those who, while agreeing with Aristotle, tried to extend and amplify his views. They were called *commentators* because, ostensibly, they were only commenting on, or interpreting, Aristotle. In fact, a certain number of new ideas were smuggled in in the process, although all were attributed to Aristotle. We sometimes see the same process today with those who use the Bible in a similar way to prove it says what they want it to say.

12.3 THE SEVENTEENTH CENTURY

We shall omit further reference to developments before the seventeenth century and refer the reader to the Selected Reading at the end of this chapter. During the seventeenth century events began to occur more quickly. In particular, this century saw a transition of major importance. Those associated with the development of science ceased being philosophers and began to be men we can call physical scientists. The rise of experimental science obviously had a major impact, since now the consequences of various theories of matter began to receive experimental tests. No longer was philosophic harmony or agreement with the Church the criterion for truth in science. The final arbiter was now an observation or an experiment, which is a controlled observation. Of course, not everyone accepted this point of view. There were those who refused to look through Galileo's telescope at the moons of Jupiter. One supposes this was in order not to be forced to accept the fact that such things could exist. This ostrichlike point of view became less and less common as time passed, although even today we find such things as the celebrated "monkey trial" in the state of Tennessee in 1925, involving the teaching of Darwin's theory of evolution.

A few of the men who contributed to the development of the atomic theory during the seventeenth century deserve mention. Peter Gassendi (1592–1665) was a Catholic priest who sought to revive the atomic theory of Democritus but to cleanse it of the elements found offensive to the Church. Essentially, without adding much that was new, Gassendi succeeded in making the atomic theory again respectable and worthy of consideration.

The French mathematician-philosopher René Descartes (1596–1650) contributed an atomic theory which was different from Democritus' but has not had much influence on what followed. One aspect of Descartes' theory, however, was far ahead of its time and a major original contribution. Descartes believed that all the particles of matter were constantly in motion and that, when an object is hot, the particles of which it is composed move faster. The sense experience of heat then occurs, because the body is able to sense this faster motion. Two centuries passed before this idea was finally fully developed into our modern ideas of heat.

The first man we shall discuss who was truly a physical scientist, an observer rather than a philosopher, was Robert Boyle (1626–1691). In his major works, *The Sceptical Chymist* (1661) and *Origin of Forms and Qualities* (1666), Boyle gave a critique of the chemistry of his day. He concluded that earth, air, fire, and water were not elements, because they were not the products of analysis. An element, to Boyle, was the end result of chemical analysis. By this he meant the decomposition by chemical processes into simpler substances. The limit beyond which this process cannot continue is an element. He did not attempt to name specific materials as elements, for the chemistry of his day was too primitive. Most of the chemical information then available had been obtained during the alchemists' search for a method to change lead into gold. In Boyle's view, the particles of the elements are the true building blocks of chemical compounds. So Boyle contributed both an empirical definition of a chemical element and the idea that atoms are the building blocks of matter.

We should mention one other scientist in this survey of the seventeenth century and state his often quoted view on atoms. Isaac Newton (1642–1727), in his treatise on optics, said in 1704:

> *. . . it seems probable to me, that God in the Beginning formed Matter in solid, massy, hard, impenetrable, moveable Particles, of such Sizes and Figures, and with such other Properties, and in such Proportion to Space, as most conduced to the End for which he formed them; and that these primitive Particles being Solids, are incomparably harder than any porous Bodies compounded of them; even so very hard, as never to wear or break in pieces; . . .*

So we see from these examples that the idea of atoms was in the air in a much more positive way than it had been in earlier times, because now the development of real physical science demanded a clearer picture of the nature of matter. Real solutions were needed, and not theoretical proposals without experimental backing.

12.4 DEVELOPMENTS BEFORE DALTON

Antoine Lavoisier (1743–1794) was a French chemist who was also a tax collector in the government of Louis XVI—a role that led to his execution during the turmoil of the French Revolution. Lavoisier made precise the concept of chemical element by providing a good working definition. This definition is:

> *If we apply the term elements . . . to express our idea of the last point that [chemical] analysis is capable of reaching, we must admit, as elements, all substances into which, by any means, we can decompose bodies.*

FIGURE 12.4 Antoine Laurent Lavoisier (1743–1794). (*Courtesy University of Pennsylvania Library.*)

This definition is what is called an *operational* definition, in that it gives us a well-defined procedure for testing whether or not a given substance is an element. It is not an element if it can be decomposed into other substances. A substance we cannot decompose *may* be an element, or perhaps we have just not yet learned how to decompose it. This was the case with a number of materials of Lavoisier's day, such as table salt, which was not decomposed into the elements sodium and chlorine until 1800. Some of the materials considered elements in Lavoisier's time were metals such as iron, silver, gold, copper, and mercury, and other materials such as carbon, oxygen, hydrogen, nitrogen, salt, lime, soda, and potash. Salt, lime, soda, and potash have since his time been shown not to be elements by the use of his own basic definition.

Lavoisier's other contribution to this story is the demonstration that mass is conserved in a chemical reaction. This statement means that the mass of material before a chemical reaction takes place is exactly the same as that after the reaction, within the accuracy of the measurement. If a log is burned in the fireplace, we know that the ashes weigh less than the log. If every bit of smoke and vapor given off is trapped, it is found that the total mass of the products (ashes, smoke, and vapor) is the same as the mass of the starting materials (log and air). The conservation-of-mass idea means that the composition by mass of compounds is information of real importance. If mass were not conserved in a chemical reaction, there would be no real reason to obtain such information.

In the late eighteenth century an important empirical law was discovered regarding the mass relations in chemical compounds. By an empirical law we mean a relationship observed in experiments to be true. We may not be able to prove that such a law ought to be the case, but we must accept the experimental evidence that this law is, in fact, true. This law is called the *law of definite, or constant, proportions*. It can be stated thus: The elements that form a given chemical compound always occur in that compound in definite proportions by mass.

As an example consider the compound carbon monoxide, a gas formed from the two elements carbon and oxygen. When pure, all samples of this compound are found to have a composition of 42% carbon and 58% oxygen, by weight or mass.

Example 12.1

How many grams of oxygen does a 140-gram sample of carbon monoxide contain? The amount of oxygen is 58 percent of the total mass; so

$$140 \text{ grams} \times 0.58 = 81 \text{ grams of oxygen}$$

Of course, the remainder is carbon.

$$140 - 81 = 59 \text{ grams of carbon}$$

The law of definite proportions did not spring into being overnight. It was the center of a controversy between two French chemists, J. L. Proust and C. L. Berthollet. (Anyone who thinks that scientists are dispassionate, unemotional individuals should witness the passions which can be generated by such scientific controversies until the facts of the situation are established.) The law was first stated by Proust just before 1800. The debate lasted until about 1808. One of the key points of contention in this debate revolved around the material formed from the reaction of copper with oxygen. Berthollet pointed out that copper oxide occurred with a range of compositions between 89% copper and 11% oxygen and 80% copper and 20% oxygen. He therefore concluded that the compound formed from copper and oxygen violated the law of definite proportions. Very careful experimental work was necessary to show that, in fact, there are two compounds of copper and oxygen. Their compositions are 89% copper and 11% oxygen for one, and 80% copper and 20% oxygen for the other. The materials Berthollet used to disprove the law of definite proportions were mixtures of these two compounds in differing amounts. However, when these two oxides of copper are completely separated, they *always* occur with exactly the percentage compositions given above.

12.5 JOHN DALTON AND THE ATOMIC THEORY

An English schoolteacher, John Dalton (1766–1844), proposed in 1808 an atomic theory of matter close to the one that we accept today. Dalton's theory accounted for the observations of his day and, most importantly, predicted a new law, which was then verified experimentally. We can summarize Dalton's theory as follows:

1 All matter consists of atoms. The atom is the smallest unit into which matter can be divided.

2 Each chemical element consists of characteristic atoms. Every atom of a given element is identical with every other atom of the same element. The elements differ from one another because their atoms differ from one another. In particular, the atoms of different elements differ in mass.

3 Atoms are unchangeable. No one can change an atom of one element, such as lead, into an atom of another element, such as gold.

4 Atoms are indestructible. They are neither created nor destroyed, but merely rearranged in any chemical process.

FIGURE 12.5 John Dalton (1766–1844.) *(Courtesy University of Pennsylvania Library.)*

Most of these ideas were not new. Some of them had been around for 2,000 years. Dalton's achievement was the synthesis of these ideas into a working model of how chemical compounds are formed.

Dalton proposed that atoms are the building blocks of all matter. A piece of any pure chemical element, such as lead, consists, on this basis, of a large number of *identical* atoms of lead. According to Dalton, the difference between lead and gold lies in the fact that their atoms are different. The determination of the number of atoms in a pound of lead requires the knowledge of the weight of one lead atom. Dalton did not have this knowledge.

To complete his atomic theory Dalton described the means by which atoms can form a chemical compound. He proposed that a compound such as carbon monoxide is an association of atoms of carbon and oxygen. The atoms are not just mixed together, but connected into units called *molecules*. A molecule is a structure consisting of two or more atoms joined together in a well-defined way. A chemical compound consists of a large number of identical molecules.

Dalton's theory accounts directly for the law of definite proportions. If every molecule of a given compound is exactly the same, that is, consists of exactly the same number of atoms of each kind, the law of definite proportions follows naturally. Suppose, for example, that carbon monoxide molecules each consist of seven atoms of carbon and five atoms of oxygen. Then, since all carbon monoxide molecules are identical, a large quantity of carbon monoxide will have seven atoms of carbon to each five atoms of oxygen. Since each atom of carbon or oxygen is identical in mass with all others, the definite-mass relationship follows. (Note that carbon monoxide does *not* have seven carbons and five oxygens. Dalton did not know, for sure, the correct chemical structures for various compounds. For the law of definite proportions to hold it is only necessary that every molecule of a compound be identical to every other molecule.)

The new law predicted by Dalton is called the *law of multiple proportions*. It involves the fact that there may be more than one compound involving the same elements, for example, copper and oxygen. Dalton guessed that such compounds would have formulas like CuO (one copper atom and one oxygen atom) and CuO_2 (one copper atom and two oxygen atoms). If so, the mass of oxygen that combines with a given mass of copper would be twice as great in the second compound as the first. This simple 2 to 1, whole number, ratio of masses was sought and found for a number of pairs of compounds.

12.6 THE MASS OF ATOMS

Dalton's theory made possible the first statements about the mass of atoms. The basic idea is contained in his proposals about molecules. He believed that the molecules actually found in nature are the simplest possible. For water, a compound of hydrogen and oxygen, he proposed the formula HO. This means that a molecule of water has one atom of hydrogen and one atom of oxygen. (The true formula is H_2O, two atoms of hydrogen and one of oxygen.) By chemical analysis 100 grams of water was found in Dalton's time to contain 86 grams of oxygen and 14 grams of hydrogen. If Dalton's formula HO is correct, the ratio of the mass of one atom of oxygen to one atom of hydrogen is 86 to 14, or about 6 to 1. The point here is that a great number of molecules of the formula HO have a mass ratio of 86 to 14. Therefore the ratio of the individual atoms must be 86 to 14. This proposal of Dalton's, although it gave wrong numerical answers, was the first step in setting up a quantitative scale for atomic masses.

12.7 THE PERIODIC CHART

Dalton's theory showed the way to develop a list of atomic masses. These are relative masses, which means that masses were developed relative to one element whose mass was arbitrarily chosen. Originally this was oxygen, but now it is carbon. It was not until after 1900 that reliable numbers existed for the mass of a single atom.

Although Dalton's theory paved the way, there was confusion for a number of years. There were a number of reasons for this, which we will not develop further. Some scientists went so far as to believe that the atomic theory would never be good for anything and that atoms, as real objects, probably do not exist. The first International Chemical Congress was convened in Karlsruhe, Germany, in 1860, in the hope that open discussion of the question by all concerned might resolve some of the contradictions among the various theories and the observed facts. This august gathering broke up in more confusion than ever. Before it was adjourned, however, a copy of a paper written by an Italian chemist, Stanislao Cannizzaro (1826–1919), was circulated among the delegates. This paper demonstrated in a very clear and convincing way how to handle the question of atomic masses, and how to determine them from chemical data. As a result of this work, good values for atomic masses became available.

The scale of atomic masses has been set up in the following manner. Oxygen was originally chosen as the element to which all masses would be referred. Its mass was set arbitrarily at 16.0000. The choice of 16 means that hydrogen comes out near 1, and all other masses are larger than 1. Most atomic masses come out near whole numbers on this scale. The units here are called *atomic mass units* (amu). Oxygen has, by definition, a mass of 16 atomic mass units. Because of changes in the kind of experiments used for atomic-mass determinations, the present standard is carbon, at 12.0000 atomic mass units. The atomic mass unit is a unit of mass whose value cannot be determined until the mass of a single atom is determined.

THE MOLE CONCEPT

A useful concept has developed here. An amount of matter equal in kilograms mass to the numerical value of the atomic mass is called a *kilogram-mole*. Sixteen kilograms of oxygen is 1 kilogram-mole. Frequently in chemistry, the *gram-mole*, or just the *mole*, is also encountered. By the definition of a mole, a mole of one element has the same number of atoms as a mole of any other element. Twelve kilograms of carbon has the same number of atoms as 16 kilograms of oxygen, since their atomic masses are in the ratio 16 to 12. To determine the number of atoms in a mole is not easy. But by 1909 there existed a number of ways to make this determination. The number of atoms in a mole is called Avogadro's number. Avogadro's number for a kilogram-mole is

$$6.0 \times 10^{26} \text{ atoms per kilogram-mole}$$

For a gram-mole the number of atoms is 6×10^{23}. Once Avogadro's number is known, the mass of an individual atom can be determined. If 12 kilograms of carbon has 6×10^{26} atoms, the mass of one carbon atom is

$$\frac{12.0 \text{ kilograms}}{6 \times 10^{26}} = 2 \times 10^{-26} \text{ kilogram}$$

Since the mass of one carbon atom is 12 atomic mass units, the value of 1 atomic mass unit is

$$1 \text{ atomic mass unit} = \frac{2 \times 10^{-26} \text{ kilogram}}{12} = 1.6 \times 10^{-27} \text{ kilogram}$$

Once the atomic-mass scale was sorted out, it became possible to search for regularities to try to find some system in the atomic masses. It was known that there are elements that undergo similar chemical reactions, such as the alkali metals lithium, sodium, potassium, rubidium, and cesium. These are all silvery metals that react violently with water to form alkaline solutions. Many of their other properties are similar as well. A number of physicists and chemists tried to correlate the chemical and physical properties of the elements with their atomic weight in order to understand these similarities.

FIGURE 12.6 Dmitri Mendeleev (1834–1907). *(Courtesy University of Pennsylvania Library.)*

In 1871 a Russian chemist, Dmitri Mendeleev, and a German scientist, Lothar Meyer, working independently, proposed tabulations of the elements which showed regularities involving these chemical and physical similarities. Mendeleev arranged the elements in a table, in order of increasing atomic mass and arranged in columns in such a way that the elements in each vertical column had similar chemical and physical properties. For example, after hydrogen, the first seven elements and their atomic masses are

Lithium	Beryllium	Boron	Carbon	Nitrogen	Oxygen	Fluorine
7	9.4	11	12	14	16	19

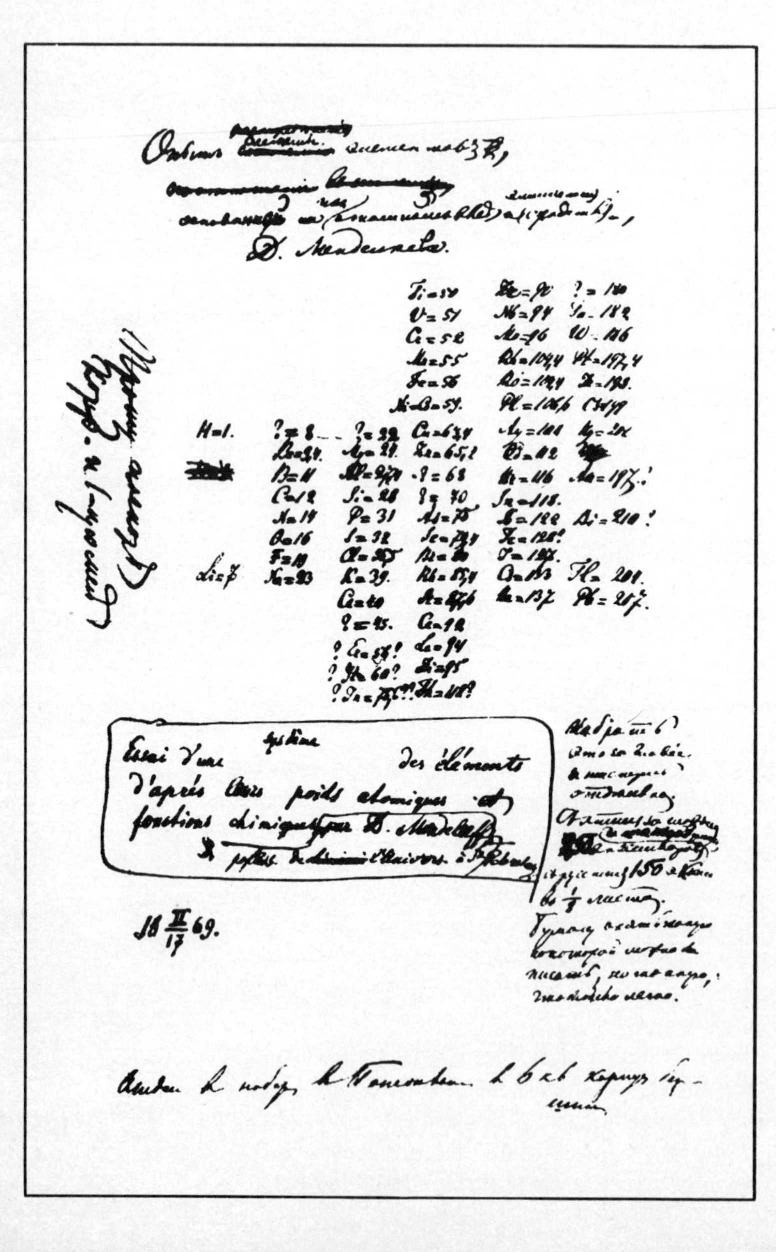

Essai d'une système des éléments d'après leurs poids atomiques et fonctions chimiques par D. Mendeleeff

FIGURE 12.7 Mendeleev's periodic table in manuscript form as presented for publication in 1869. (*Courtesy The New York Public Library.*)

	Group							
Series	I	II	III	IV	V	VI	VII	VIII
1	H (1)							
2	Li (7)	Be (9.4)	B (11)	C (12)	N (14)	O (16)	F (19)	
3	Na (23)	Mg (24)	Al (27.3)	Si (28)	P (31)	S (32)	Cl (35.5)	
4	K (39)	Ca (40)	___ (44)	Ti (48)	V (51)	Cr (52)	Mn (55)	Fe (56), Co (59), Ni (59), Cu (63)
5	[Cu (63)]	Zn (65)	___ (68)	___ (72)	As (75)	Se (78)	Br (80)	
6	Rb (85)	Sr (87)	?Yt (88)	Zr (90)	Nb (94)	Mo (96)	___ (100)	Ru (104), Rh (104), Pd (106), Ag (108)
7	[Ag (108)]	Cd (112)	In (113)	Sn (118)	Sb (122)	Te (125)	I (127)	
8	Cs (133)	Ba (137)	?Di (138)	?Ce (140)	___	___	___	
9	___	___	___	___	___	___	___	
10	___	___	?Er (178)	?La (180)	Ta (182)	W (184)	___	Os (195), Ir (197), Pt (198), Au (199)
11	[Au (199)]	Hg (200)	Tl (204)	Pb (207)	Bi (208)	___	___	
12	___	___	___	Th (231)	___	U (240)		

TABLE 12.1 Periodic classification of the elements; Mendeleev, 1872

The next element of increasing atomic mass known to Mendeleev was sodium, which behaves chemically much like lithium. So the second row of the table begins with sodium and is

Sodium	Magnesium	Aluminum	Silicon	Phosphorus	Sulfur	Chlorine
23	24	27.3	28	31	32	35.5

Each of these is chemically similar to the one directly above it. The next element known to Mendeleev was potassium, which is similar to sodium and lithium and so starts a new row, or period. The complete table, using only the elements known to Mendeleev, is given in Table 12.1.

Each entry contains the chemical symbol for the element. In parentheses is the atomic mass. Among other features we see a number of conspicuous blank spaces. These were left by Mendeleev because, when he put each element beneath those chemically similar to it, these blanks occurred. He predicted that there were as yet undiscovered elements whose atomic mass and chemical properties were correct for the blank spaces. Using the properties of the other elements in a vertical column, or group, Mendeleev predicted the properties of the missing elements. In the spot under carbon and silicon at atomic mass 72, Mendeleev predicted an element later discovered and named germanium. He also predicted its properties with astonishing accuracy. These are shown in Table 12.2.

Another example of the periodic aspect of the properties of the elements is shown in Fig. 12.8, which is a plot of the atomic volume

I	II							
1 H 1.00797								
3 Li 6.939	4 Be 9.0122							
11 Na 22.9898	12 Mg 24.312							
19 K 39.102	20 Ca 40.08	21 Sc 44.956		22 Ti 47.90	23 V 50.942	24 Cr 51.996	25 Mn 54.9380	26 Fe 55.847
37 Rb 85.47	38 Sr 87.62	39 Y 88.905		40 Zr 91.22	41 Nb 92.906	42 Mo 95.94	43 Tc (99)	44 Ru 101.07
55 Cs 132.905	56 Ba 137.34	57 La 138.91	58 to 71	72 Hf 178.49	73 Ta 180.948	74 W 183.85	75 Re 186.2	76 Os 190.2
87 Fr (223)	88 Ra (226)	89 Ac (227)	90 to 103	104 (257)	105 (262)			

Lanthanides	58 Ce 140.12	59 Pr 140.907	60 Nd 144.24	61 Pm (147)
Actinides	90 Th 232.038	91 Pa (231)	92 U 238.03	93 Np (237)

Values in parentheses are atomic masses of longest-lived or best-known isotopes.

TABLE 12.2 Prediction of the element germanium and its properties by Mendeleev in 1871. (mp = melting point; bp = boiling point)

	Predicted, element X	Found (1886), element Ge
Atomic mass	72	72.6
Properties	Grey, high mp, density = 5.5	Grey, mp = 958°C, density = 5.36
Oxide	XO_2, high mp, density = 4.7	GeO_2, mp = 1100°C, density = 4.70
Sulfide	XS_2, insoluble in water, soluble in yellow ammonium sulfide solution	GeS_2, insoluble in water, very soluble in yellow ammonium sulfide solution
Chloride	XCl_4, volatile liquid, bp < 100°C, density = 1.9	$GeCl_4$, volatile liquid, bp = 83°C, density = 1.88

TABLE 12.3 Periodic table of the elements. The number above the symbol is the atomic number; the number below the symbol is the atomic mass

				III	IV	V	VI	VII	2 He 4.0026
				5 B 10.811	6 C 12.01115	7 N 14.0067	8 O 15.9994	9 F 18.9984	10 Ne 20.183
				13 Al 26.9815	14 Si 28.086	15 P 30.9738	16 S 32.064	17 Cl 35.453	18 Ar 39.948
27 Co 58.9332	28 Ni 58.71	29 Cu 63.54	30 Zn 65.37	31 Ga 69.72	32 Ge 72.59	33 As 74.9216	34 Se 78.96	35 Br 79.909	36 Kr 83.80
45 Rh 102.905	46 Pd 106.4	47 Ag 107.870	48 Cd 112.40	49 In 114.82	50 Sn 118.69	51 Sb 121.75	52 Te 127.60	53 I 126.9044	54 Xe 131.30
77 Ir 192.2	78 Pt 195.09	79 Au 196.967	80 Hg 200.59	81 Tl 204.37	82 Pb 207.19	83 Bi 208.980	84 Po (210)	85 At (210)	86 Rn (222)

62 Sm 150.35	63 Eu 151.96	64 Gd 157.25	65 Tb 158.924	66 Dy 162.50	67 Ho 164.930	68 Er 167.26	69 Tm 168.934	70 Yb 173.04	71 Lu 174.97
94 Pu (242)	95 Am (243)	96 Cm (247)	97 Bk (247)	98 Cf (249)	99 Es (254)	100 Fm (253)	101 Md (256)	102 No (254)	103 Lw (257)

versus the atomic mass. The atomic volume is the volume occupied by one atom in the solid state of the element. It is obtained by dividing the atomic mass by the density. The periodic repetition of the peaks at the alkali metals is very striking. Many of the other properties of atoms show a similar periodicity when plotted against the atomic mass.

Until the 1940s the last element in the periodic table was uranium, atomic mass 238. Since that time a significant number of elements have been artificially created in nuclear reactors and accelerators, and the present table contains 105 elements. The modern version of the table is shown in Table 12.3. As in Mendeleev's table, elements in vertical columns, or groups, are chemically similar. Each entry has the chemical symbol for the element, the atomic mass below the symbol and, above the symbol, a number which gives the numerical position of the element in the table and is called the *atomic number*. In Chapter 19 we shall find additional physical significance for this number.

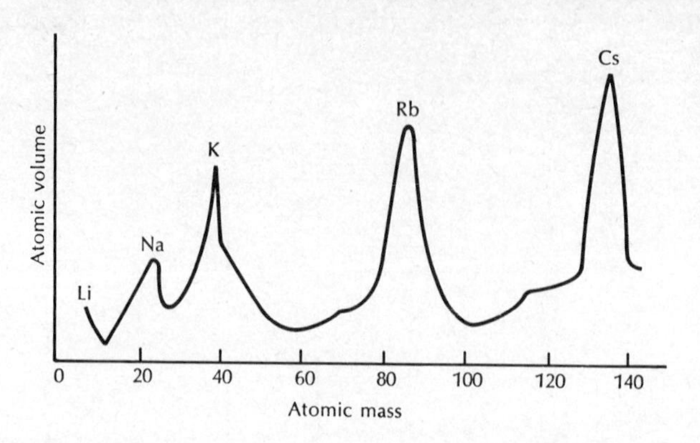

FIGURE 12.8 A plot of atomic volume as a function of atomic mass, showing the periodicity of this property. The atomic volume is essentially the volume occupied by a single atom of the element in question. The elements lithium, sodium, potassium, rubidium, and cesium all occupy one vertical column in the periodic table.

We now close this part of the story of the atom. Knowledge of the atom in Mendeleev's time was limited to its atomic mass, an idea of its size (about 10^{-10} meter), and the chemical and physical properties of the elements. But there was very little information regarding the structure, if any, of the atom. Our discussion of this must wait until we have acquired more background.

SUMMARY

The earliest atomic theory was that of Democritus. Opposing his point of view was the idea that earth, air, fire, and water are the four elementary materials. Aristotle opposed Democritus and was supported by the early church. Aristotle's ideas dominated science until the seventeenth century.

In the seventeenth century the atomic theory was revived by Gassendi, Boyle, and Newton.

Lavoisier provided the first working definition of an element, and also showed that mass is conserved in chemical reactions.

The law of definite proportions: Elements that form a chemical compound always occur in that compound in definite proportions by mass.

Dalton's atomic theory allowed predictions that could be tested by experiment. One of the key ideas was that all atoms of a given element are identical, particularly in mass. Dalton showed how to obtain the relative masses of atoms.

Accurate values of atomic masses were not available until after 1860. An amount of matter numerically equal in kilogram mass to the atomic weight is called a kilogram-mole. A mole of any element has the same number of atoms as a mole of any other element.

Mendeleev and Meyer developed the periodic chart by using the periodic repetition of certain properties as one ascends the scale of atomic masses.

SELECTED READING

Van Melsen, A. G.: "From Atomos to Atom: The History of the Concept Atom," Harper Torchbook TB 517H, Harper and Row, Publishers, New York, 1960. A detailed history of the atomic concept from the early Greeks to modern times. This concept is one of the central ideas of modern science, and the history of its evolution through 2,000 years is fascinating reading.

QUESTIONS

1 How did Robert Boyle define an element?
2 What is an empirical law?
3 Why does the law of definite proportions hold for both the mass and the weight of the constituent elements? Would this be true on the moon?
4 What is a molecule?
5 Are the statements of Dalton's theory given in the text consistent with modern theories of the structure of matter?
6 How is the kilogram-mole defined?
7 Does a mole of water contain more or fewer oxygen atoms than a mole of atomic oxygen?
8 Is the mass of the exhaust gases of your car larger than, equal to, or less than the mass of gasoline burned?
9 What is the mass of N_2O in atomic mass units? What is the mass of 1 kg-mole of N_2O?

SELF-TEST

__________ Thales
__________ Heraclitus
__________ Anaximenes
__________ Parmenides
__________ Democritus
__________ Empedocles
__________ Anaxagoras
__________ Aristotle
__________ Lucretius
__________ Gassendi
__________ Descartes
__________ Boyle
__________ Lavoisier

__________ Dalton
__________ Cannizzaro
__________ Mendeleev
__________ Meyer
__________ element
__________ law of definite proportions
__________ law of mass conservation
__________ molecule
__________ atomic mass
__________ mole
__________ kilogram-mole
__________ periodic chart

1 Developed the first real atomic theory
2 Defined a chemical element
3 Greek who believed in the earth, air, fire, and water concept
4 Greek with an atomic theory
5 Proposed water as the primary matter
6 Materialist philosopher following Democritus
7 Believed particles of the elements are the true building blocks of matter
8 Catholic priest who made atomic theory "respectable"
9 Proposed air as the primary matter
10 Believed change is impossible
11 Believed matter consists of seeds that can be subdivided
12 Leading scientist of the Greeks
13 Believed particles of matter are in constant motion
14 Finally cleared up the confusion about the atomic theory
15 Predicted the element germanium
16 Suggested fire as the primary constituent of matter
17 Also proposed a periodic chart
18 An amount of matter whose mass is numerically equal in grams to its atomic mass
19 The arrangement of elements in order of their mass
20 Several atoms joined together to form a specific chemical species
21 The composition by mass of chemical compounds is always the same
22 The end result of chemical analysis
23 Mass of an element, based on carbon = 12.0000
24 Mass of an element numerically equal in kilograms to the atomic mass of the element
25 The mass of products is equal to the mass of reactants in a chemical process

PROBLEMS

(Use masses from the periodic chart, Table 12.3)

1 The analysis of 50 g of water yields 5.5 g of hydrogen and 44.5 g of oxygen. Find the percentage composition of water.
2 Carbon dioxide is 27% carbon and 73% oxygen. Calculate the masses of carbon and oxygen in 1.5 kg of carbon dioxide.
3 One of the oxides of copper is 89% copper and 11% oxygen. Calculate the masses of copper and oxygen in 250 g of this oxide.
4 The analysis of 58.5 g of table salt yields 23 g of sodium and 35.5 g of chlorine. Calculate the percentage composition of table salt.
5 48 grams of sulfur and 171 g of fluorine react to form 219 g of sulfur hexafluoride. Calculate the percentage composition of this compound.
6 One of the oxides of nitrogen is 53.5% oxygen and 46.5% nitrogen by mass. How much oxygen and nitrogen are in 350 g of this oxide?
7 Propane gas is 82% carbon and 18% hydrogen by mass. How much carbon and hydrogen are in 150 g of propane?
8 28 grams of hydrogen reacts with 56 g of carbon to form methane (natural gas). What is the percentage composition, by mass, of methane?
9 480 grams of oxygen reacts completely with 372 grams of phosphorus to make P_2O_5, phosphorus pentoxide. What is the percentage composition, by mass, of P_2O_5?
10 Calculate the mass of *one* atom of lead.
11 Calculate the number of atoms in 6 kg of carbon.
12 Calculate the number of atoms in 35 kg of nitrogen.
13 Calculate the mass of *one* atom of silver.
14 Calculate the number of atoms in 10 g of oxygen.
15 Calculate the number of atoms in 150 g of sulfur.

The Greeks believed that all matter was composed of four elements—earth, air, water, and fire—in various mixtures. These correspond roughly to the four states in which modern scientists agree that matter is found—solids, gases, liquids, and plasmas. In this chapter we will discuss a few of the properties of each of these states of matter.

13 STATES OF MATTER

13.1 SOLIDS

A *solid* is most easily characterized as an object that maintains its shape regardless of its container (Fig. 13.1). Its shape is maintained because the atoms that make up the solid remain pretty much in the same place and are not free to move about. Solids can be divided into two types—crystalline and amorphous. The atoms in a crystalline solid are arranged in a regular geometrical array. One of the simplest of these is ordinary table salt. Sodium and chlorine atoms are arrayed as shown in Fig. 13.2. Diamond is another material with a simple atomic arrangement as shown in Fig. 13.3. All the chemical elements, and most compounds, form regular crystals. Figure 13.4 shows some crystals. The shape of these crystals is a reflection of the regular arrangement of the atoms in the material. The existence of crystals such as those shown in Fig. 13.4 was thus one of the first clues that matter consists of atoms.

CRYSTALS

The atoms of a crystal are constantly in motion. These motions are mostly vibrations back and forth that do not change the average position of the atoms. The magnitude of these vibrations depends on the temperature. As the temperature is lowered, the magnitude of the vibrations decreases. At absolute zero (0 K) there is only a very small amount of vibration. Frequently scientists study crystals at very low temperatures in order to remove the confusing effects of all the vibrating motion.

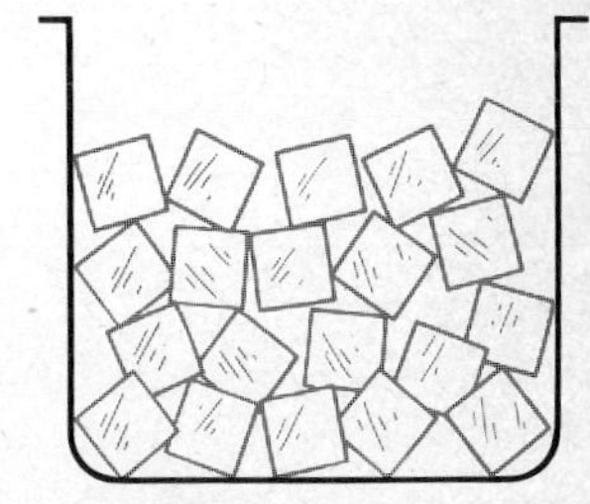

FIGURE 13.1 Ice cubes maintain their shape independent of their container.

The properties of crystalline materials are very sensitive to minor defects in the crystal structure. It is very difficult to prepare crystals in

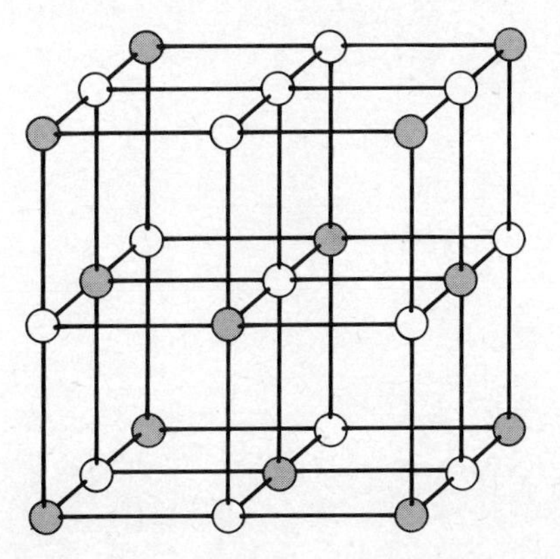

FIGURE 13.2 The sodium chloride crystal lattice. The dark circles represent sodium ions, and the light circles chloride ions.

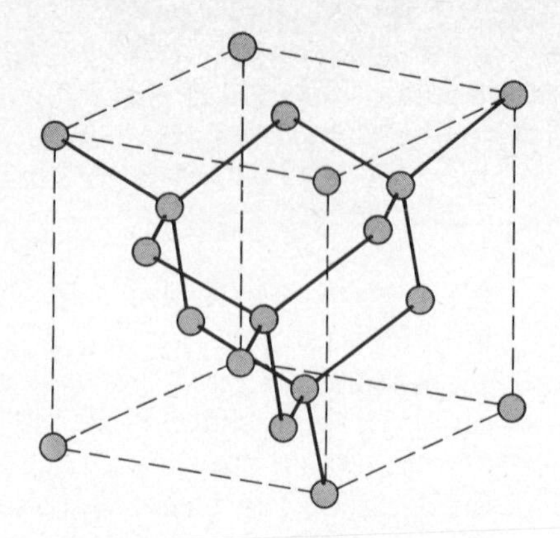

FIGURE 13.3 In the crystal of diamond, each carbon has a bond to four other carbons.

which the rows of atoms extend for large distances with no defects whatsoever. An example of a simple defect is shown in Fig. 13.5. In most cases such defects reduce the strength of materials markedly.

One area where this is of vital importance is in the construction of high-energy nuclear reactors. The constant stream of high-energy nuclear particles encountered causes all structural materials to be significantly damaged. Such items as fuel rods, and pipes carrying cooling materials, are weakened. They also swell, sometimes as much as 10 percent. This creates very severe problems for reactor construction, operation, and safety.

FIGURE 13.4 A group of natural quartz crystals. (*Photograph by Richard Eichelberger.*)

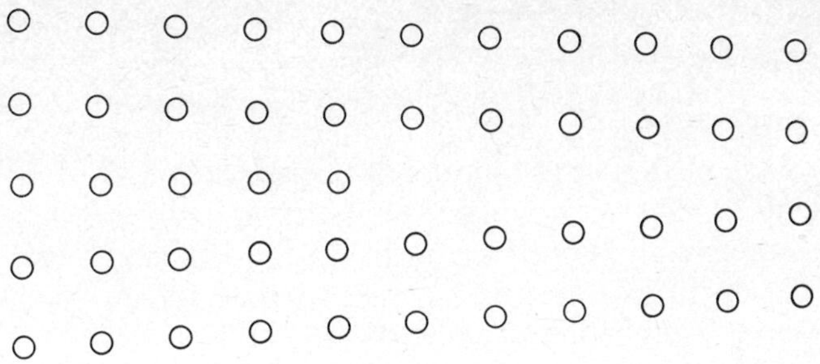

FIGURE 13.5 A crystal defect in which one row of atoms comes to an end. This is called *dislocation*.

Among the crystalline materials we distinguish between metals and nonmetals. Metals have good electric conductivity, and that of nonmetals is poor. (A few elements are in between, such as silicon and germanium, called *semiconductors*, and arsenic, antimony, and bismuth, called *semimetals*.)

Among the nonmetals we distinguish three kinds of crystals called *ionic*, *covalent*, and *molecular* crystals. The distinction is based on the kind of bond that holds the atoms in place.

In ionic crystals the basic structural units are ions—atoms that have lost or gained one or more electrons. An example is sodium chloride, ordinary table salt. The crystal lattice consists of sodium ions (Na^+) and chloride ions (Cl^-). These ions are held together by an electric force between positive and negative ions. In Fig. 13.2 we showed a sodium chloride lattice.

Covalent crystals are held together by ordinary covalent chemical bonds. Here electrons are shared by a pair of atoms, so that no ions are formed. Covalent bonds are strong, so covalent crystals are very strong and hard. The best example of a covalent crystal is the diamond, which is composed of carbon. Each carbon has a bond to four other carbons, as shown in Fig. 13.3.

Molecular crystals are crystals held together by very weak forces between molecules. Therefore they tend to be soft, and melt at low temperatures. Ordinary sugar is a molecular crystal.

AMORPHOUS MATERIALS OR "GLASSES"

Amorphous, or glassy, solids have no regular, well-defined crystalline arrangement. Glass is an amorphous solid, as are most plastics. The easiest way to obtain an amorphous solid is to cool a liquid substance quickly, so that the atoms have no time to rearrange themselves into regular crystals.

13.2 SOLID-STATE TECHNOLOGY

You have probably heard of solid-state technology, perhaps in an advertisement for a color TV or a new stereo music system. What this means is that the devices which control the electric currents in the TV or sound system are devices made from silicon or germanium, called *transistors*, rather than old-fashioned vacuum tubes. Silicon and germanium are called semiconductors because of their unusual electrical properties. It is found that small amounts of impurities produce very great changes in these electrical properties. Some impurities of particular interest are phosphorus, arsenic, and antimony (one column to the right of silicon and germanium on the periodic chart, Table 12.3), and boron, gallium, and indium (one column to the left on the periodic chart). The impurity concentrations used are less than .01 percent. The impurity atoms replace a silicon or germanium atom in the crystal. This is shown in Fig. 13.6.

The crystal lattice of silicon and germanium is similar to that of diamond. When an atom of phosphorus, arsenic, or antimony is added to germanium or silicon, there is one leftover electron, which then can wander through the lattice. An atom of boron, gallium, or indium is shy one electron of perfect bonding, so that an electron vacancy is created. This vacancy is called a *hole*. Holes can wander through the lattice, as well as electrons. The electrical properties of electrons and holes are used to give transistors their useful characteristics.

One of the most dramatic applications of solid-state technology is in the electronic computer. The first computer using vacuum tubes, built in 1945, was called ENIAC (*e*lectronic *n*umerical *i*ntegrator and *c*alculator). It weighed 30 tons and generated enough heat to nearly melt itself. The invention of the transistor dramatically reduced the size and heat output and increased the speed and reliability of computers. Originally, individual transistors, made from pieces of silicon or germanium, were connected with other circuit elements by the wires shown in Fig. 13.7. It has now become possible to create many transistors on a single chip of silicon, connecting them with tiny bands of evaporated metal. Such a device is called an integrated circuit. This has made possible very powerful computers of quite modest size, and also the newly de-

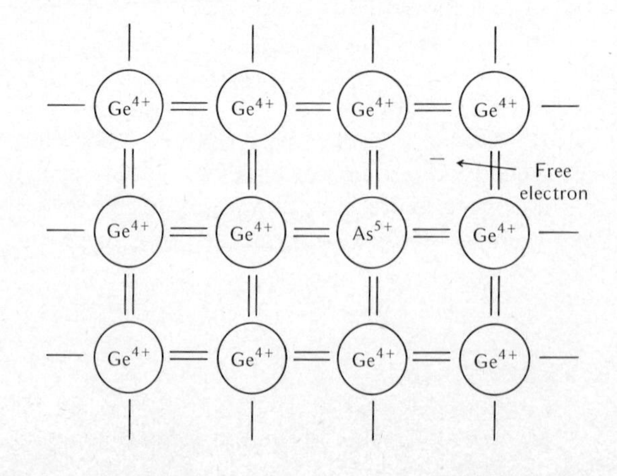

FIGURE 13.6 An arsenic impurity in a germanium lattice. One electron is left over, and is free to move about in the lattice.

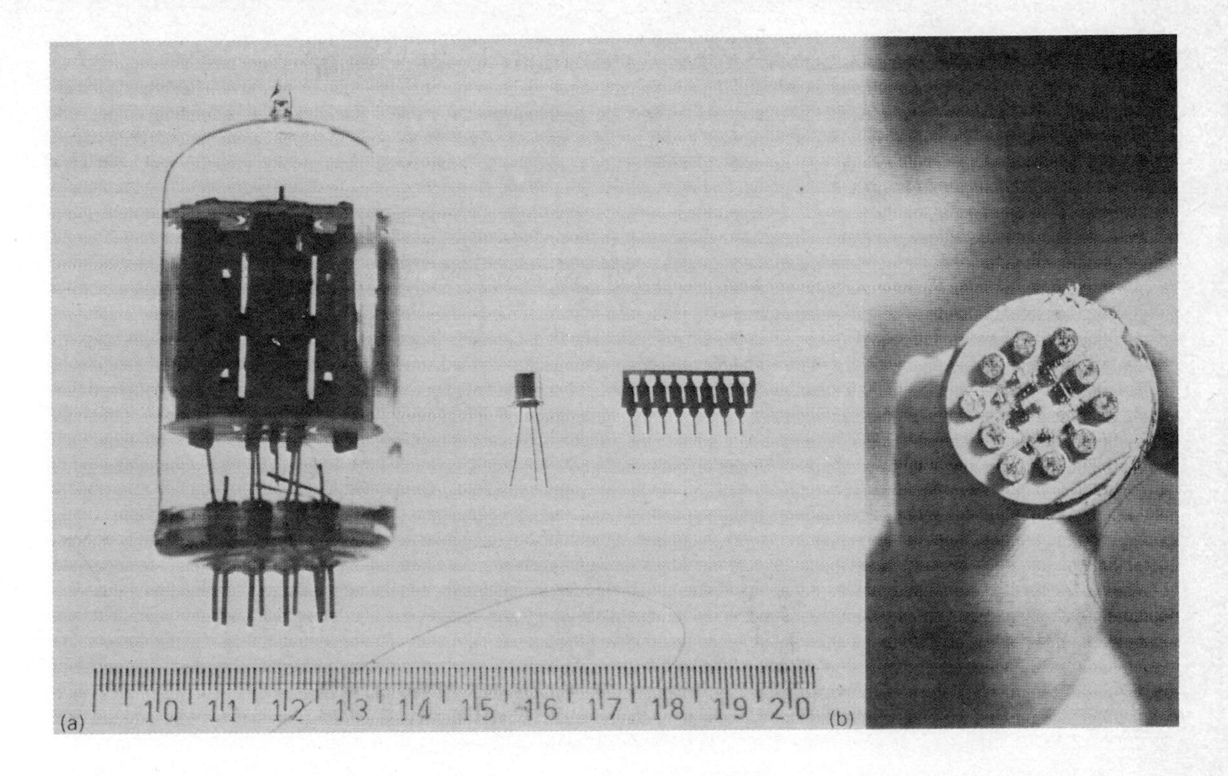

FIGURE 13.7 (*a*) A vacuum tube, left, a transistor, center, and an integrated circuit, right. The integrated circuit can do the work of 600 transistors. (*b*) This photograph shows the size of an integrated circuit package.

veloped pocket computers and calculators. (A computer is more than a calculator. It remembers the structure of a calculation, and will repeat the same calculation on another set of input numbers. A computer also can make decisions, for example, if the answer is greater than zero do this, if less than zero do that.) Figure 13.8 shows one of the newest of the pocket computers and the integrated circuits it uses. Note that the size of this device is not determined by the electronics, but by the surrounding hardware.

Another situation in which impurities are important is in the crystals used in certain types of lasers. A ruby is a crystal of aluminum oxide with a small amount of chromium impurity. When used in a ruby laser, the chromium atoms are the active element. One of the most powerful of modern lasers uses a glass with neodymium (a rare-earth element) as an impurity. Neodymium lasers have been suggested for use in the new laser-fusion devices to create a controlled hydrogen bomb reaction. We will have more to say about lasers in Chapter 18, and fusion in Chapter 21.

MELTING AND SUBLIMATION

One of the striking things that happens to crystalline solids is that they *melt*. They change from the solid state to the liquid state abruptly

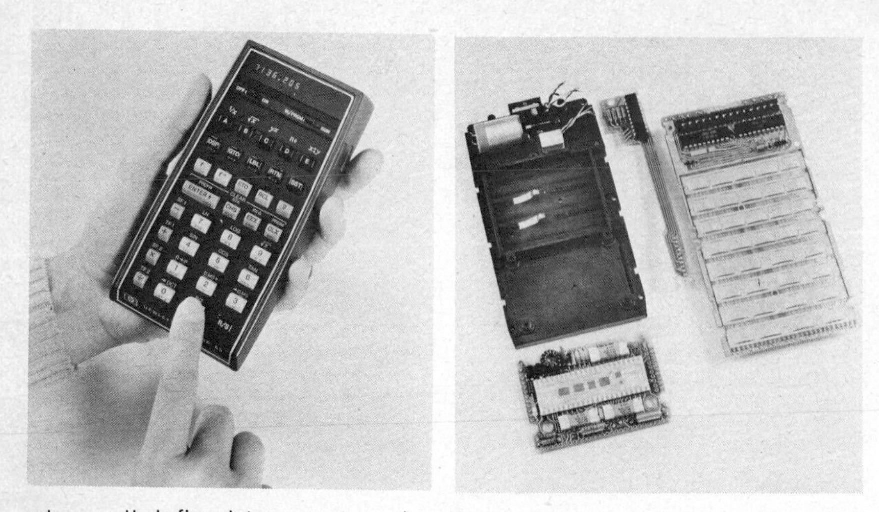

FIGURE 13.8 (*a*) A hand-held pocket computer. (*b*) The inside of the pocket computer. The six small grey areas in the center of the bottom left part are the integrated circuits. All timing control, memory, and arithmetic functions are provided by these circuits. (*Courtesy Hewlett-Packard.*)

at a well-defined temperature. In almost every case heat is absorbed during the melting process, but the temperature does not rise. This heat is necessary for the transition between the solid and the liquid to occur. When the liquid freezes back to a solid, the same amount of heat must be removed.

The heat of melting can be very large. For ice it is 80 calories for each gram melted. So 1 gram of water can be cooled from 80 to 0°C by 1 gram of ice melting. (Remember, 1 gram of water changes temperature by 1°C for each calorie added or removed.)

It has been proposed to use the heat of melting of certain crystalline solids for energy storage in solar heating schemes. Heat from some kind of solar energy collector is used to melt the material during the day. At night, the heat is removed, causing the material to resolidify. The advantage of this process is that a lot of energy is stored without having to deal with a wide temperature variation in the storage material.

A less obvious fact is that many solids *sublime*. That is, they change directly from the solid state to the gas state without becoming liquid. Some examples are dry ice (solid carbon dioxide), moth balls, iodine, and even ice (frost frequently disappears, leaving no water). A large amount of energy, more than is needed for melting, is required for sublimation.

When a liquid is slowly cooled and frozen, most of the impurities remain behind in the liquid portion. The regular crystal lattice tends to "squeeze out" foreign atoms. This fact has led to a process for purifying elements called *zone refining*. All solid-state technology requires very pure materials, so the zone refining process has been invaluable.

DENSITY

An important physical property of all materials is their density. Density is defined as the mass of material in a specified volume. One consistent

TABLE 13.1 Density of several materials

Material	Density, grams/centimeter^3
Lithium metal	0.53
Ice	0.92
Water	1.00
Diamond	3.51
Aluminum	2.70
Iron	7.86
Nickel	8.90
Copper	8.92
Silver	10.5
Platinum	21.4
Gold	19.3
Mercury	13.6
Lead	11.3
Uranium	18.7

use of units gives densities as kilograms per cubic meter. A more common use, however, is the unit grams per cubic centimeter. The density of water is 1 gram/centimeter^3. Table 13.1 lists the densities of some elements and common materials. Note that the densities of the elements increase as one goes down the periodic chart, although not completely regularly. Thus some idea of the chemical composition of an object can be inferred from its density.

Most materials are denser in the solid than in the liquid form. An exception is water. Solid water, ice, is less dense than the liquid. Therefore ice floats on water, and lakes freeze from the top down. The effect on life on earth of the reverse situation would be dramatic. Lakes and oceans would freeze from the bottom up, and would thaw only at the top in summer. Very little life could exist in lakes or oceans.

The average density of the planets can be used to infer something about their composition. The average density of the earth is about 5.5 grams/centimeter^3. Rocks near the surface have a density less than this. Therefore the core of the earth is believed to consist of mostly very dense iron and nickel. But the density of Jupiter is 1.3 grams/centimeter^3, and that of Saturn 0.7. Therefore these planets are believed to consist mostly of hydrogen and other light elements. The differences in composition, indicated by their densities, between Earth and the outer planets must be accounted for in any model of how the solar system was formed.

13.3 LIQUIDS

In the *liquid* state the atoms are no longer arranged in neat rows, as in the solid state, but are disordered. However, the distances between atoms in liquids are not much different than in solids. Liquids flow from point to point, and take the shape of their container, as shown in Fig. 13.9.

Liquids can change to the gaseous state. This process is called *boiling* or *vaporization*. Like the state change between solid and liquid, this occurs, for pure materials, at a well-defined temperature, and requires that heat be added. For water, this amount of heat, called the *heat of vaporization*, is very large—540 calories/gram. Above a certain critical temperature, characteristic of each material, it is no longer possible to make a distinction between liquid and gas. The material changes continuously, and no distinct phase change can be observed. This is shown in Fig. 13.10. Liquids and gases are both sometimes called *fluids*. The distinction between solids and fluids is that solids resist sideways or shear forces, and fluids do not. As an example of shear forces, put an ice cube on top of a large sheet of ice. Place a little water on top of the ice sheet. The ice cube can easily be moved across the ice sheet as long as the water is liquid. If the water freezes, the cube can no longer be moved without breaking the ice. In this example the liquid water does not resist shear forces, but the solid ice does (Fig. 13.11).

FIGURE 13.9 Liquids take the shape of their container.

PRESSURE

One of the important quantities used to discuss both liquids and gases is *pressure*. Pressure is related to force, and is defined as force per unit area. A liquid exerts pressure on the sides and bottom of its container. A gas exerts pressure on all sides. The pressure at the bottom of a liquid depends directly on the depth of the liquid above. Essentially it is caused by the weight of the liquid above. But it is independent of the shape of the vessel holding the liquid, and depends only on depth. A common way to demonstrate this is a device called Pascal's vases, shown in Fig. 13.12. The water level is the same in each vase, regardless of its size or shape. If the weight of the liquid in vase *A* caused a greater pressure than in vase *B*, the excess pressure would cause the liquid in *B* to rise to a higher level. This is not observed to occur.

Related to these ideas is Archimedes' principle. This can be stated in several ways.

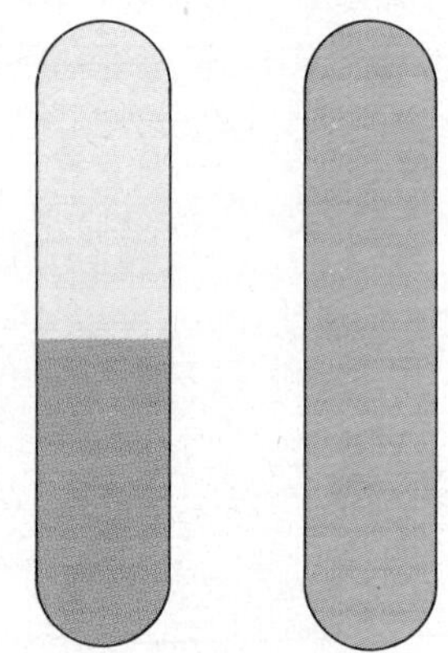

FIGURE 13.10 On the left there is a distinct boundary between water and vapor at 370°C. On the right the boundary completely disappears above 374°C.

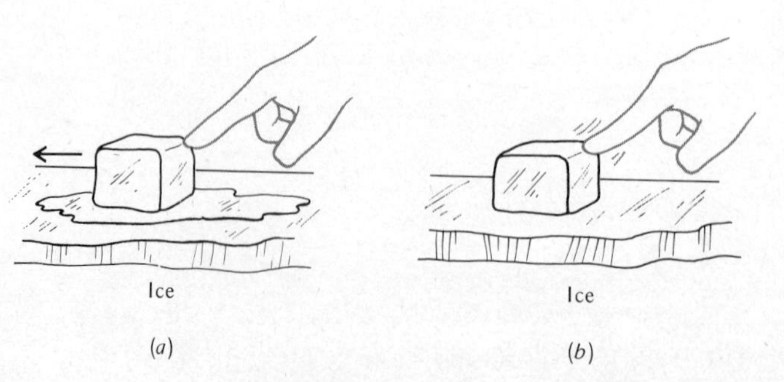

FIGURE 13.11 Water does not resist shear forces; ice does.

FIGURE 13.12 Water, or any liquid, reaches the same level in a container of any shape.

An object floating in a liquid displaces (occupies the space of) an amount of liquid whose weight equals the weight of the floating object.

Or

An object immersed in a liquid has an upward force on it equal to the weight of the volume of liquid displaced.

As an example, consider an ice cube floating in water. The forces on the ice cube are shown in Fig. 13.13. The upward force is due to the water displaced and is called the *buoyant force*. We can see from Archimedes' principle that the weight of the ice cube is equal to the weight of a volume of water smaller than the ice cube. That is, the density of ice is less than that of water. For this reason it floats.

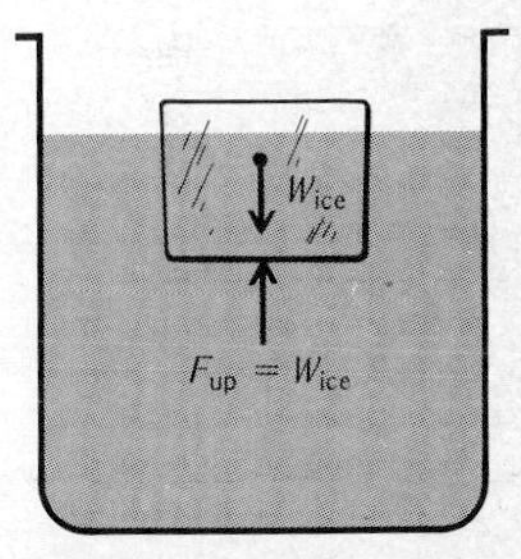

FIGURE 13.13 The ice displaces a volume of water just equal to its own weight. The weight down of the ice is just balanced by the buoyant upward force, due to the water displaced.

These ideas have an interesting application in geophysics. It is now believed that the continents on earth "float" like giant rafts on denser rock beneath them. It is also believed that the continents, or plates, drift about on these denser rocks. The rocks of the mantle, below the crust, are sufficiently fluid to permit this motion. (Many solids will flow slowly if sufficient force is applied. This is called *plastic flow*. Ice in a glacier flows in this fashion.) It is now believed that at one time all the continents were next to each other and have drifted to their present positions. The close "fit" between eastern South America and western Africa is one piece of evidence for this. The motion of these continental rafts, or plates, continues today.

Most liquids are not very compressible. If pressure is applied to them, their volume and density change only slightly. So if pressure is applied, it is then transmitted directly to other parts of the system. This leads to a class of machines called *hydraulic* devices, that act as force multipliers. (Energy is still conserved, and we will see that the work in equals the work out, except for friction.) Figure 13.14 shows a simple hydraulic lifting arrangement. Force is applied to the small piston on the left. The pressure is transmitted to the large piston on the right. Since pressure is force divided by area, the total force on a piston is given by pressure times area. The total force on the right is larger because the area is larger.

In terms of work, however, which is force times distance, the piston

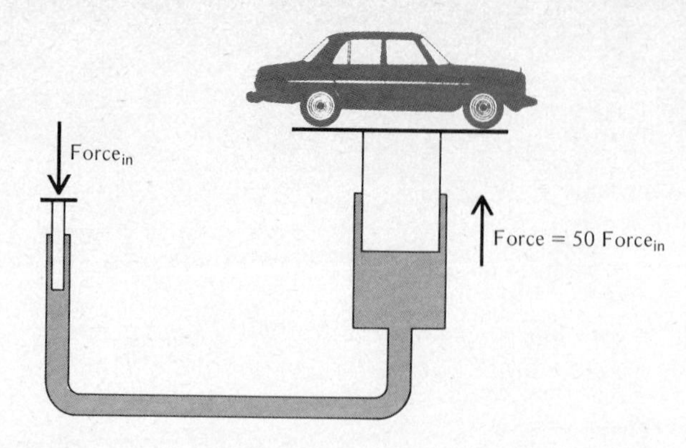

FIGURE 13.14 The pressure on each side is the same. The force holding the car up is 50 times the force in on the left, because the piston on the right has 50 times the area of the one on the left.

on the left has to move much further than the one on the right, since the volume of fluid remains the same. If the area on the right is 50 times the area on the left, the force on the right is 50 times the force on the left. To compensate, the piston on the left must be moved 50 times as far as the one on the right. In many hydraulic systems this much motion would become awkward, so provision is made to add more fluid on the left.

Another interesting property of liquids is called *surface tension*. In Chapter 11 we spoke of floating a needle on the surface of a bowl of water. Surface tension is what makes this possible. The surface atoms of a liquid are attracted to the molecules below them, but are not attracted from above. So the surface atoms or molecules behave differently from the rest because of this unbalanced force. The result is called surface tension. Certain water bugs float on the surface of the water as a result of surface tension. Droplets of moisture in the air become spherical in shape because of surface tension (a sphere has the smallest surface for its volume of any shape).

13.4 GASES

In many ways gases are the simplest of the states of matter. We can justify this statement by the following observation. When a liquid such as water is boiled, the resulting gas (water vapor) occupies a volume about 1,000 times greater than the original liquid. This must mean that the atoms or molecules of a gas are roughly 10 times further apart in the gas than in the liquid. So we can look at a gas as a collection of atoms or molecules that are relatively far apart. Therefore they do not interact very strongly. As a result gases can be considered a collection of individual objects (atoms or molecules), and the properties of the gas do not depend strongly on the nature of the particular atoms or molecules of which it is composed.

The atoms or molecules of a gas are in constant motion. How do we know this? For one reason, if they were not in motion, they would fall to the bottom of their container, behaving more like a liquid than a gas

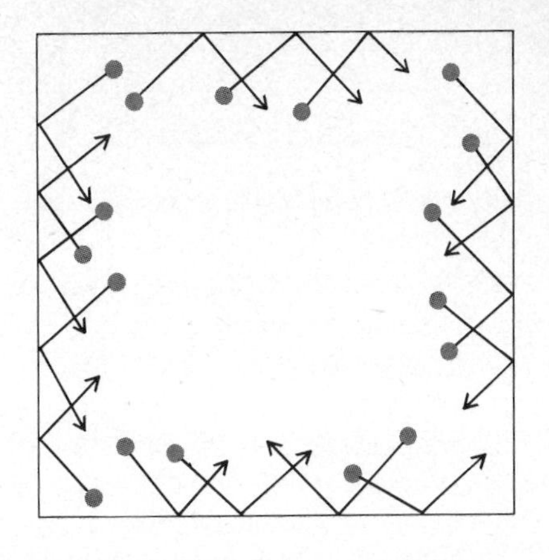

FIGURE 13.15 The collisions of atoms off the wall of a container cause the pressure of a gas.

(the molecules of a liquid are also in motion, but less violently). Second, a gas exerts pressure on all the walls of its container. We interpret this pressure as being caused by the collisions of the molecules with the container walls. Figure 13.15 shows the idea. The constant bombardment of the wall by atoms gives a net force, and this force is expressed as the pressure.

How can we show that a gas exerts pressure? There are several experiments. One of the simplest is to take a 1- or 2-gallon metal can and remove the air from it. This can be done with a vacuum pump, or

FIGURE 13.16 A can is completely crushed if the air is removed. The total force on the can is several tons. (*Photograph by Alexis Kelner.*)

by boiling some water in the can and then corking it. When the can is cooled, the water vapor condenses and the can collapses. This experiment shows that the pressure of the air is sufficient to crush a can completely.

Another experiment is related to an observation made by Galileo. Water can be pumped from a well only from a depth of 34 feet. If the well is any deeper, the water cannot be pumped by a pump at the top of the well. (Deeper wells are pumped using electric pumps down in the well.) In Galileo's day pumps were regarded as operating according to the principle that "nature abhors a vacuum." When air was pumped out of the top of the pipe, the water rushed in to fill its place. But Galileo was puzzled by the fact that the water would fill the pipe only to a height of 34 feet! (Figure 13.17.)

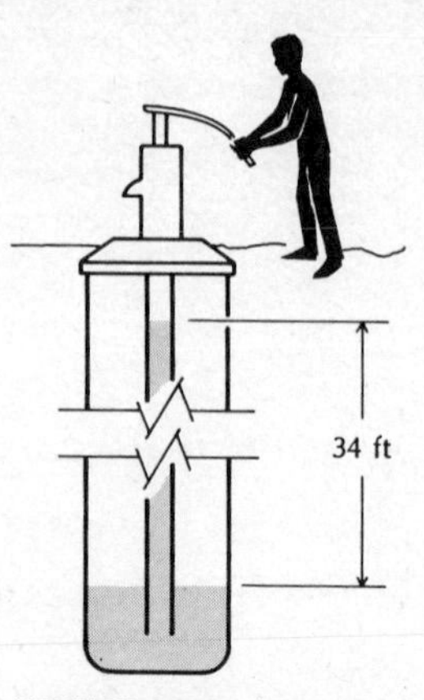

FIGURE 13.17 A pump at sea level can only pump water to a height of approximately 34 feet.

A student of Galileo, Evangelista Torricelli, invented a device—the *barometer*—which helps to solve the problem. He filled a glass tube with the heavy liquid metal mercury and, putting a finger over the open end, inverted it in a dish of mercury (Fig. 13.18). If the tube was longer than 76 centimeters, the surface of the mercury fell until it was 76 centimeters above the mercury level in the dish. Because nothing could pass through the glass walls of the tube, the space above the mercury must have been a vacuum. Again, nature abhors a vacuum, but in this case only to a height of 76 centimeters of mercury.

We can see the relation between Galileo's well and Torricelli's experiment by comparing the mass of a column of water with a 1-centimeter cross section 34 feet high and the mass of a column of mercury with the same cross section, but 76 centimeters high. Since 34 feet is approximately 1,030 centimeters, the volume of water is

$$V = 1 \text{ centimeter}^2 \times 1{,}030 \text{ centimeters} = 1{,}030 \text{ centimeter}^3$$

The unit of mass, the gram, was defined as the mass of 1 centimeter³ of water. Therefore this column of water whose volume is 1,030 centimeter³ has a mass of 1,030 grams, or 1.03 kilograms.

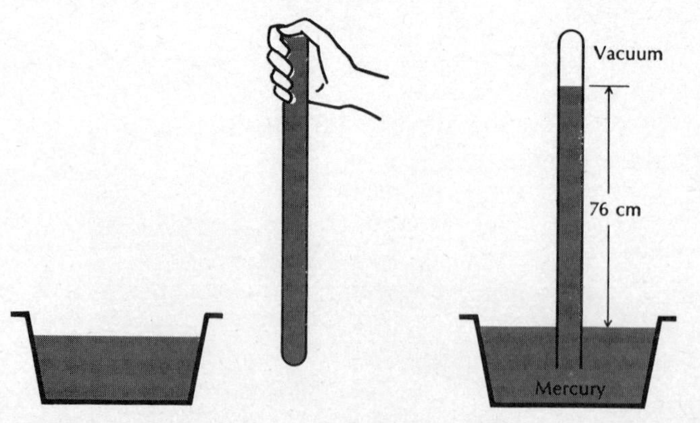

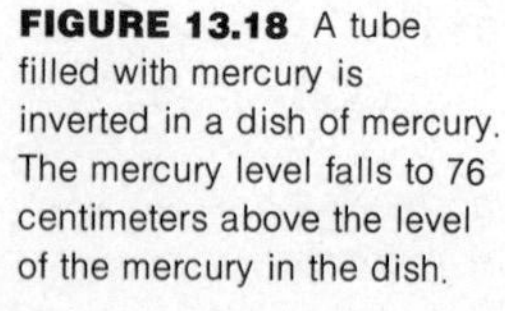

FIGURE 13.18 A tube filled with mercury is inverted in a dish of mercury. The mercury level falls to 76 centimeters above the level of the mercury in the dish.

The volume of the mercury column 76 centimeters high with a cross section of 1 centimeter^2 is

$$V = 1 \text{ centimeter}^2 \times 76 \text{ centimeters} = 76 \text{ centimeters}^3$$

Mercury is known to have a density of 13.6 grams/centimeter^3. Therefore the mass of mercury in this column is

$$m = 76 \text{ centimeter}^3 \times 13.6 \text{ grams/centimeter}^3$$
$$m = 1{,}030 \text{ grams} = 1.03 \text{ kilograms}$$

So we see that there is a relationship between the column of water 34 feet high which a pump will raise and the column of mercury which can be supported in a closed tube. If the cross-sectional area of the two columns is the same, they have exactly the same mass, and therefore exactly the same weight. The weight of each of these columns is

$$\begin{aligned} \textit{Weight} &= \text{Mass} \times g \\ &= (1.03 \text{ kilograms})(9.8 \text{ meters/second}^2) \\ W &= 10.1 \text{ newtons} \end{aligned}$$

Notice that this result does not depend on the cross-sectional area of the column measured. If the cross section had been 2 centimeter^2, the weight of both mercury and water columns would have been 20.2 newtons, still identical.

Now let us ask why a column of mercury 76 centimeters high, or a column of water 34 feet high, should stand in a closed tube. It is obvious that, if we take a tube open at both ends and pour water into it, the water will flow out the bottom. The same thing is true for mercury. What is the difference between this situation and that of the closed tube of Torricelli? There must be some force which acts to support the mercury column in Torricelli's tube or the water column in a pump. This force is exactly the same in both cases and is equal to the weight of each column. We have calculated this as 10.1 newtons for each square centimeter of cross section of the tubes. The amount of force necessary to support the mercury or water in these tubes depends on the size of the tube. We are led to describe the force which supports the mercury column in terms of *pressure*. Pressure is a measure of *force per unit area*. Its units are newtons per square meter, or pounds per square inch. Armed with this idea, we can say that the column of mercury in the closed tube is supported because there is a pressure of 10.1 newtons/centimeter^2 upward on the mercury at the bottom of the tube. This balances the weight of the mercury (Fig. 13.19).

Now suppose we take a tube in the shape of a U which has one arm 34 feet high and one arm 76 centimeters high, as in Fig. 13.20. If we first put mercury in the tube and then fill the long side with water, we shall find that 34 feet of water exactly balances 76 centimeters of mercury.

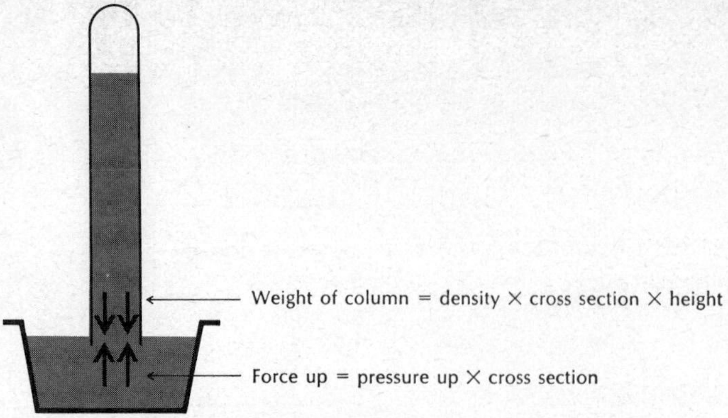

Weight of column = density × cross section × height

Force up = pressure up × cross section

FIGURE 13.19 The weight of the column of mercury must be supported by an upward pressure from the mercury in the dish.

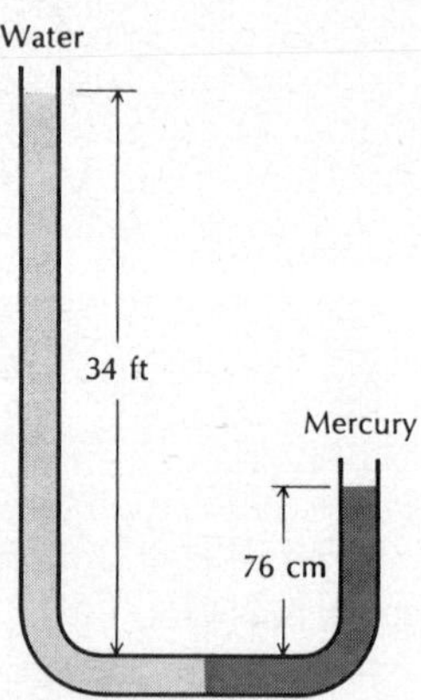

FIGURE 13.20 A column of water 34 feet high balances a column of mercury 76 centimeters high.

This is not surprising, since the weight of 34 feet of water is the same as that of 76 centimeters of mercury. (There are practical difficulties in carrying out this experiment exactly as described, but let us ignore them and treat it as a thought experiment, since all we wish to show is that a column of mercury 76 centimeters high exactly balances a column of water 34 feet high.) Now construct a tube, as in Fig. 13.21, which is larger in diameter on the long side. Now the weight of the water on the left is more than the weight of the mercury on the right. But in doing the experiment, we still find that 34 feet of water balances 76 centimeters of mercury, even though the weight of water is several times greater than the weight of mercury. Perhaps we can explain this result with another tube of a similar sort. Suppose we make it in the shape shown in Fig. 13.22 and fill both sides with water. No matter how deep the water is, it will come to exactly the same level on each side of the tube. Let us take a look at point *A*. The water on the right is held in position by a pressure on it at *A*. The same pressure acts on every part of the wall at *A*. So the extra weight of the water on the left yields a pressure, part of which pushes on the walls as shown, and part of which pushes on the water on the right to hold it up. Because the water level does not fall, there must be an equal pressure at *A* due to the water in the right column. So in dealing with liquids, the pressure, and not the actual volume or weight, is the important quantity. This experiment is equivalent to Pascal's vases, discussed in Section 13.3.

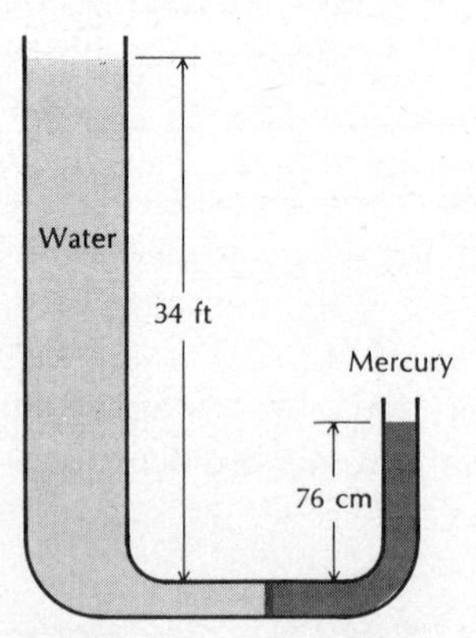

FIGURE 13.21 Even when the size of the tubes on the two sides differs, we find that 34 feet of water balances 76 centimeters of mercury.

13.5 THE PRESSURE OF THE ATMOSPHERE

In Section 13.3 we made a distinction between solids and fluids. (Both liquids and gases are fluids.) The particular distinction that is important here is that a force downward on the upper surface of a solid is transmitted only to the bottom. But a force on the upper surface of a fluid is transmitted in *all* directions. Therefore the pressure of a column of water or mercury is directed both downward and outward against the walls of the tube.

The similarity between liquids and gases was used by Torricelli to explain the apparently unsupported mercury column in the closed tube. He proposed that we live in a sea of air, as shown in Fig. 13.23, and that the weight of 76 centimeters of mercury represents the weight of mercury that can be supported by the pressure of this sea of air on the open surface of the mercury in the dish. The pressure of the air is then, using our previous results, the weight of a column of air extending to the top of the atmosphere. Since we have shown that a mercury column 76 centimeters high can be supported by the pressure of the air, and that the weight of a mercury column 1 centimeter^2 and 76 centimeters high is 10.1 newtons, it must be true that the *weight* of a column of air 1 centimeter^2 and extending to the top of the atmosphere is 10.1 newtons. The pressure of the air is the weight of air above every square centimeter of the earth's surface. In mks units the pressure of the air is 1.01×10^5 newtons/meter^2, and in English units it is 14.7 pounds/inch^2.

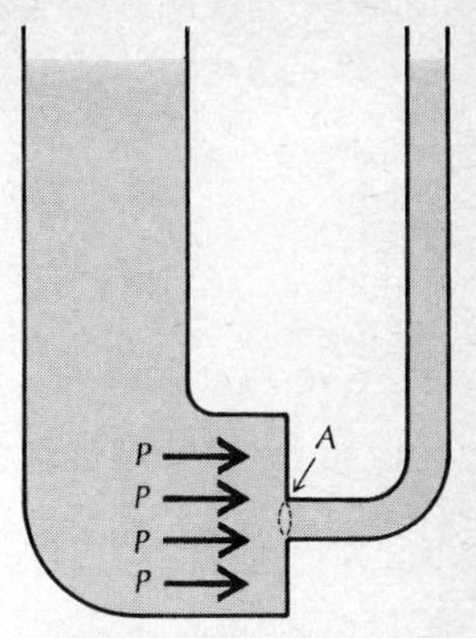

FIGURE 13.22 Columns of water which are unequal in volume still come to the same level.

Torricelli further pointed out that if the sea-of-air analogy were correct, the height of the mercury column which could be supported by the atmosphere would *decrease* if one climbed a mountain, because the weight of the column of air which supported the mercury would decrease. This would occur because on a mountain one would be above part of the sea of air. This experiment was performed by Blaise Pascal (1623–1662), and Torricelli was proven right. The height of mercury which can be supported on a closed tube is then a measure of the pressure of the air, and Torricelli's instrument is the barometer, now used as an essential tool in weather forecasting, since changes in the pressure of the air of about 2.5 centimeters of mercury are associated

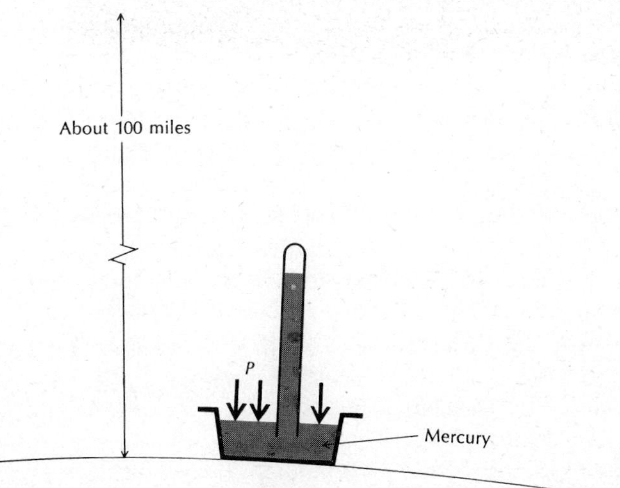

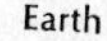

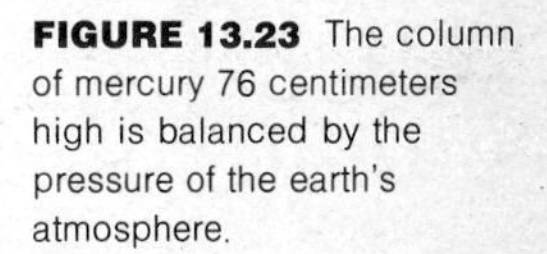

FIGURE 13.23 The column of mercury 76 centimeters high is balanced by the pressure of the earth's atmosphere.

with different weather systems. The measurement of the pressure of the air is also used to determine the altitude of airplanes above sea level.

Since air pressure is measured using a barometer, it has become customary to report pressure in terms of the height of the mercury column it will support. At sea level this column is about 76 centimeters, or just under 30 inches, and therefore the barometric pressure is reported at 29.90 inches of mercury, or 76 centimeters of mercury. What is meant, in fact, is that the pressure will support a column of mercury that high.

FIGURE 13.24 Robert Boyle (1627–1691). *(Courtesy University of Pennsylvania Library.)*

13.6 BOYLE'S LAW

Now we can proceed to experiments that tell us something about the properties of gases. The first of these was performed by Robert Boyle. Boyle studied the volume a gas occupies as the pressure on it is changed. If we put a small amount of mercury in the tube, as shown in Fig. 13.25*a*, and open the valve, the height of the mercury will be the same on both sides. Then we close the valve. The pressure of the air in region A must be exactly what it was before the valve was closed, that is, atmospheric pressure of 1.01×10^5 newtons/meter^2. If the tube is of uniform cross section, the distance A is an exact measure of the space the gas between the mercury and the valve occupies. Volume = cross section $\times A =$ constant $\times A$. This is what is called the volume of the gas sample.

Now suppose we add some mercury to the left side of the tube, as in Fig. 13.25*b*. We shall find that the mercury level is not the same on both sides. If we designate the difference in levels by B, we will have the following result. The gas trapped in A is supporting a column of mercury of height B, the difference between the mercury levels in the two arms. But the gas trapped in A would have supported a column 76 centimeters high *before* adding the extra mercury if there had been a vacuum above the mercury on the left side. That is, the pressure of the gas trapped in A was initially equivalent to a column of mercury 76 centimeters high. After adding the extra mercury to the left side, the pressure on the gas in the volume A' (which is less than the original volume A) must be 76 centimeters plus the difference in the height of the mercury columns B. The pressure is $B + 76$ centimeters of mercury. As B is increased by adding more mercury, the volume of gas, as measured by the distance A', decreases. If we add enough mercury to make B exactly 76 centimeters, the pressure will be twice what it was initially (Fig. 13.25*c*): 76 centimeters + 76 centimeters = 152 centimeters of mercury. If the volume A' is now measured, it is found to be reduced to one-half its original value.

This experimental result can be generalized, and the result is *Boyle's law for gases*, which states that, if the pressure on a gas is doubled, its volume is cut in half; if the pressure is tripled, its volume

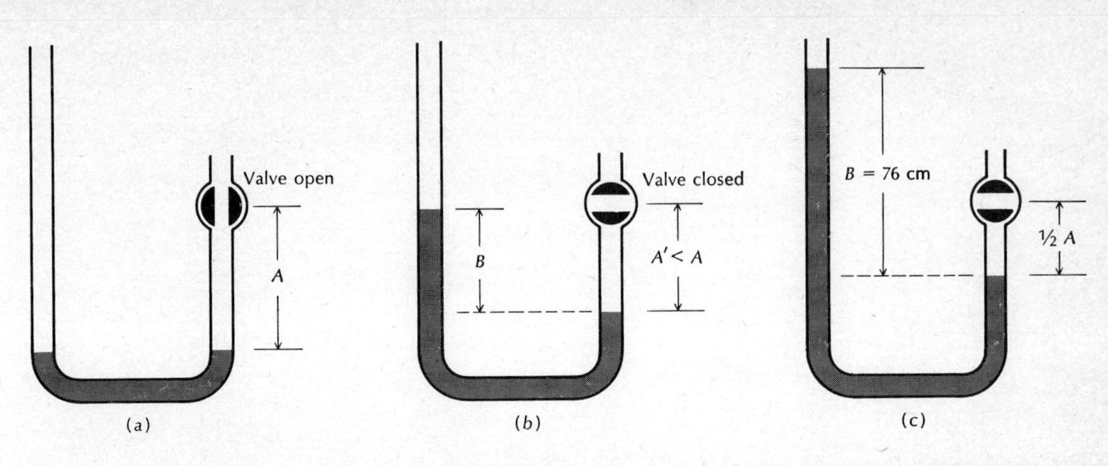

FIGURE 13.25 (*a*) With the valve open the mercury level is the same in both sides of the tube. (*b*) With the valve closed and some mercury added to the left, the gas trapped on the right is squeezed into a smaller volume. (*c*) When the difference *B* between the two mercury columns is 76 centimeters (at sea level), the trapped gas is reduced to one-half its original volume.

is reduced to one-third of its original volume. Mathematically, this is stated by saying that the product of pressure and volume is constant. This is

$$\text{Pressure} \times \text{volume} = \text{constant}$$
$$PV = \text{constant}$$

or

$$P_1V_1 = P_2V_2 = P_3V_3, \quad \text{etc.} \tag{13.1}$$

where P_1 and V_1 are the pressure and volume measured under one set of conditions, and P_2 and V_2 are the pressure and volume after the conditions have been changed, perhaps by adding more mercury. We should make one qualification: All these measurements must be made at the same temperature. Boyle was aware of this qualification but never put it into quantitative terms. Not until 150 years later was the effect of temperature change expressed quantitatively by Charles and Gay-Lussac.

13.7 THE LAWS OF CHARLES AND GAY-LUSSAC

In this section we discuss how the volume of gases changes with temperature. We use the Celsius and Kelvin temperature scales first discussed in Chapter 7. Let the volume of a sample of air be measured as the temperature is changed. To make things simple, the pressure is kept constant by adjusting the height of a mercury column supported by the sample. The easiest way to show these results is to make a graphical plot of the measured volume as a function of temperature. The solid black points in Fig. 13.26 represent measured values of the volume at various temperature. Air becomes a liquid at −196°C and, because of the peculiarities in its behavior near this temperature, only data for higher temperatures are included in Fig. 13.26. The graph shows a steady decrease in volume as the temperature is decreased. All the data points fall on a straight line. If we were to extend or extrapolate the

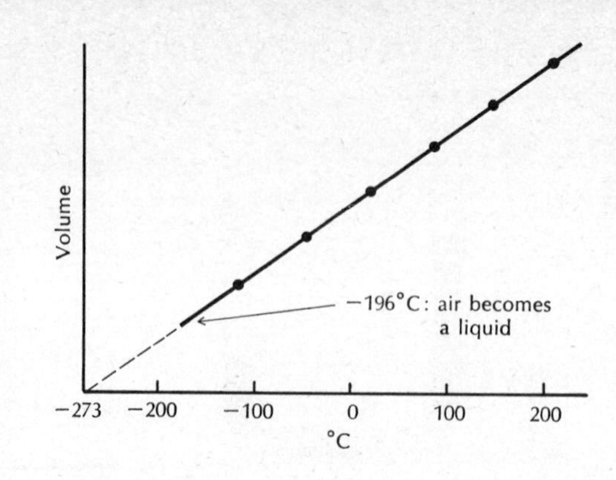

FIGURE 13.26 The volume of a sample of air as the temperature is decreased. The data points follow a straight line to the point where air becomes a liquid (−196°C). The extension of this straight line shows that the volume would be reduced to zero at −273°C if the straight-line relationship were obeyed to that temperature.

straight line beyond the point where air becomes a liquid, as shown by the dashed line, it would appear as though the volume would become zero at −273°C. If the same experiment is performed with other gases, the same result will be found: All gases change volume with temperature in the same way. Their volumes decrease with temperature in such a way that their volume would be zero at −273°C if the gas did not liquefy first.

We can express mathematically the results of this experiment. At constant pressure, the volume of *any* gas depends on the temperature as follows:

$$V \propto t + 273 \tag{13.2}$$

Here t is the temperature in degrees Celsius. In words, Equation 13.2 says that the volume of any gas is directly proportional to the temperature in degrees Celsius plus 273°. (If the student has difficulties with the graphical presentation or its interpretation as a proportionality, he should review Section A.5 in the appendix.)

For a change in temperature from 0 to 1°C, the volume of a gas increases by $\frac{1}{273}$ of its original value. This is *Charles' law*.

If we hold the volume of a gas constant and vary the temperature, a similar result will be found. At constant volume the pressure of a gas varies with temperature as

$$\text{Pressure} \propto t + 273 \tag{13.3}$$

This is *Gay-Lussac's law*.

There is something very suggestive about the number 273 which appears in both of these laws. Since these laws are found experimentally to hold for all gases, it seems that there may be something unique about the temperature −273°C at which the volumes of gases seem to vanish. This fact is used as the basis for the Kelvin temperature scale. In this scale the size of the unit is the same as the Celsius degree, but 0 K is defined as −273°C. We first encountered this scale in Chapter 7. Experiments on gases, as well as on heat engines, show that 0 K is the absolute zero of temperature. This temperature can never be attained, although temperatures have been achieved in the laboratory as low as 0.000001 K for short periods of time.

If we express the laws of Charles and Gay-Lussac using the Kelvin scale, we find

$$V \propto T \qquad \text{Charles' law} \tag{13.4}$$

and

$$P \propto T \qquad \text{Gay-Lussac's law} \tag{13.5}$$

where T is the temperature in kelvins.

13.8 THE IDEAL GAS LAW

If we consider allowing both pressure and volume to change as we vary the temperature, we find experimentally that the product of pressure and volume is proportional to the temperature in kelvins. This can be written as

$$PV \propto T$$

or

$$\frac{PV}{T} = \text{a constant} \tag{13.6}$$

Equation 13.6 states that the product of pressure and volume divided by the temperature in kelvins is constant for a given sample of gas. This is a combination of Boyle's, Charles', and Gay-Lussac's laws. If, for a given sample of gas, the pressure P_1, volume V_1, and temperature T_1 are measured, and then some change made in one or another of these quantities, the following must be true:

$$\frac{P_1V_1}{T_1} = \frac{P_2V_2}{T_2} \tag{13.7}$$

where P_2, V_2, and T_2 are the pressure, volume, and temperature after the change has been made. Equation 13.7 is called the *ideal gas law*. It is called this because it describes the behavior of gases under the simplest assumptions. Most gases near the point where they become liquid will show deviations from ideal gas behavior.

Example 13.1

The temperature of 2 meters^3 of air at 27°C and a pressure of 76 centimeters of mercury is raised to 127°C, and the pressure is decreased to 38 centimeters of mercury. What is the new volume?

$$\frac{P_1V_1}{T_1} = \frac{P_2V_2}{T_2}$$

$$\frac{(76 \text{ centimeters})(2 \text{ meters}^3)}{(0+273) \text{ kelvins}} = \frac{(38 \text{ centimeters})\, V_2}{(100+273) \text{ kelvins}}$$

$$\frac{76 \times 2}{273}\,\frac{373}{38} = V_2$$

$$5.46 \text{ meters}^3 = V_2$$

The student should note that the ideal gas law holds *only* for temperatures expressed in kelvins. It should also be clear that any unit of pressure can be used, with the restriction that the same unit be used for P_1 and P_2. This is also true for the units of volume.

The density of gases is clearly related to the volume that a sample of gas occupies. If the same mass of gas is squeezed into a smaller volume, its density will be increased. Therefore we can write for density ρ (Greek rho)

$$\frac{\rho_2}{\rho_1} = \frac{P_2 T_1}{P_1 T_2} \tag{13.8}$$

since the equation for volumes is (from Equation 13.7)

$$\frac{V_1}{V_2} = \frac{P_2 T_1}{P_1 T_2} \tag{13.9}$$

STANDARD CONDITIONS

It is very convenient, for comparison purposes, to agree to a set of conditions of temperature and pressure, which are called *standard conditions*, or STP (standard temperature and pressure). The agreed conditions are those corresponding to the average pressure at sea level and a temperature of 0°C, or 273 K. Standard pressure is 76 centimeters of mercury, or 10.1 newtons/meter². Therefore, if the volume of a gas is said to be reduced to standard conditions, this means the volume that sample of gas would occupy at 76 centimeters of mercury and 273 K.

13.9 PLASMAS

Recently there has been much interest in an area called *plasma physics*. This interest is a result of attempts to achieve controlled thermonuclear fusion (the controlled release of the energy of the hydrogen bomb). A plasma is a gas that has been heated to a high enough temperature that the atoms have lost one or more electrons. Therefore the gas becomes electrically conducting. To fully ionize a gas so that all atoms have lost all electrons requires very high temperatures. But it is possible to have partially ionized plasmas in ordinary flames. Because a plasma is electrically conducting, its properties are very different from those of ordinary gases. This has led to the idea that plasmas are a "fourth state of matter."

The importance of the plasma state lies in the fact that this is the material of which stars are made, and this is the state that must be achieved if we are to have thermonuclear fusion as a source of energy. To make the fusion process work requires temperatures in excess of 10 million kelvins. To maintain these temperatures requires that the plasma be kept away from the walls of the container. Magnetic fields have been proposed as a means for doing this, because moving charged particles experience a force when moving through a magnetic field. This has led to containment proposals called "magnetic bottles." We will have more to say about thermonuclear fusion in Chapter 21.

Another practical reason for the interest in plasmas is a means of electric energy generation called magnetohydrodynamics (MHD). In a MHD generator a hot plasma (created either by burning coal or oil, or from a fusion device) is passed through a magnetic field. Positively charged atoms curve one way in the magnetic field, and negatively charged electrons curve the other way. Electrodes inserted in the gas stream take advantage of this charge separation, and a current flows through an external circuit. Since no heat engine is involved, the energy efficiency of a MHD generator is expected to be quite high. This may allow more efficient use of our coal resources, for instance.

SUMMARY

A solid maintains its shape regardless of its container. The atoms of a crystalline solid are arranged in a regular array. The atoms of an amorphous solid are disordered. Minor defects and impurities in crystals can have a major effect on their properties. Semiconductors are one practical result of this.

In ionic crystals, the basic unit is an ion, an atom that has gained or lost an electron. In covalent crystals, like diamond, the atoms are held together by ordinary covalent chemical bonds. In molecular crystals molecules are held together by very weak intermolecular forces.

The heat of melting is the heat that must be added to a solid to convert it to the liquid state, without any temperature change. The heat of sublimation is the heat needed to go directly from the solid to the vapor state.

The density of a substance is defined as the mass per unit volume, such as kilograms per cubic meter, or grams per cubic centimeter.

The atoms in a liquid are free to flow about. Therefore, liquid takes the shape of its container, and has a flat upper surface. The heat of vaporization is the heat required to go from the liquid to the vapor state.

Pressure is defined as force per unit area. A gas exerts pressure on all sides of its container. A liquid exerts pressure on the bottom and sides of its container.

Archimedes' principle: (1) An object floating on a liquid displaces an amount of liquid whose weight equals the weight of the object. (2) An object immersed in a liquid has an upward force on it equal to the weight of the volume of liquid displaced.

In a hydraulic system, force can be multiplied, but the work on one side equals the work on the other.

The fact that water can be pumped only 34 feet and that mercury stands in a closed tube 76 centimeters is a reflection of the pressure of the atmosphere. The barometer is an instrument using this property. An altimeter is a barometer which uses the pressure of the air to determine altitude above sea level.

In two systems of units, 1 atmosphere of pressure is

1.01×10^5 newtons/meter^2

14.7 pounds/inch^2

Boyle's law: The product of pressure times volume of a gas is constant (at constant temperature). $PV = \text{constant}$.

Charles' law: The volume of a gas at constant pressure is directly proportional to the temperature in kelvins. $V \propto T$.

Gay-Lussac's law: The pressure of a gas at constant volume is proportional to the temperature in kelvins. $P \propto T$.

The ideal gas law: Combining these results,

$$\frac{P_1V_1}{T_1} = \frac{P_2V_2}{T_2}$$

or

$$\frac{V_1}{V_2} = \frac{P_2T_1}{P_1T_2}$$

The density ρ of gases obeys a similar relationship:

$$\frac{\rho_2}{\rho_1} = \frac{P_2T_1}{P_1T_2}$$

Standard conditions are 273 K and 1 atmosphere of pressure.

Plasmas are gases in which the atoms have lost one or more electrons as a result of the high temperature. Therefore, plasmas are electrically conducting.

SELECTED READING

Holden, Alan, and Phylis Singer: "Crystals and Crystal Growing," Science Study Series S-7, Doubleday and Company, Garden City, New York, 1960. A variety of experiments with crystals, and instructions on how to grow crystals.

QUESTIONS

1 Would you expect glass to melt at a well-defined temperature, or to soften over a range of temperatures? Why?
2 The statement was made that liquids are nearly incompressible. Why should this be true?
3 Covalent crystals, like diamond, are among the hardest substances known to man. Why?
4 Archimedes is supposed to have developed his principle when he was ordered by the king to determine whether a crown was pure gold, or had been adulterated. How did this work?
5 A ship is spoken of as having a displacement of 30,000 tons. What does this mean?
6 How does an air pressure of 30 to 40 lb/in.^{2} in a tire support a 2,000-lb car?
7 Why should you *not* measure the air pressure in your tires after driving a long distance?
8 If atmospheric pressure at sea level is 14.7 lb/in.^{2}, why are we not crushed?
9 Why is the pressure of the atmosphere less during stormy weather?
10 What is the difference between a computer and a calculator?
11 How many crystalline materials can you name?

SELF-TEST

____________ solid
____________ liquid
____________ gas
____________ plasma
____________ crystal
____________ ionic crystal
____________ covalent crystal
____________ molecular crystal
____________ metal

____________ nonmetal
____________ amorphous solid
____________ sublimation
____________ vaporization
____________ zone refining
____________ density
____________ heat of vaporization
____________ pressure
____________ Archimedes' principle
____________ hydraulic device
____________ barometer
____________ vacuum
____________ 1 atmosphere
____________ Boyle's law
____________ Charles' law
____________ Gay-Lussac's law
____________ ideal gas law

1 A poor electric conductor
2 Purification, taking advantage of the fact that impurities are prevented from being incorporated in a crystal
3 Force per unit area
4 Takes the shape of its container; exerts force on all surfaces of the container
5 Variation in pressure of a gas with temperature
6 Gas in which atoms have lost one or more electrons as a result of high temperature
7 An object in a liquid has an upward force on it equal to the weight of the displaced liquid
8 Passing directly from the solid to the vapor state
9 Solid that conducts electricity well
10 Pressure of the air around us
11 Device for measuring gas pressure
12 Variation in volume of a gas with temperature
13 Mass per unit volume
14 A force multiplier using the incompressibility of liquids
15 Passing from the liquid to the gaseous state
16 Solid in which the atoms are arranged in a regular array
17 Relationship between pressure and volume for a gas
18 The absence of nearly any gas
19 The combination of Boyle's, Charles', and Gay-Lussac's laws

20 Material that takes the shape of its container; exerts pressure on the walls and sides of the container

21 Material in which the entire solid is joined by chemical bonds among its atoms

22 Material that maintains its shape independent of its container

23 Material in which weak bonds among molecules hold the solid together

24 Energy associated with the change between liquid and gas

25 Material in which the individual atoms have lost or gained one or more electrons

26 Solid material in which the atoms are disordered

PROBLEMS

1 Calculate the volume of 3 m^3 of gas whose temperature is raised from 100 to 200°C, maintaining 1 atmosphere pressure.

2 Calculate the pressure on 6 m^3 of gas whose temperature is raised from 27 to 177°C, keeping the volume constant. The original pressure was 2 atmospheres.

3 Calculate the pressure on 0.5 m^3 of gas cooled from 1 atmosphere and 2°C to −148°C, without changing the volume.

4 Calculate the volume of 0.1 m^3 of gas cooled from 373 to 273°C, keeping the pressure at 5 atmospheres.

5 The pressure on 1 m^3 of gas at 1 atmosphere and 0°C is increased to 3 atmospheres. The temperature remains constant. What is the new volume?

6 It is desired to reduce the volume of 2 m^3 of gas at 2 atmospheres pressure to $\frac{1}{2}$ m^3. If the temperature is constant, what is the new pressure?

7 The pressure on 75 m^3 of air is increased from 1 atmosphere to 3 atmospheres. If the initial temperature was 0°C, what must the final temperature be if the volume is to remain 75 m^3?

8 The density of air at 0°C and 1 atmosphere is 1.29 g/liter. What is the density of air at 100°C and 1 atmosphere?

9 The volume of a sample of gas whose density is 1.5 g/liter is reduced from 3,000 to 1,000 cm^3. The temperature is kept the same. What is the new density?

10 The temperature of 3.0 m^3 of helium gas is raised from −73 to +127°C. The pressure is decreased from 2 to 1 atmosphere. Calculate the final volume.

11 The temperature of 0.03 m^3 of gas at 0°C is raised to 273°C. The initial pressure was 65 cm mercury, and the final pressure is 195 cm mercury. What is the final volume?

12 10 m^3 of gas has a temperature of 0°C and a pressure of 1 atmosphere. The temperature is raised to 200°C, and the pressure increased to 4 atmospheres. What is the final volume?

13 The pressure on 5 m^3 of gas is raised from 1×10^5 newtons/m^2 to 4×10^5 newtons/m^2. The initial temperature was -173°C. What must the final temperature be if the volume is to remain at 5 m^3?

14 The density of helium gas at 300 K and 1 atmosphere is 0.16 g/liter. What is it at 400 K and 3 atmospheres pressure?

15 The density of helium gas at 300 K and 1 atmosphere is 0.16 g/liter. What is its density at 25 K and 3 atmospheres pressure?

14

THE ELECTRON

Modern society would be very different if it were not for our knowledge of the properties of the electron. Radio, TV, telephones, computers—the list of everyday devices which depend on the use of the electron is nearly endless. The proof for the electron's existence comes from two kinds of evidence: one chemical, the other physical.

14.1 FARADAY AND THE LAWS OF ELECTROLYSIS

Electrolysis is the production of chemical change by passing an electric current through a chemical compound. The two most common types of experiments using electrolysis involve solutions in water of many materials and the electrolysis of molten materials. In general, the process leads to chemical decomposition of the material under study into other materials that are frequently, but not always, its constituent elements.

As an example of this process we can consider the passage of an electric current through water which has been made electrically conducting by adding some common salt or acid. The process releases the gases hydrogen and oxygen, as shown in Fig. 14.1. Hydrogen appears at the negative terminal, called the *cathode*, and oxygen at the positive terminal, called the *anode*. The two gases can be mixed again and reacted or burned. The product is water.

The passage of electricity through molten materials is typified by the electrolysis of molten common salt by Humphrey Davy in 1807. In this experiment the metal sodium is found at the cathode, and the gas chlorine is liberated at the anode. These two materials can be reacted to form salt, showing that the salt has indeed been decomposed into its elements by the electric current. Thus salt is sodium chloride, a compound of sodium and chlorine.

Faraday was a student of Davy's and carried on this work after Davy. The first law Faraday discovered relates the mass of an element deposited in electrolysis to the amount of current passed through the apparatus. He found the mass of any material deposited always was proportional to the total amount of electric charge passed through the apparatus. This is his *first law of electrolysis*. Doubling the total charge

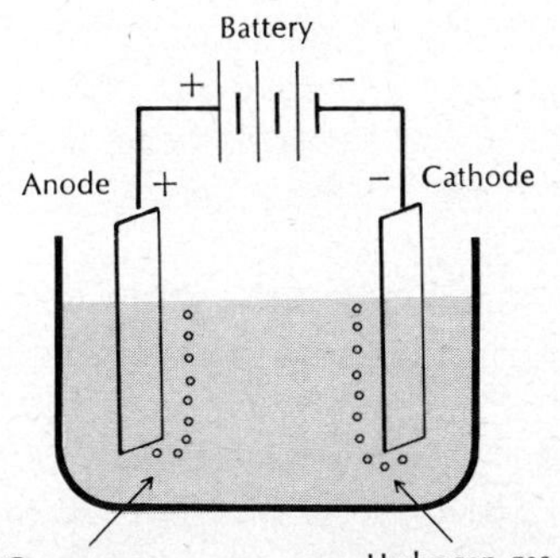

FIGURE 14.1 Electric current passed through water decomposes it into the elements hydrogen and oxygen.

doubles the mass of the material. [Remember that current in amperes is coulombs per second, so that the total charge in coulombs is given by the current in amperes multiplied by the time ($Q = It$).] Expressed mathematically, Faraday's first law of electrolysis is

$$\text{Mass deposited} \propto \text{charge} \qquad (14.1)$$

A physical model for this result involves the idea that perhaps electric charge is carried by atoms and that each atom which arrives at an electrode delivers a certain amount of charge. Since the mass of material is proportional to the number of atoms, this model will account for the proportionality between mass and charge.

The second of Faraday's results relates the masses of *different* elements deposited. Suppose that a current of electricity is passed through molten sodium chloride for a long enough time to deposit exactly 23 grams of sodium metal. If we now measure the mass of chlorine liberated, we shall find 35.45 grams. From the table of atomic masses, Table 12.3, we see that 23.0 grams is 1 gram-mole of sodium, and 35.45 grams is 1 gram-mole of chlorine. Remember that 1 gram-mole consists of Avodagro's number of atoms. Such a result might be coincidence, but if we investigate the electrolysis of many substances, we will find a large number (not all, however) for which this simple relationship holds.

To a scientist, a relationship of this kind suggests something about the underlying physical processes of the phenomenon. The scientific problem is to devise a model of the processes occurring at the atomic level which will account for the large-scale, or macroscopic, observations. It is, of course, possible to make a very simple proposal to account partially for these observations. If exactly one atom of sodium is deposited for each atom of chlorine, then the masses experimentally found can be understood. That is, the masses experimentally found show that exactly the same number of sodium and chlorine atoms is deposited. A further conclusion is also possible. In this experiment a certain amount of electric charge enters the apparatus at the positive terminal (anode), and the same amount leaves at the negative terminal (cathode). This implies that, in whatever manner the electric charge enters into the process, the sodium atom and the chlorine atom carry the same amount of electric charge. Otherwise there would occur an accumulation of positive or negative charge in the apparatus, since we have just concluded that one sodium atom is deposited for each chlorine atom. We shall return to this idea shortly.

A slightly different result occurs in the electrolysis of a material such as water. In the electrolysis of water, if 1 gram of hydrogen is evolved at the cathode, 8 grams of oxygen is found at the anode. For 1 gram-mole of hydrogen, $\frac{1}{2}$ gram-mole of oxygen is liberated. Remember that a gram-mole is a mass of an element in grams numerically equal to its atomic mass. The atomic mass of hydrogen is 1, and of oxygen, 16. On the basis of the results for sodium chloride, one would tend to predict 16 grams of oxygen for 1 gram of hydrogen. In fact, the mass of oxygen deposited is exactly half this much. Similarly, if a compound of aluminum and chlorine, aluminum chloride, is electrolyzed and enough current is passed to deposit 35.5 grams of chlorine, the amount of aluminum is found to be 9 grams. Since the atomic mass of aluminum is 27, this is $\frac{1}{3}$ gram-mole, and one-third the amount expected on the basis of the experiment with sodium chloride. To summarize these results, we can say that an amount of charge which will deposit a gram-mole of chlorine sometimes deposits a gram-mole of other elements (sodium, in the example above), sometimes deposits $\frac{1}{2}$ gram-mole (oxygen, in the example above), and sometimes deposits $\frac{1}{3}$ gram-mole (aluminum, in the example).

THE PROPOSAL OF STONEY AND HELMHOLTZ

In 1874, G. Johnstone Stoney proposed a model for the nature of electric charge which he based on these observations of electrolysis. This idea was independently suggested by Hermann von Helmholtz in 1881 (and Stoney was piqued for years because Helmholtz seemed to have received most of the credit for it). The proposal was that electricity must come in distinct bundles, or lumps, in order to account for the laws of electrolysis.

The basic reasoning can be seen from a simple analogy. In Fig. 14.2 we see a group of six Indians crossing three rivers. They cross the first river in six canoes, the second river in three canoes, and the last in two canoes. It is easy to see that this is done by putting one Indian in each canoe at the first river, two in each at the second, and three in each at the third. Only whole Indians are transported.

The Indians represent the bundle of charge. The canoes represent atoms. The idea is that some atoms carry exactly twice the charge of

FIGURE 14.2 The canoes can carry one, two, or three Indians—but never part of one.

others. If we pass an electric current through the three solutions shown in Fig. 14.3, the mass of each metal plated can be measured. From this the number of atoms of each of the metals can be calculated. If the experiment were done as shown, it would be found that twice as many sodium atoms plated as calcium atoms, and three times as many sodium atoms as aluminum atoms. Stoney interpreted this kind of experiment to mean that sodium atoms carry one bundle of electric charge, calcium atoms two bundles, and aluminum atoms three bundles, because the amount of electric charge that passes through each part of the apparatus is the same.

14.2 CATHODE RAYS

The proposal of Stoney and Helmholtz was highly suggestive, but did not constitute proof that electric charge comes in bundles. The proof came just before 1900 from work on the passage of electricity through gases. The spectacular glow produced by passing an electric current through gases at low pressures had been known for years. Michael Faraday, in 1838, made qualitative observations on some aspects of the phenomenon. Perhaps because of its spectacular visual nature, this work attracted a lot of interest, and an extraordinary number of significant discoveries have occurred through these studies. Among them are the discovery of x-rays and isotopes and much information about the structure of the atom.

The basic experimental arrangement for cathode-ray experiments is shown in Fig. 14.4. An electric field is created in a glass tube between the cathode and the anode by a high-voltage source. Then the air is pumped out. At first nothing happens. Suddenly, at a pressure about one-hundredth that of atmospheric pressure, the remaining air in the tube glows a beautiful pink, and electric current flows between the cathode and the anode. This bright glow was described in 1748 by its discoverer as an "arch of lambent flame." The color in the tube depends on the gas used. Neon, for instance, glows red, as in a neon sign.

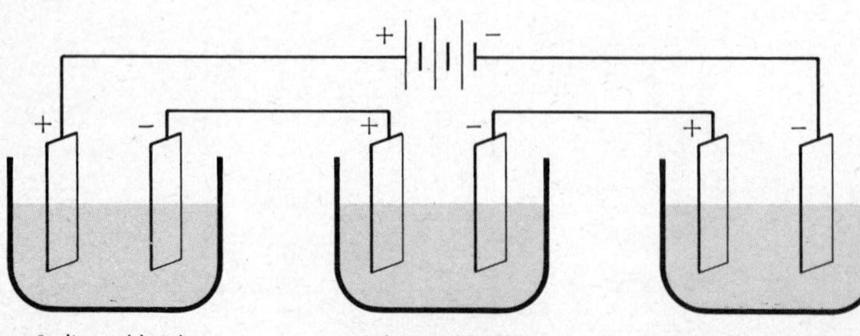
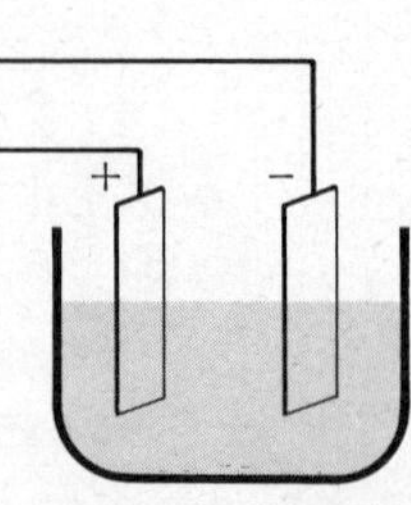

FIGURE 14.3 The same amount of electricity deposits 1 gram-mole of sodium, $\frac{1}{2}$ gram-mole of calcium, and $\frac{1}{3}$ gram-mole of aluminum from the three compounds shown.

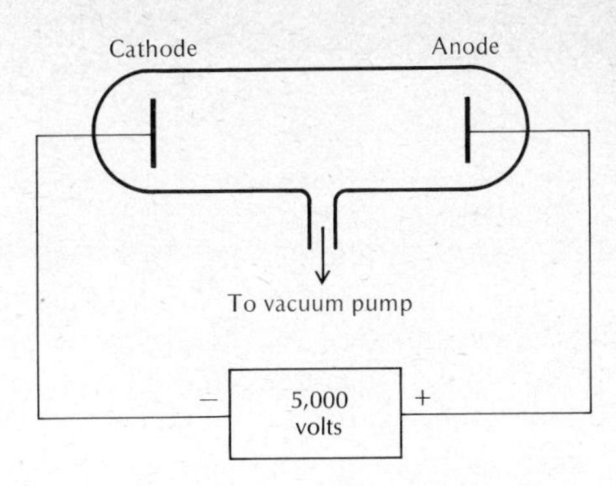

FIGURE 14.4 When the air pressure is reduced to about one-hundredth of atmospheric pressure, the gas glows brightly, and conducts electricity. This is the principle of the neon sign.

If the pressure of the gas is reduced still further, the bright glow fades, and the only effect still observed is a pale-blue or greenish glow, called *fluorescence*, that appears on some parts of the glass tube. In 1869, a solid object was placed between the cathode of such a tube and the glass wall, and a shadow was seen in the fluorescence of the glass. Therefore, whatever causes the fluorescence is emitted from the cathode and travels in straight lines. One immediate hypothesis was that these "rays" were a form of light invisible to the eye, ultraviolet light, since ultraviolet light was known to cause various objects to fluoresce. The name cathode rays became associated with these rays, since they are emitted from the cathode, or negative electric terminal. Ultraviolet light will be discussed again in Chapter 15.

A major flaw in the idea that cathode rays are a form of light lay in the fact that their path was observed to be curved in a magnetic field. Light is not known to act in this fashion but, as we have seen in Chapter 11, a moving, charged object does move in a curved path in a magnetic field. It was a number of years, however, before enough evidence accumulated to show that cathode rays are indeed particles. The protagonists of the particle viewpoint generally believed the cathode rays to be a "torrent of molecules" which had acquired electric charge at the cathode. In 1891, however, it was found that cathode rays could pass through a thin metal foil which was sufficiently thick to hold a good vacuum. Thus cathode rays could pass through a metal film which an atom could not. The implication is that cathode rays, if they are particles, are much smaller than the size of an atom. By this time the size of an atom was known to be about 10^{-10} meter.

In 1884 a magnetic field was first used to determine some of the properties of cathode rays. As we previously showed, charged particles moving in a magnetic field experience a force moving at right angles to the direction of motion. The accuracy of these early experiments was insufficient to completely determine the nature of the electron. In particular the velocity of the particles was not measured in these experiments.

14.3 THOMSON AND THE "DISCOVERY" OF THE ELECTRON

In 1894 an Englishman, J. J. Thomson, made his first studies of cathode rays. At about this time it was shown conclusively that cathode rays are negatively charged. Thomson's procedure was to use the effects of both electric and magnetic fields on the cathode rays. Both had been tried before, but Thomson used them together in a way which gave him a value for the velocity of the cathode rays. Thomson's apparatus is shown schematically in Fig. 14.6. The central region has a magnetic field directed perpendicular to the direction of motion of the cathode rays. The cathode rays were detected by their impact on a fluorescent screen. (A TV picture tube is just such a device.) The two plates can have an electric voltage placed on them so as to create an electric field in the region between. The force on a charged particle, with charge $e^{\dagger}$ between the plates, is given by

FIGURE 14.5 J. J. Thomson (1856–1940). (*Courtesy University of Pennsylvania Library.*)

$$\text{Force}_{\text{electric}} = eE \tag{14.2}$$

The magnetic force on a moving, charged particle was discussed briefly in Chapter 11. The force increases with the velocity of the particle, and increases with the magnetic field. If the magnetic field is perpendicular to the direction of motion, the magnetic force is given by

$$\text{Force}_{\text{magnetic}} = \text{charge} \times \text{velocity} \times \text{magnetic field}$$

$$\text{Force}_{\text{magnetic}} = evB \tag{14.3}$$

Thomson adjusted the values of E and B so that the two forces were identical but in opposite directions. Experimentally, this was determined by finding the adjustment which left the cathode rays undeflected from the path they would follow in the absence of both fields. That is, if the electric force upward and the magnetic force downward are equal, there will be no net force on the particle.

This measurement gave a value for the velocity of the cathode rays. Thomson combined this result with the results of experiments with a magnetic field alone. This gave a numerical value for the ratio of charge

†We shall use the letter e in this chapter, and in later chapters, to designate the charge of the electron. It is the specific value of electric charge which is the charge of the electron. This is much like the use of g to indicate the specific acceleration due to gravity, instead of a for some general acceleration.

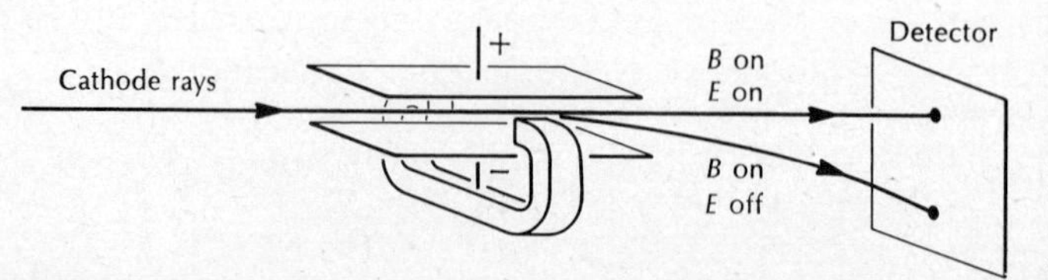

FIGURE 14.6 Schematic drawing of Thomson's apparatus for measuring e/m for the electron. The electric field is created by the parallel plates and deflects the charged particles up or down. The magnetic field is perpendicular to the paper, and also deflects the moving particles up or down. When E and B are properly adjusted, the cathode rays are undeflected, as shown by the upper line.

FIGURE 14.7 A photograph of Thomson's apparatus for measuring *e/m* of the electron. (*Courtesy the Science Museum, London.*)

to mass for the cathode rays. This value was about 2,000 times larger than one would expect for even the lightest atom. This meant that cathode rays are either very small and have very little mass, or have a very large electric charge. Thomson believed that they were small mass objects, with less than one two-thousandth the mass of an atom. Because he was the first to make a conclusive measurement of the properties of cathode-ray particles, Thomson is credited with "discovering" the electron, which is the name this particle was given by Stoney.

To find the individual values of charge and mass for the electron, a separate measurement of one of them is necessary. This will be the subject of Section 14.4.

To J. J. Thomson is attributed the following toast at a scientific meeting: "To the electron; may it never be of any use to anyone." The number of different ways in which the electron is of use to us today would fill many books. All modern electronics, TV, radio, computers, telephones, radar, and so on, would not exist except for the electron and its properties. This is simply one more in a long chain of examples of the fact that any scientific discovery may eventually have an impact on the lives of all, and that no one can predict what discovery will lead to which invention.

14.4 THE CHARGE OF THE ELECTRON

Thomson's work showed clearly that there exists a distinct object, or particle, which has a well-defined mass and electric charge. Thomson measured the ratio *e/m* of the charge to the mass of this particle and found a result of about 10^{11} coulombs/kilogram. Because electrons (cathode rays) were observed to pass through thin metal foils which would not allow an atom to pass, the conclusion was reached that an electron is much smaller than an atom. Thomson believed, in fact, that the mass of an electron was much smaller than that of an atom, and that the amount of electric charge on an electron was the same as the charge on an electrically charged atom as measured in electrolysis. (The charge-to-mass ratio for the hydrogen atom in electrolysis is about 10^8 coulombs/kilogram. If one supposed the mass of an electron to be about one two-thousandth that of an atom, the charge-to-mass ratio of 10^{11} coulombs/kilogram measured for the electron implies that its charge is about the same as that of a hydrogen atom in electrolysis.) Clearly, a direct measurement of either the charge or the mass of an electron was necessary in order to test this idea.

Stoney and Helmholtz inferred an approximate value for the unit of electric charge e from experiments in electrolysis. Stoney's result was approximately 10^{-20} coulomb. At this point we want to elaborate on the way he reached this value. The basic idea is that, if N atoms are deposited by electrolysis at one electrode, and each carries a charge e, the total charge which appears at that electrode must be given by

$$\begin{aligned} \text{Charge} &= \text{number of atoms} \times \text{charge per atom} \\ Q &= Ne \end{aligned} \tag{14.4}$$

Experimentally 96,500 coulombs of charge is found to deposit 1 gram of hydrogen, 23 grams of sodium, and 35.5 grams of chlorine. Each of these is an amount, in grams, numerically equal to the atomic mass of the particular element; it is a gram-atomic weight, or mole, of the element. Since the masses of individual atoms are in the ratio of their atomic masses, 1 gram of hydrogen, 23 grams of sodium, and 35.5 grams of chlorine must contain the same number of atoms. The number of atoms of an element whose total mass is numerically equal to the atomic mass of the element in question is called Avogadro's number, as was discussed in Chapter 12.

From the amount of charge (96,500 coulombs) necessary to deposit Avogadro's number of atoms (23 grams of sodium, for example), we could obtain the amount of charge carried by each atom if we knew Avogadro's number. If we assume that the charge on each atom is the basic unit of charge, the charge on the electron e, we could obtain a value for e from the following. (This assumption needs proof, of course.)

$$\text{Charge per atom } e = \frac{Q}{N} = \frac{96{,}500 \text{ coulombs/gram-mole}}{N \text{ atoms/gram-mole}} \tag{14.5}$$

Therefore one way of obtaining a value for the charge e is from a knowledge of Avogadro's number. If the atom in question carries two units of electric charge, an appropriate factor of 2 needs to be inserted in the last equation. A material such as oxygen or copper requires $2 \times 96{,}500 =$

193,000 coulombs of charge to deposit Avogadro's number of atoms (16.0 grams of oxygen, 63.5 grams of copper). A direct application of the reasoning above gives for these materials

$$\text{Charge per atom} = \frac{193{,}000 \text{ coulombs}}{N} \tag{14.6}$$

Equation 14.6 yields a value of the charge twice as large as that in Equation 14.5; therefore we conclude that oxygen and copper carry two of the basic units of charge.

By 1909 a French physicist, J. B. Perrin, was able to list 14 different ways of experimentally obtaining Avogadro's number. All gave a value near 6×10^{23} or 7×10^{23} atoms/gram-mole. From this, one obtains a value of e:

$$e = \frac{96{,}500 \text{ coulombs/gram-mole}}{6 \text{ or } 7 \times 10^{23} \text{ atoms/gram-mole}}$$

$$e = 1.4 \text{ to } 1.6 \times 10^{-19} \text{ coulomb/atom}$$

This method leads to an average value for the charge on an atom, but does not prove that the charge on all atoms is the same.

14.5 THE MILLIKAN OIL-DROP EXPERIMENT

In the period around the start of the twentieth century a number of experimenters began work in a wholly different direction. This work led to the measurement of the magnitude of individual electric charges.

We begin with a slight digression. This concerns the mechanism involved in the formation of tiny droplets of moisture (fog) in moist air. It turns out that the formation of droplets is in some way triggered by tiny particles of dust in the air. In fact, it has been shown that particles of dust in the earth's atmosphere have a very strong bearing on whether rain will or will not occur under conditions which are otherwise identical. One source of this dust is debris from meteors whose sources are comets which have long since disappeared. In about 1897, C. T. R. Wilson discovered that electrically charged atoms, or ions, in the air can also cause condensation into droplets of moisture if conditions are right. This observation became the basis of the Wilson cloud chamber for the detection of radioactive particles, which we shall discuss later.

FIGURE 14.8 R. A. Millikan (1868–1953.) (*Courtesy University of Pennsylvania Library.*)

J. J. Thomson used this fact to attempt to measure the value of the fundamental electric charge. The method is based on the idea that, if a charged ion is the nucleus on which a droplet of fog condenses, then the droplet will carry an electric charge exactly equal to the charge of the ion. If an electric field is applied, there will then be an electric force on each droplet, and a cloud of such droplets will move in response to this force. Thomson produced ions in the air by directing a beam of x-rays through his apparatus. We shall discuss x-rays further in Chapter 16. For the moment, the only property of interest is the fact that they create electrically charged atoms or molecules (ions) in the air.

An American physicist, R. A. Millikan, worked out a variation of this scheme which allowed a precise and unequivocal determination of the value of the fundamental electric charge. Instead of studying clouds of drops, Millikan arranged to study an individual droplet. Suppose we consider a small droplet of oil. (Millikan used mineral oil because the evaporation of water from his drops created experimental difficulties.) If the drop has a mass m and carries an electric charge Q, there can be two forces on the drop, as shown in Figs. 14.9 and 14.10. There will be a force downward on the drop equal to its weight mg. In addition, if we arrange an electric field E so that the electric force is upward, there will be an upward force given by QE, where Q is the charge on the drop of oil. Since the electric field can be varied, we can adjust these two forces to be exactly equal. If the electric force and the weight are exactly equal but in opposite directions, the drop will neither fall nor rise. When this occurs, we can write

Electric force = QE

Weight = mg

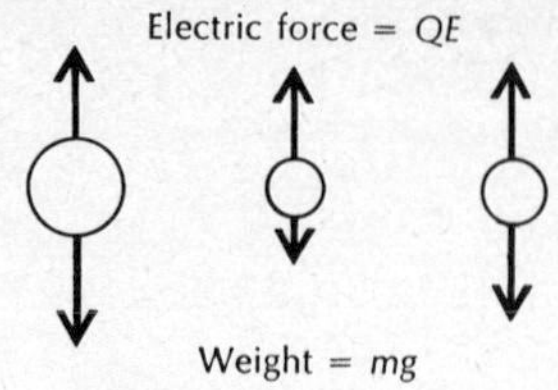

FIGURE 14.9 Electric and gravitational forces on a drop of oil in Millikan's experiment. The drop on the left has a weight greater than the electric force, and so it falls. The drop in the center has a weight less than the electric force, and so it rises. The drop on the right has a weight which precisely balances the electric force, and so it neither rises nor falls.

$$QE = mg$$

or

$$Q = \frac{mg}{E} \tag{14.7}$$

A knowledge of the mass of the drop and the value of the electric field for which the two forces are equal and opposite yields a value for Q, the electric charge on the drop.

Let us estimate the magnitude of the forces involved in this experiment. If the electric field is generated by two parallel plates 1 centimeter apart, and a potential of 1,000 volts is applied to the plates, the electric field is 1,000 volts/centimeter or 10^5 volts/meter, which is 10^5 newtons/coulomb (see Chapter 10). If the charge on the drop Q is assumed to be about 10^{-19} coulomb, approximately the value given by Thomson's measurements or Stoney's estimate, the electric force on the drop will be

$$F_E = (10^{-19} \text{ coulomb})(10^5 \text{ newtons/coulomb})$$

$$F_E = 10^{-14} \text{ newton}$$

The mass of a drop of oil whose weight is equal to this electric force of 10^{-14} newton is

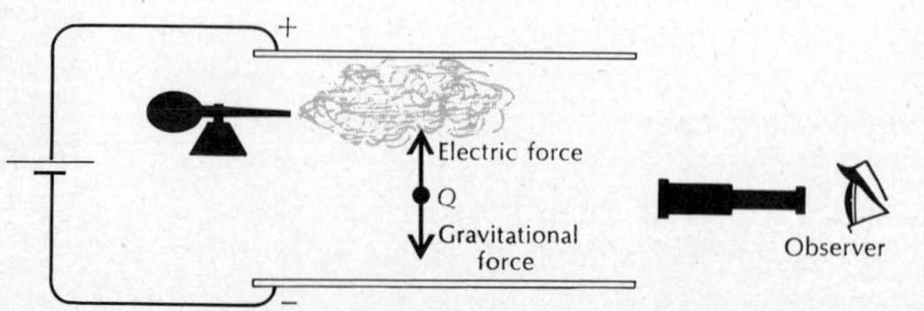

FIGURE 14.10 Schematic diagram of Millikan's experiment. The oil drops are created by an atomizer. The motion of the oil drop is followed through the telescope. The electric field is adjusted so that the drop neither rises nor falls.

$$m = \frac{W}{g}$$

$$m = \frac{10^{-14}\text{ newton}}{9.8\text{ meters/second}^2}$$

$$m \cong 10^{-15}\text{ kilogram}$$

The radius of a drop whose mass is this small is, roughly, 5×10^{-7} meter, or 5×10^{-5} centimeter. This is sufficiently small so that Millikan could not make a precise measurement, directly, of the size of the drop. Therefore he found the mass of his drops by measuring the speed with which a drop fell when the electric field was turned off. The speed of a falling drop in air depends on the properties of the air and on the size of the drop. Once the mass of his drops was determined, Millikan could find the value of the charge Q, using Equation 14.7. (We have eliminated many of the details regarding the way the measurements were carried out, but none of the essential ideas have been sacrificed.)

Millikan found that he could change the value of the electric charge on a droplet of oil by exposing the droplets to x-rays. He was thus able to make measurements for many hours on the same drop of oil with many different values for the charge on that drop. All his measurements gave the following result: The amount of electric charge found on a drop always was a whole-number multiple of a certain minimum quantity of charge.

The logic of Millikan's experiment can be seen in the following analogy. Suppose a survey of a class is made, and each student is asked how much money he or she has. The answers come in: 45¢, 35¢, 70¢, 65¢, 40¢, 55¢. What can you prove from this data as to the smallest coin in the monetary system? You could prove that there must exist a coin as small as 5¢, because each number is a whole-number multiple of 5¢, and 5¢ is the smallest difference between values. Millikan used his data in a similar way. The charge on his drops of oil was always found to be a whole number times some minimum quantity. Millikan took this quantity as the basic unit of electric charge, the charge on a single electron. The value he found was 1.6×10^{-19} coulomb. His experiment does not prove that electric charge cannot come in smaller quantities. He just never observed anything smaller. Many other lines of evidence have shown that in fact the Millikan result is the charge of the electron and is the basic quantity of electric charge. No smaller quantity of charge has ever been observed. Millikan's own words probably say this more clearly than any others:†

Relationships of exactly this sort have been found to hold absolutely without exception, no matter in what gas the drops have been suspended or what sort of droplets were used upon which to catch the

†R. A. Millikan: "The Electron," 2d ed., University of Chicago Press, Chicago, 1924, p. 71.

ions. In many cases a given drop has been held under observation for five or six hours at a time and has been seen to catch not eight or ten ions, . . . but hundreds of them. Indeed I have observed, all told, the capture of many thousands of ions in this way, and in no case have I ever found one the charge of which, when tested as above, did not have either exactly the value of the smallest charge ever captured, or else a very small multiple of that value. Here, then, is direct unimpeachable proof that the electron is not a "statistical mean," but that rather the electrical charges found in ions all have either exactly the same value or else small exact multiples of that value.

As other atomic particles have been discovered (Chapter 22), the same basic unit of charge turns up. The quantity of charge e seems to be the basic unit of charge in the building of any particle. It thus has much greater significance than merely the value of the charge of the electron.

There has been a very recent suggestion that electric charge might exist in quantities equal to one-third or two-thirds of e. The object carrying this amount of charge has been named a *quark*. An experiment at Stanford University, similar to Millikan's, has been undertaken to attempt direct observations of quarks. At the time of this writing, the result is unclear; so, as of now, no amount of electric charge less than e has been observed.

SUMMARY

The experimental results of electrolysis can be summarized in Faraday's two laws:

1 The mass deposited is proportional to the amount of charge passed through the apparatus.
2 The mass deposited is proportional to the atomic mass of the element deposited divided by a small number (1, 2, or 3).

Stoney and Helmholtz argued that the small whole number in the second statement could be interpreted as meaning that each atom carries one, two, or three basic units of electric charge.

Cathode rays are found to be moving charged particles.

The discovery of the electron is credited to Thomson, who measured the ratio of its charge to its mass, e/m. The magnitude of the unit of electric charge can be inferred from electrolysis by dividing the amount of charge necessary to deposit Avogadro's number of atoms by Avogadro's number. This amount of charge is found, experimentally, to be 96,500 coulombs, giving

$$e = \frac{96{,}500}{\text{Avogadro's number}}$$

This yields a result which is as accurate as the value of Avogadro's number used.

Millikan showed that the amount of charge on drops of oil is always given by *ne*, when *n* is a whole number and *e* is the minimum quantity of electric charge. The magnitude of *e* is 1.6×10^{-19} coulomb.

SELECTED READING

Anderson, David L.: "The Discovery of the Electron," Momentum Book 3, D. Van Nostrand Company, Inc., Princeton, N.J. (1964). A detailed history of the experiments and ideas leading to the discovery of the electron.

QUESTIONS

1 Why does the fact that a solid object cast a shadow when placed in the path of cathode rays imply that the cathode rays travel in straight lines?

2 Does the operation of an automobile battery have any relationship to the phenomenon of electrolysis?

3 Does the mass deposited in an electrolysis experiment depend on how long the experiment is carried on?

4 Sometimes the mass of one of the elements deposited in an electrolysis is not a gram-mole, while the mass of the other is a gram-mole. Why?

5 Why must water be made conducting in order to electrolyze it? Why are hydrogen and oxygen generated rather than the material put into the water to make it conducting?

6 Must the effect of gravity be included in the analysis of the Thomson experiment?

7 Show that in the Thomson experiment, when the electric and magnetic forces just balance each other, $v = E/B$.

8 What evidence supported Thomson's belief that cathode rays were of very small mass?

9 Why do only some parts of a glass discharge tube exhibit fluorescence?

10 Did Millikan make any change in the mass of his oil drop when he changed its charge?

11 Assume that the charge on the oil drop described in the text of mass 10^{-15} kg changes by e. How much does the voltage needed to keep it suspended change?

SELF-TEST

__________ Thomson
__________ Millikan
__________ Faraday
__________ electrolysis
__________ cathode rays
__________ fluorescence
__________ e/m
__________ electron
__________ Avogadro's number
__________ oil-drop experiment
__________ ion

1 Glow of visible light from the glass of a cathode-ray tube
2 Number of atoms in a mole of material
3 An atom that has lost or gained an electron
4 Measured the charge of an electron
5 "Discovered" the electron
6 Studied the laws of electrolysis
7 Light object carrying the smallest known charge
8 The method of measuring the charge of an electron
9 The beam of electrons from the negative terminal of an evacuated glass tube with high voltage on it
10 Process caused by the passage of electricity through conducting solutions
11 Ratio of charge to mass

PROBLEMS (Use Table 12.3 for atomic masses.)

1 A charge of 9,650 coulombs deposits 3.2 g of copper in an electrolysis experiment. How many units of charge does each copper atom carry?
2 A charge of 200,000 coulombs is required to deposit 23 g of sodium in an electrolysis experiment. Find the value of the charge of the electron implied by these data. (It is not the correct value.)
3 In a Millikan oil-drop experiment an oil drop of mass 1×10^{-16} kg

is found to be exactly balanced (it does not fall or rise) by an electric field of 102 newtons/coulomb. How many electronic charges whose charge is equal to the charge on the electron are on the drop? ($e = 1.6 \times 10^{-19}$ coulomb.)

4 In a Millikan experiment a drop carrying three elementary charges neither rises nor falls in an electric field of 1,000 newtons/coulomb. Calculate the mass of the oil drop.

5 In Thomson's experiment, an electric field of 1,000 newtons/coulomb and a magnetic field of 0.1 newton/ampere-m are simultaneously applied to a beam of electrons. The electrons move in a straight line. What is their velocity?

6 In a Thomson experiment, an electric field of 3×10^4 volts/m just balances the effects of a magnetic field of 0.15 newton/ampere-m. What is the velocity of these electrons?

7 In a Millikan oil-drop experiment the mass of the drop is 3×10^{-16} kg. If the drop neither rises nor falls at the values of the electric field given, what value do you infer for the charge of the electron?

$E = 294$ newtons/coulomb

$E = 118$ newtons/coulomb

$E = 196$ newtons/coulomb

$E = 98$ newtons/coulomb

(*Hint:* Calculate the charge on the drop for each value of the electric field, and then use the logic described on page 263.)

8 In a Millikan experiment a drop of oil whose mass is 2×10^{-16} kg is observed to neither rise nor fall at the values of the electric field given. (*a*) What value do you infer for the charge of the electron? (*b*) How many elementary charges are on the drop in each situation?

$E_1 = 1{,}000$ newtons/coulomb

$E_2 = 1{,}250$ newtons/coulomb

$E_3 = 1{,}430$ newtons/coulomb

$E_4 = 1{,}670$ newtons/coulomb

9 In a Millikan oil-drop experiment, a drop of oil of mass 10^{-14} kg is observed. The gravitational force is exactly balanced at values of the electric field given. What value for the charge of the electron do you infer from these data? (It is not the correct value.)

$E = 10 \times 10^4$ newtons/coulomb

$E = 6 \times 10^4$ newtons/coulomb

$E = 4.2 \times 10^4$ newtons/coulomb

$E = 3.7 \times 10^4$ newtons/coulomb

THE WAVE NATURE OF LIGHT

The nature of light and the interaction of light with matter are so central to the development of twentieth-century physics that it would be nearly impossible to discuss modern developments without some understanding of light. Historically, there have been two important theories of the nature of light: first, it consists of a stream of particles, and second, it consists of some form of wave motion. The classical view of light as developed in the experimental work of Thomas Young and the theoretical work of James Clerk Maxwell is that light is a wave phenomenon. The next two chapters will develop this idea and show how it accounts for most of the experimental properties of light. This will set the stage for an understanding of the photoelectric effect, which led to a major change in the accepted view of light.

Chapter 15 is a discussion of most of the more obvious properties of light, such as its speed and its composition in terms of colors. Chapter 16 discusses the specifically wavelike properties of light.

THE ORDINARY PROPERTIES OF LIGHT

This chapter has been titled the ordinary properties of light because it deals with those properties of light one is most likely to encounter in everyday experience. These are the properties the scientifically uneducated person might list if you asked for a list of the properties of light. In the next chapter, and in Chapter 18, we will discuss some more specialized properties.

15.1 THE BASIC PROPERTIES OF LIGHT

There are some properties of light which are immediately obvious to the most casual observer. We can see that light casts shadows and, since the shape of a shadow is the same shape as the object which casts it, *light must travel in straight lines*.

A second evident property of light is that it *carries energy*. We can see this because it heats objects it strikes. The light of the sun, in fact, brings 2 calories/minute to each square centimeter of the earth's surface that receives its direct rays (this includes some energy not visible to the eye). Since we have shown that heat is a form of energy, any model we construct for the nature of light must allow for the fact that it can carry energy over long distances.

A third property of light which is perhaps slightly less obvious is that it *travels with great speed*. When we turn a light switch, there is no delay before we see the light. A flashbulb on a camera illuminates a whole room at one time; the light does not appear to arrive at one part of the room before another. From these observations and others like them we can conclude that light travels very fast, possibly infinitely fast. Certainly its speed is much greater than the speed of sound, for we can see a flash of lightning and sometimes wait many seconds before the arrival of the thunderclap. The speed of sound is known to be about 1,000 feet/second.

A fourth property is that there seem to be different kinds of lignt. The difference lies in the observation that light *comes in different colors*. Any model we create for light must account for the existence of color.

The Greek atomists, following Democritus, believed that light consisted of particles that flowed out from the source of the light. This view made it easy to understand that light travels in straight lines, since a stream of fast-moving particles would travel in straight lines like bullets. And a stream of fast-moving particles would certainly carry energy, although the Greeks themselves had no concept of energy. If light consists of particles, it is difficult to conceive of extremely high velocities for these particles.

The same thing that happened to the atomists' view of the nature of matter happened as well to their point of view on light. Aristotle believed that light was a phenomenon occurring in the transparent medium between the eye and the source of the light, and not a stream of particles.

Since the views of Aristotle were accepted by the early Church, and others were rejected, there the matter rested until the seventeenth century.

15.2 THE SPEED OF LIGHT

During the seventeenth century additional information accumulated about the properties of light that had to be incorporated in any theory concerning its nature. To build up our ideas regarding the nature of light in as simple a fashion as possible, we shall discuss these developments slightly out of chronological order.

We have already stated that light moves fast, if not instantaneously, from point to point. Aristotle believed the latter point of view. Galileo made an attempt to determine the velocity of light. He stationed an assistant at some distance from himself and flashed a lantern pointed at the assistant. The assistant was supposed to flash his own lantern when he saw the light from Galileo's lantern. The distance the light traveled, divided by the observed time delay, would give the speed of light. About all Galileo could say from these crude experiments was that the speed of light was too great to be measured by this technique.

The first observations to lead to a value for the velocity of light were made by a Danish astronomer named Ole Roemer, working for the French Royal Academy of Science in 1767. His work was motivated by intensely practical considerations. One of the primary needs of navigation is a good clock. The position of a ship in latitude, or north-south direction, can be determined from the maximum height of the sun above the horizon. But to determine longitude, or east-west position, the time must be known precisely. An error of 1 minute leads to an error in the position of the ship of 17 miles at the equator. Since sailing ships remain at sea for months at a time, a good clock is an essential item. Roemer was involved in finding an astronomical observation that could be used to check a ship's clock (or chronometer).

One astronomical observation which might fill the need is the eclipse of one of the large moons of Jupiter by the planet itself. The time of such an eclipse can be precisely measured from any point on earth where the planet is visible, and the time when the eclipse ought to occur can be calculated, since the period of the moon can be measured. Therefore a table predicting the time of these eclipses in the future could be published, and a navigator at sea could check a ship's clock by comparing it with the observations of the eclipses of Jupiter's moon.

The period of the innermost bright moon, Io, is about $42\frac{1}{2}$ hours. The time when it will be eclipsed can be predicted at various times in the future. Roemer observed these eclipses at various times of the year, and noted that under certain circumstances the eclipses occurred as

OPTICAL SPECTRA

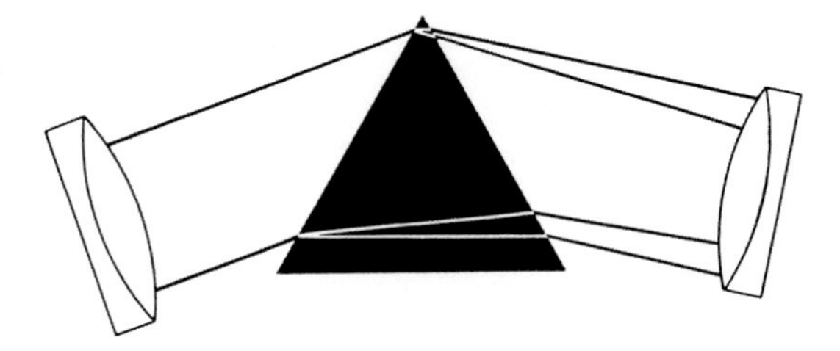

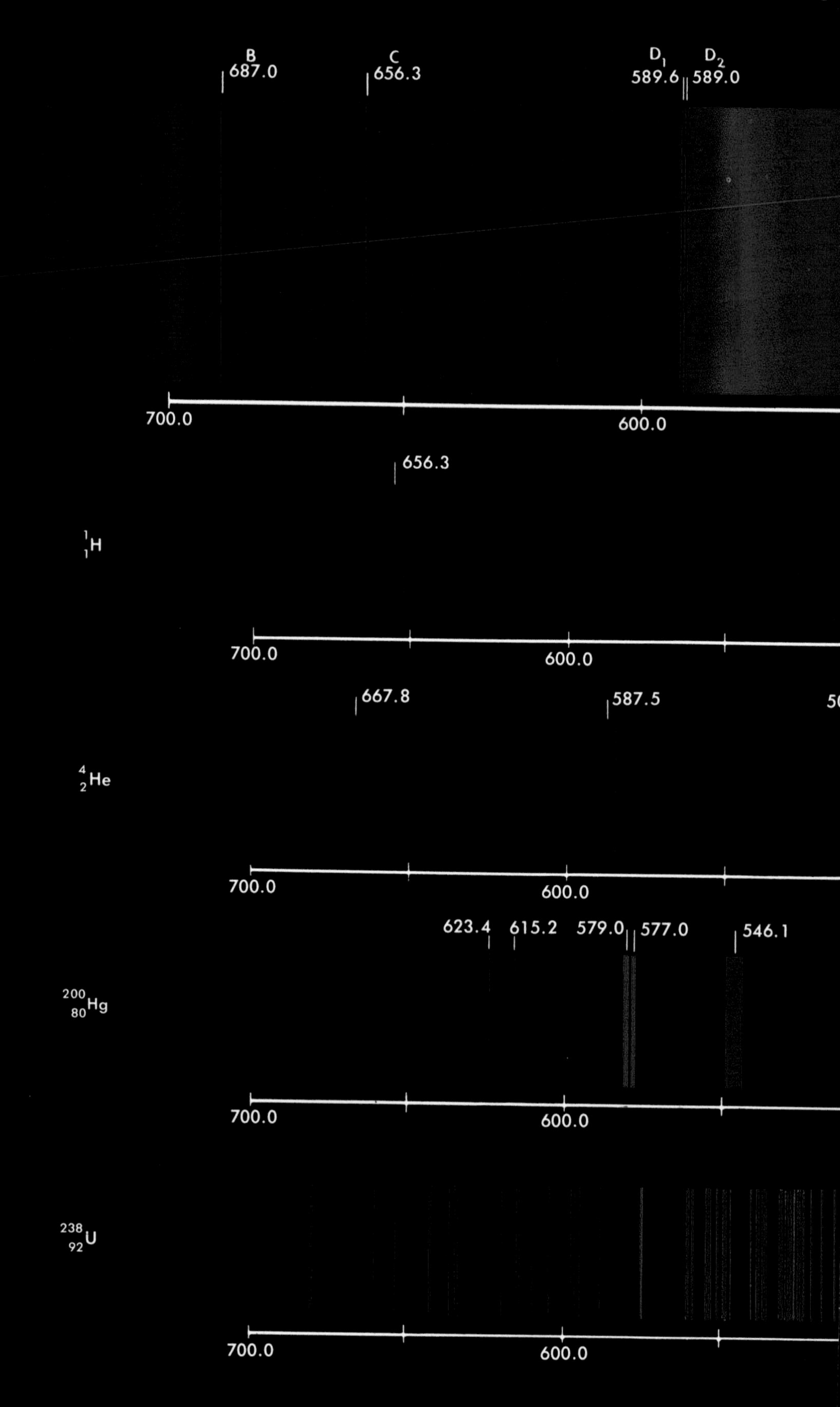

B
687.0
C
656.3
D1
589.6
D2
589.0
700.0
600.0
656.3
1H
700.0
600.0
667.8
587.5
4He
700.0
600.0
623.4
615.2
579.0
577.0
546.1
200Hg
700.0
600.0
238U
700.0
600.0

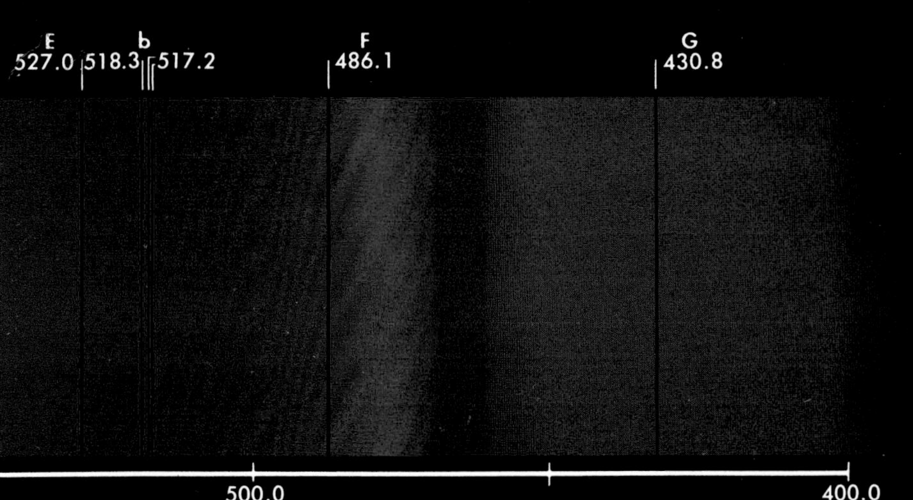

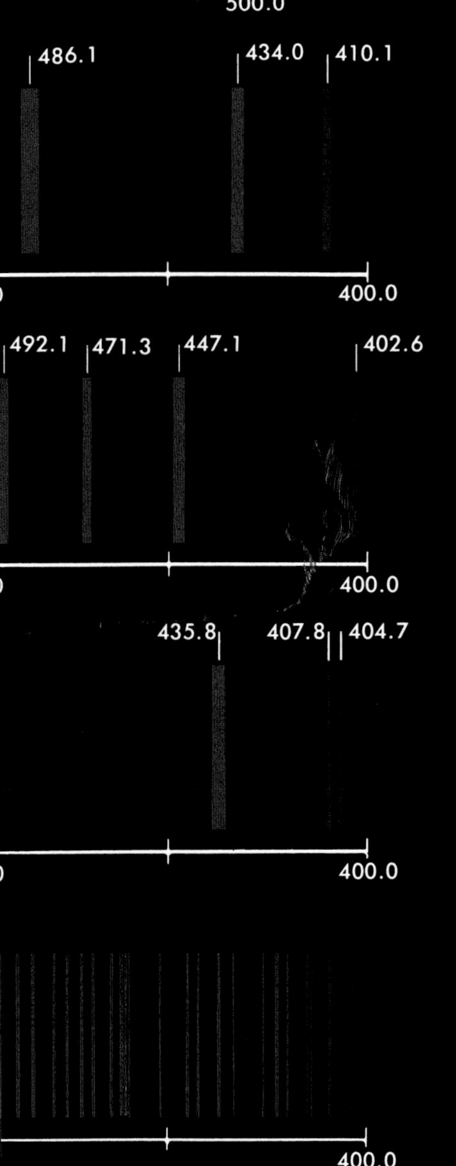

Such diverse and fundamental information on the nature of matter as the composition of distant stars and the structure of atoms and molecules has been obtained by analysis of the light emitted from substances heated to incandescence.

In the **SPECTROSCOPE**, such light, passed through a slit and a prism, is broken up into its component wavelengths, which are observed as colored lines (i.e., light of different energies) characteristic of the differences between the various electron energy levels of the atoms. This **EMISSION SPECTRUM** is **CONTINUOUS** when the images of the wavelengths are uninterruptedly overlapping; it is a **LINE SPECTRUM** when only certain specific wavelengths are emitted, as shown here for the elements hydrogen, helium, mercury, and uranium.

On the solar spectrum across the top of this plate appears a series of dark lines—**FRAUNHOFER LINES**—forming an **ABSORPTION SPECTRUM**. Some of the light from the intensely hot interior of the sun is absorbed by the cooler gases of its outer layers as the light energies raise the atoms in the cooler layers to higher energy states; bright lines are not, therefore, seen for these changes.

The spectra are calibrated in nanometers (1 nm = 10^{-9} m); the letters are arbitrary designations introduced by Fraunhofer for lines important in spectroscopy.

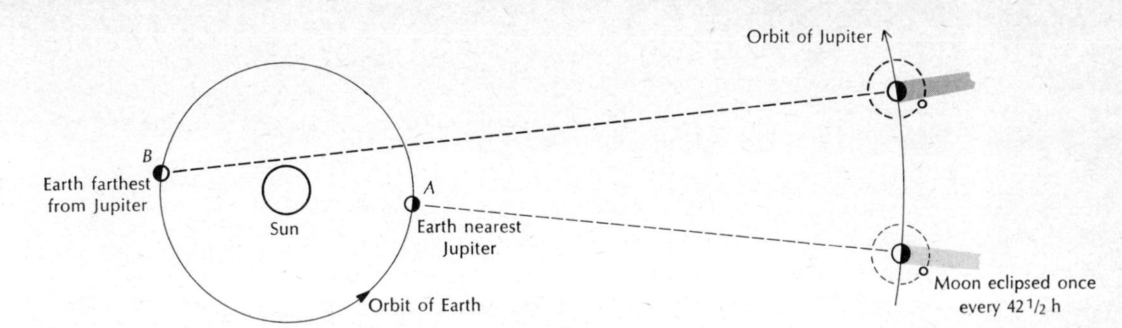

FIGURE 15.1 Eclipses of Jupiter's moon occur *late* when the earth is on the opposite side of the sun from Jupiter, because of the longer distance the light must travel.

much as 22 minutes later than predicted. This is an intolerable error for navigational purposes, and he sought an explanation.

Figure 15.1 shows the general situation. The schedule of the moon eclipses is set up when the earth is nearest to Jupiter, as at *A*. Six months later, when the earth is at *B*, the moon was found by Roemer to be 22 minutes behind schedule. (Note that the period of revolution has not changed by 22 minutes, just the time when eclipses appear to occur as observed from earth).

Roemer interpreted these results as indicating that the eclipses actually occurred on schedule, but that it took 22 minutes longer for the light from Jupiter to reach the earth when the earth was at *B* than when it was at *A*. Roemer believed the distance across the earth's orbit (twice the distance from the earth to the sun) to be about 182 million miles, or 1.82×10^8 miles, which is 2.93×10^{11} meters. He concluded from his data that it took 22 minutes, or 1.32×10^3 seconds, for light to travel this distance. Therefore the velocity of light is calculated to be

$$c = \frac{d}{t} = \frac{2.93 \times 10^{11} \text{ meters}}{1.32 \times 10^3 \text{ seconds}}$$

$$c = 2.22 \times 10^8 \text{ meters/second or about } 138{,}000 \text{ miles/second}$$

The symbol c is universally used to indicate the velocity of light.

Roemer's value is quite low compared with the modern value of 186,000 miles/second or 3×10^8 meters/second. Nevertheless, it represents the first measurement of the velocity of light which gave a result that was not infinite and also gave an answer of the correct magnitude. The time delay of the eclipses is now known to be about 16 minutes (this involves making accurate time measurements over a period of 6 months), and the distance across the earth's orbit is 186 million miles. Using these two numbers yields the modern value for the velocity of light.

MICHELSON'S MEASUREMENTS

In the two centuries since Roemer's calculation, the speed of light has been measured using a number of sophisticated techniques. Until quite recently the best values were obtained using variants of an elegant and simple technique originally proposed by Jean Foucault (1819–1868) and later improved on by A. A. Michelson (1852–1931). The measurements of the velocity of light Michelson made eventually earned him a Nobel Prize. The basic idea is shown in Fig. 15.3.

In this experiment a light source is located so that it can shine on face *A* of the eight-sided mirror M_1 and follow the path indicated by the solid line. It is reflected by a mirror M_2, several miles distant, and returns to face *C* of the octagonal mirror. From there it is observed. If the eight-sided mirror is rotated in a clockwise direction and observed from the distant mirror M_2, one will see a series of flashes as the sides of the mirror M_1 come into position to reflect the light in the proper direction. If the octagonal mirror is rotated only slightly when the flash of light is reflected from M_2 to M_1, face *C* will have turned so that the returning flash is not reflected to the observer. But if the speed of rotation of the mirror is properly adjusted so that the mirror turns by exactly one-eighth in the time the light travels to M_2 and back, then face *B* will be in the proper position to reflect the light to the observer. From the speed of rotation necessary to achieve this, it is possible to obtain the velocity of light. The round-trip distance is known, and the time to make this round trip is one-eighth of the time required for the mirror to make one complete revolution.

FIGURE 15.2 A. A. Michelson (1852–1931). (*Courtesy American Institute of Physics.*)

Michelson carried out this experiment several times during his life. Between 1924 and 1927 mirrors were set up on Mt. Wilson and Mt. San Antonio in southern California. The 22-mile distance between them was surveyed to an accuracy of about 1 inch. The result was a value for the speed of light of $c = 2.99798 \times 10^8$ meters/second. Beginning in 1930, shortly before Michelson's death, and continuing afterward until 1933, a measurement of light in an evacuated tube 1 mile long was carried out. This gave a result of $c = 2.99774 \times 10^8$ meters/second. These two results should be compared with a value obtained by Michelson in 1882, using the same method. He had invented this method while an instructor at the United States Naval Academy a few years earlier. This value was $c = 2.99853 \times 10^8$ meters/second. Michelson was, throughout his life, a superlative experimental scientist.

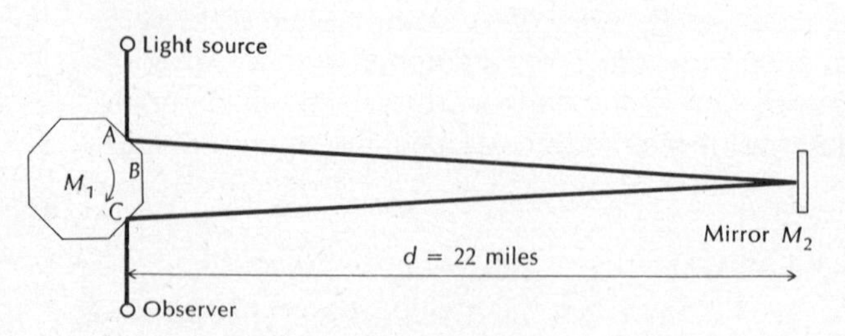

FIGURE 15.3 Schematic diagram of Michelson's measurement of the velocity of light. The mirror M_1 rotates clockwise; and when the speed is correctly adjusted, the light returning from M_2 is reflected from face *B* to the observer.

15.3 REFLECTION AND REFRACTION

One of the properties of light in most common use is *reflection*. The simplest example of this involves an ordinary plane mirror. When you observe yourself or some object in a mirror, there is an illusion that an object exists at some point behind the mirror, when in fact this is clearly not the case. Understanding this apparent object, or image, will help in understanding both reflection and refraction.

The law of reflection for light from a plane mirror is relatively simple. It is illustrated in Fig. 15.4. If we call the angle between the incident beam of light and a line perpendicular to the surface the angle of incidence, and the similar angle between the perpendicular and the reflected beam the angle of reflection, experiment with a flat mirror shows that *these two angles are always equal.*

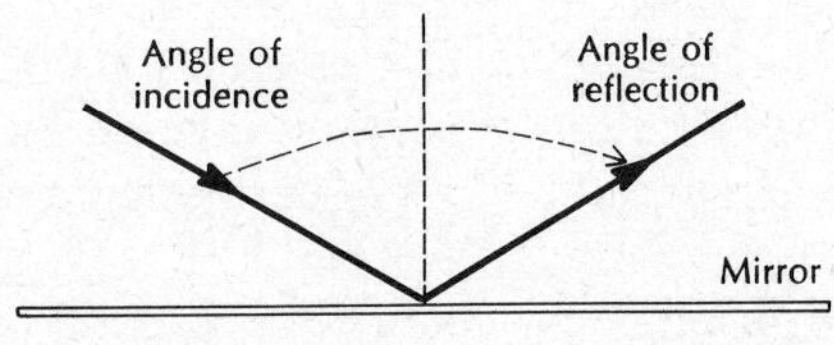

FIGURE 15.4 For reflection from a plane surface the angle of incidence is equal to the angle of reflection.

Figure 15.5 shows how the image of the candle appears to be behind the mirror. The image of the candle appears to be in the direction from which the light arrives at our eye. This result will always occur. The image will appear to be in a position indicated by the direction from which the light arrives at the eye.

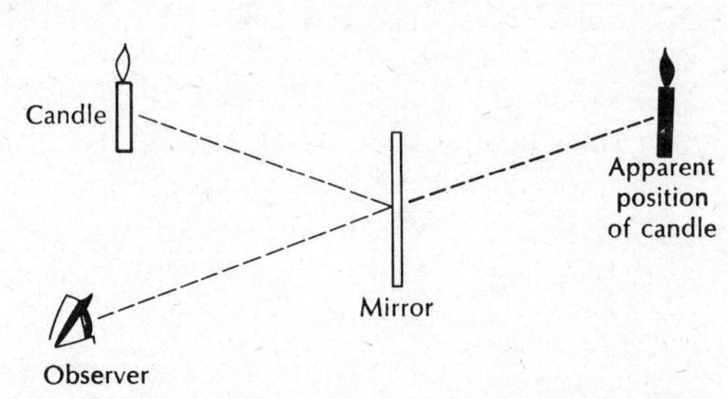

FIGURE 15.5 The candle appears to be behind the mirror, because the light arrives at our eye from that direction.

The second of the seventeenth-century developments involving the properties of light is the phenomenon of *refraction*, or the bending of light as it passes from one medium to another, as from air to glass. This is of importance for two reasons—one practical, the other theoretical. The practical importance of refraction lies in the fact that it is the basis of all optical instruments: telescopes, eyeglasses, microscopes, and the like. The second reason for its importance is the bearing it has on whether light is a wavelike or particlelike phenomenon.

We can observe refraction when we look at the oar of a boat, which is partly above and partly below the water surface. The oar appears to

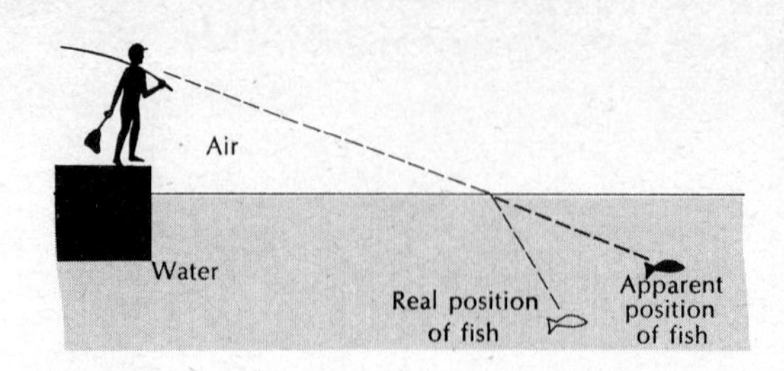

FIGURE 15.6 The apparent position of the fish is not as deep as its actual position, because of refraction at the air-water surface. The apparent position is determined by the direction from which the light arrives at the eye.

be bent! A fisherman is familiar with the same phenomenon. If he sees a fish that appears to be 4 feet below the water surface, he knows the fish is, in fact, closer to 6 feet deep. If he does not allow for this, he may catch no fish. This is shown in Fig. 15.6. A ray of light from the fish to the eye of the fisherman travels in straight lines, except at the surface of the water, where it is bent. This refraction occurs wherever light passes from one transparent medium to another, except when the path of the light is exactly perpendicular to the boundary between the two materials.

Figure 15.7 shows the path of a beam of light from air into glass (or glass into air—the direction of the beam of light along the path does not affect the angle of bending). The angle of incidence is the angle between the line perpendicular to the surface and the path of light. The angle of refraction is the angle between the perpendicular and the path of light in the second medium, in this case glass. The angle through which light is bent is observed to increase as the angle of incidence increases.

There were many attempts to describe mathematically the phenomenon of refraction before it was finally achieved by Willebrord Snell in 1621. We can describe what is going on in terms of the angles of incidence and refraction θ_1 and θ_2. Or we can go a distance d away from the surface in each direction and draw two triangles, shown in Fig. 15.8, which have sides d, ℓ_1, and h_1 and d, ℓ_2, and h_2. Snell described the behavior of the light crossing the boundary in terms of a quantity called the *index of refraction* for the two materials in question. We shall indicate the index of refraction by μ (greek mu). Snell's law says that

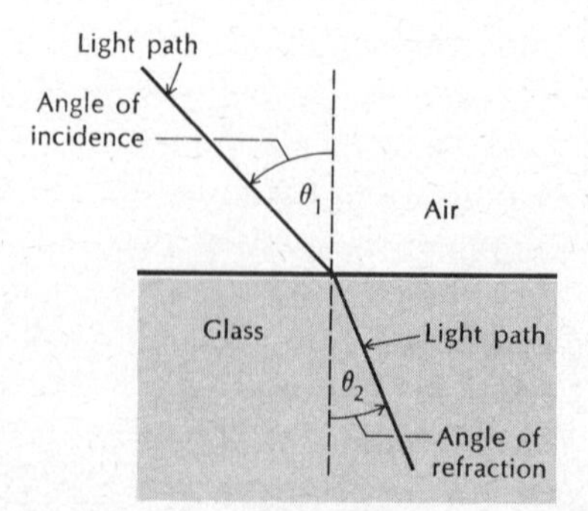

FIGURE 15.7 The light path is bent in going from air into glass.

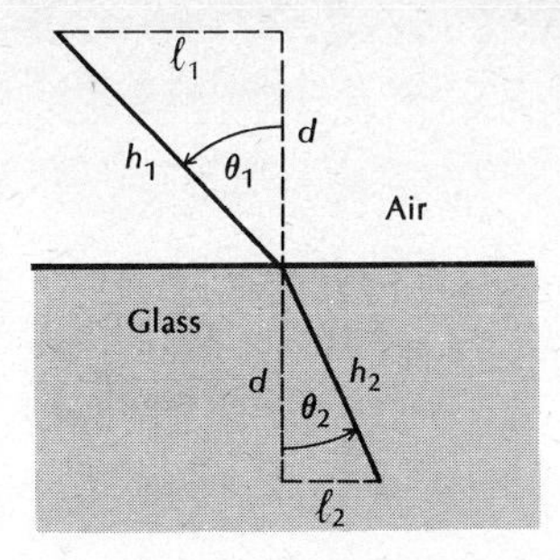

FIGURE 15.8 Snell's law relates the quantities ℓ_1, h_1, ℓ_2, and h_2 to the refractive index of glass with respect to air. θ_1 and θ_2 are the angles of incidence and refraction, respectively.

for a particular pair of materials, such as air and glass, the following relationship holds, no matter at what angle the light arrives at the surface.

$$\frac{\ell_1}{h_1} \times \frac{h_2}{\ell_2} = \mu \qquad (15.1)$$

Students who know a little trigonometry will be interested to know that the index of refraction and Snell's law are more usually discussed in terms of trigonometric functions of the angles θ_1 and θ_2 in Fig. 15.8. In terms of these, we can write Snell's law, if μ is the refractive index,

$$\mu = \frac{\sin\,\theta_1}{\sin\,\theta_2} \qquad (15.2)$$

The derivation of this result from Equation 15.1 is a simple exercise in trigonometry.

The index of refraction is normally stated with air or vacuum as medium 1. In almost all cases this makes the numerical value of the index larger than 1. Table 15.1 gives a list of refractive indices of several common materials with respect to vacuum. The index of refraction of glass with respect to air and vacuum is about 1.5 (different kinds of glass have different values); that of water with respect to air is 1.3. This means that light is bent through a greater angle in going from air to glass than in going from air to water. This is shown in Fig. 15.9 for the two cases.

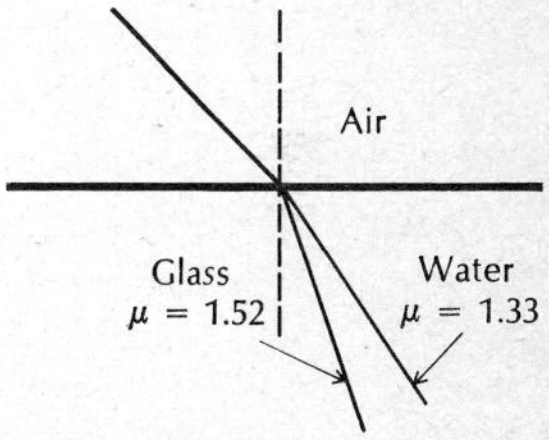

Air
Glass $\mu = 1.52$
Water $\mu = 1.33$

FIGURE 15.9 The refractive index of glass is greater than that of water; therefore the light is bent through a greater angle in going from air to glass than from air to water.

15.4 THE WAVE MODEL VERSUS THE PARTICLE MODEL FOR LIGHT

The atomists from Democritus onward believed light to be a stream of particles emitted by the source of light. Aristotle believed that it was some form of disturbance in the medium between the source and the observer. The Dutch scientist Christian Huygens (1629–1695), in 1678, proposed that light was a form of wave, or vibration, which spread out from the source like waves in water. Newton, in his treatise *Opticks* in

Material	Index of refraction
Vacuum	1.0000
Air	1.0003
Water	1.33
Glass (crown)	1.52
Glass (flint)	1.66
Diamond	2.42

TABLE 15.1 Index of refraction of various materials (compared with vacuum). Crown and flint glasses are glasses with different chemical compositions. These values change slightly with the color of the light, as discussed in Section 15.5

1704 took the side of the particle point of view, although not unequivocally.

The phenomenon of refraction offers a good opportunity to make a clear-cut distinction between the two ideas. When the implications of these two ideas are worked out, the following two predictions are made: If light is a stream of particles, the speed of light will be *faster* in glass or water than in air. If light is a wave-type phenomenon, the speed of light will be *slower* in glass or water than in air.

Unfortunately, it was many years before a direct experimental test was made of these predictions. Finally, in 1849, Armand Fizeau (1819–1896), a French physicist, in an experiment which was the predecessor of Foucault's and Michelson's, showed by direct measurement that light travels more slowly in water than in air—an apparent demonstration that light has a wave nature. By that time the debate between the particle and wave viewpoints had already been settled by Young, as we shall see in the next chapter.

It is possible to relate the refractive index between two materials directly to the velocity of light in each. The refractive index of glass with respect to air is

$$\mu_{\text{glass-air}} = \frac{\text{velocity of light in air}}{\text{velocity of light in glass}} \tag{15.3}$$

Therefore the larger the index of refraction, the greater the change in velocity for light in the two materials. A corollary is that, if the velocity of light is the *same* in the two materials, the refractive index between them is 1.00, and *no* bending of light occurs at the boundary.

Example 15.1

Calculate the velocity of light in glass having a refractive index of 1.5.

$$\mu = \frac{c}{c_{\text{glass}}}$$

$$c_{\text{glass}} = \frac{c}{\mu} = \frac{3.00 \times 10^8 \text{ meters/second}}{1.5}$$

$$= 2.00 \times 10^8 \text{ meters/second}$$

15.5 NEWTON AND COLOR

One facet of the phenomenon of refraction attracted the interest of Isaac Newton. When he arranged for sunlight from a small hole in his window-shade to fall on a triangular piece of glass, called a prism, he could produce on the opposite wall an oblong area showing the colors of the rainbow, as in Fig. 15.10. This phenomenon had long been known. The beauty of a crystal chandelier and of a cut diamond depends on it. But here Newton showed his skill as a scientist in finding a method to understand this phenomenon. He placed another prism upside-down with respect to the first and showed that the colors blended back into white light, as in Fig. 15.11. In so doing he showed that no change is made in light passing through the prism; it is just separated into its component colors. White light must be a mixture of light of all the colors of the rainbow.

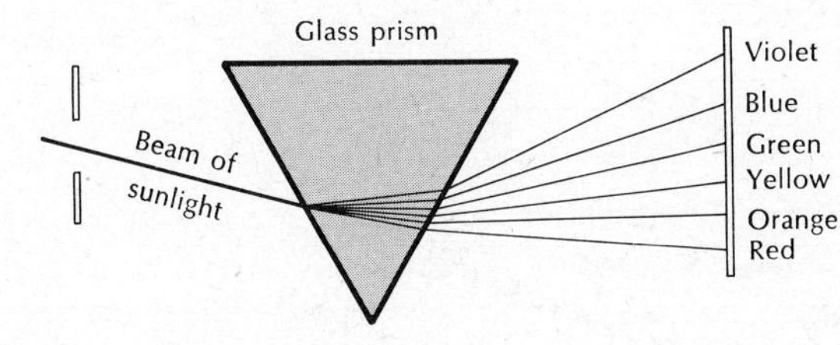

FIGURE 15.10 A prism can be used to break up white light (sunlight) into the colors of the rainbow.

To demonstrate further that light is not changed in passing through a prism, Newton performed the experiment diagramed in Fig. 15.12. A slit or pinhole is placed between the two prisms so that only one color of light, such as blue, from the first prism hits the second prism. Newton showed that the angle α (Greek alpha) through which the blue light is bent in the second prism is exactly the same as that in passing through the first prism. He repeated the experiment for all colors of light, with the same results. He concluded that the reason white light is separated into colors is that the index of refraction of glass is different for different colors. It is greatest for purple light, which is thus bent through the sharpest angle, and least for red, which is bent through the least angle. Therefore the velocity of violet light in glass is less than that of red if all colors have the same velocity in air, which they nearly do.

The change in index of refraction with color of any material is called the *dispersion* of that material. Gem stones are beautiful because they show a strong dispersion. The primary reason that glass does not look like diamond is that it has less dispersion, or separates the colors less.

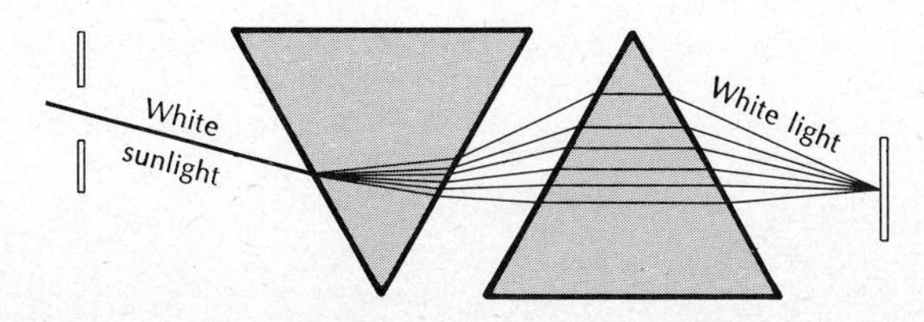

FIGURE 15.11 Newton showed that after white light is broken up into colors it can be remixed to give white light again.

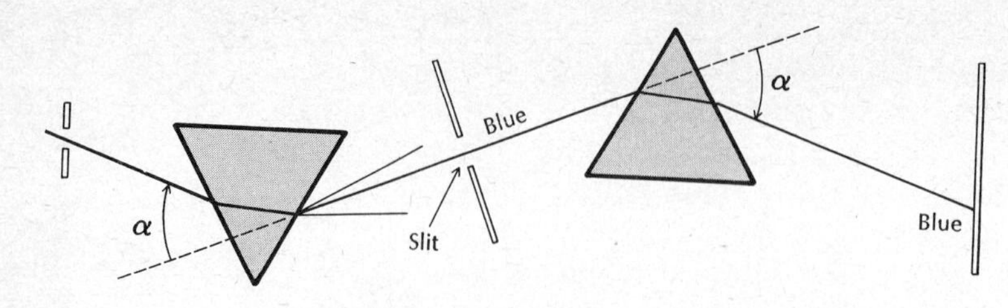

FIGURE 15.12 Blue light is bent through the same angle at each of two prisms.

Therefore a piece of glass cut like a diamond has no "fire." As an indication of the change in index of refraction for various colors, some data for crown glass and diamond are given in Table 15.2. The rather small change in the refractive index of glass as compared with diamond indicates the rather low dispersion of glass. We shall return to this subject and discuss the use of a prism to separate light into its various colors as a tool in the study of the structure of the atom.

TABLE 15.2 Index of refraction of glass (crown) and diamond for various colors

Material	Violet	Blue	Green	Yellow	Orange	Red
Glass	1.532	1.528	1.519	1.517	1.514	1.513
Diamond	2.485	2.465	2.435	2.417		2.402

15.6 THE RAINBOW

A rainbow is caused by the refraction of sunlight falling on raindrops and reflection at the interior of a drop. The path of sunlight in a raindrop is shown in Fig. 15.13. This path corresponds to the brightest, or primary, rainbow. A rainbow is always seen in the direction away from the sun, with the red color at the top of the arc and the violet inside. Each color seen in the rainbow comes from a different group of drops, situated so that the particular color of light travels toward the observer. Figure 15.14 shows how the position of the rainbow in the sky is determined. Because the light is bent at a particular angle from the path of the sunlight, any given color appears as a part of the circle, where every point of that circle is at the same angle from the direction of the sunlight.

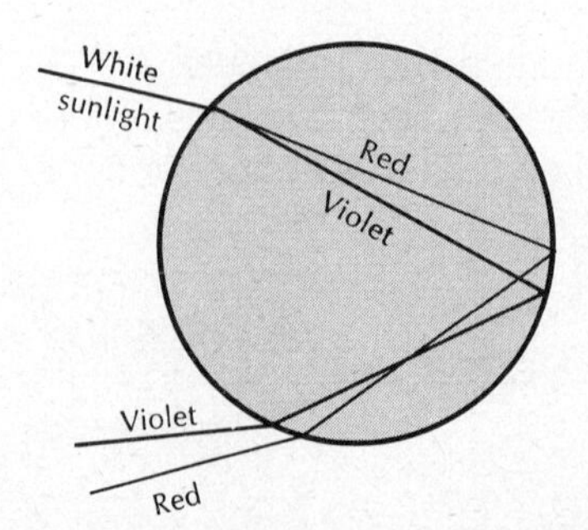

FIGURE 15.13 The path of a beam of white light in a raindrop. The beam is bent at two air-water surfaces, and reflected once inside the drop. Red appears at the top of a rainbow, because the drops which send red light to your eye are higher in the sky.

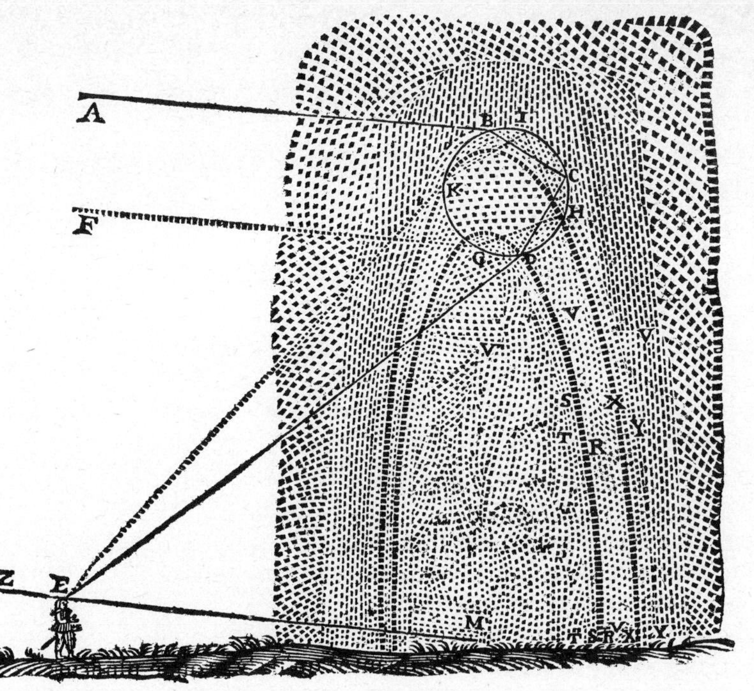

FIGURE 15.14 A figure taken from Descartes' *Dioptics*, showing the apparent position in the sky of the rainbow. The solid line *A* gives the position of the primary rainbow. The dotted line *F* gives the position of the secondary rainbow sometimes seen. For the secondary rainbow the light is reflected twice within the raindrop.

SUMMARY

The first measurements of the velocity of light were made by Roemer, using astronomical data about the eclipses of the moons of Jupiter. The modern value, due in large measure to A. A. Michelson, to sufficient accuracy for most purposes, is

$$c = 3.00 \times 10^8 \text{ meters/second}$$

Refraction is the bending of light as it passes from one medium (such as air) to another (such as water). Snell's law gives the relationship between the directions of the two light beams and the refractive index μ. In terms of the velocity of light in the two media, the refractive index of glass with respect to air is given by

$$\mu = \frac{\text{velocity of light in air}}{\text{velocity of light in glass}}$$

Isaac Newton showed that white light is composed of all the colors of the rainbow. Light of different colors can be separated by taking advantage of the different refractive indexes of light of different colors. The mechanism by which light is separated into its components is called dispersion.

SELECTED READING

Jaffe, Bernard: "Michelson and the Velocity of Light," Science Study Series S13, Doubleday and Company, Garden City, N. Y., 1959. A biography of Michelson, showing his abiding interest in measurements of the velocity of light.

QUESTIONS

1. Is there anything in this chapter that allows you to argue against a particle theory for light?
2. Explain why it is necessary to know the exact time for purposes of accurate navigation.
3. If the index of refraction were less than one, how would refraction work?
4. Explain why red appears at the top of a rainbow.
5. If violet were to appear at the top of the primary rainbow, what would this imply about the dispersion of water?
6. Modern laboratory techniques are capable of measuring time intervals of 10^{-9} second and less. How far does light travel in this time?

SELF-TEST

__________ Ole Roemer
__________ Armand Fizeau
__________ A. A. Michelson
__________ Christian Huygens
__________ velocity of light
__________ refraction
__________ dispersion
__________ index of refraction
__________ Snell's law

1. Bending of light at the boundary of two media
2. 3×10^8 m/second
3. Ratio of the velocity of light in two media
4. Variation of refractive index with color
5. Relates angle of incident beam and angle of refracted beam
6. Measured the velocity of light in water
7. Proposed that light is a form of wave

8 First measured the velocity of light
9 Measured the velocity of light between 1882 and 1930

PROBLEMS

1 Calculate the velocity of light in diamond.
2 Calculate the velocity of light in crown glass.
3 How long did a flash of light take to make the round trip of 44 miles in Michelson's experiment?
4 The moon is 240,000 miles from the earth. How long does a round trip for a beam of light take?
5 In the drawing shown, is the velocity of light greater in medium 1 or in medium 2?

16

INTERFERENCE, DIFFRACTION, AND WAVES

Newton's optical studies, in which he contributed much that was original both experimentally and theoretically, involved him in a controversy with Christian Huygens (1629–1695) over the nature of light. Because Newton did not relish controversy and debate, he delayed the publication of his treatise, *Opticks*, for several years to avoid such controversy. It was finally published in 1704.

Newton believed that light must consist of a stream of particles, primarily because all the evidence seemed to him to point that way. Huygens believed that light must be a form of wave phenomenon, like sound, and that the waves spread in all directions from their source. Both of these points of view could account for all the properties observed for light at that time. Newton rejected the wave point of view, because he could find no evidence for one particular type of effect that he knew to be characteristic of wave phenomena and that he knew would not occur if light were a stream of particles. This effect is called *diffraction* and, in order to understand Newton's point of view, we need to understand something more about waves and their behavior.

16.1 WAVES

Let us look at a type of wave which is a little more familiar to us and more easily observed than light: waves in a lake or the ocean. What do we mean by waves? The surface of the water is not flat; there are regions which are high and regions which are low, as shown in Fig. 16.1, and these regions move. The high regions we shall call peaks, or crests, and the low regions valleys, or troughs. One point should be made clear: Although the waves travel, the particles of water which make up the waves do not move in the direction of the wave motion. Like a cork sitting on the water, they bob up and down as the wave passes. In Fig. 16.1 several of the descriptive terms we shall use for waves are illustrated. The *wavelength* λ (Greek lambda) is the distance from peak to peak or valley to valley. The *amplitude* is the height of the peak, above the position of the calm sea (or the depth of a valley below it). The *velocity* of the wave is the velocity with which the disturbance moves; that is, it is the velocity with which a given peak moves through the water. The *frequency* of a wave is the number of peaks passing a given point in one second.

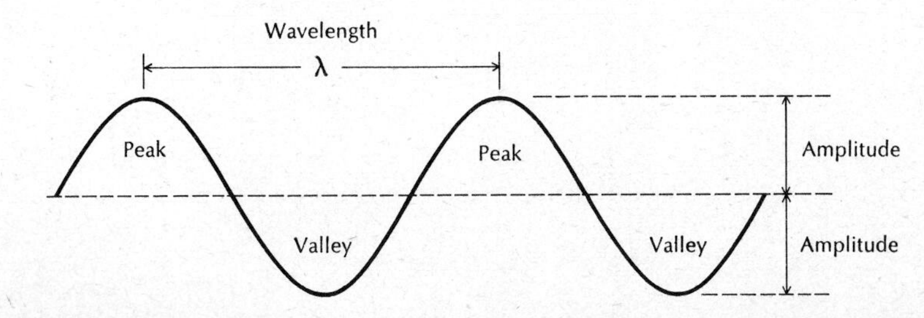

FIGURE 16.1 The terms wavelength, amplitude, peak, and valley are illustrated in this diagram of a wave.

The phenomenon of diffraction which accompanies all waves, and which Newton could not find for light, can be demonstrated with water waves in several ways. Its essential characteristic is that waves arriving at a barrier bend into the shadow region. By shadow we mean the region which would be free of any disturbance if the waves traveled only in straight lines. Figure 16.2 shows the diffraction of ocean waves into the shadow of a breakwater. Figure 16.3 shows waves in a shallow pan of water passing through an opening. The peaks and valleys show as light and dark regions in this photograph. This picture shows clearly the waves bending into regions where no waves would occur *if* waves

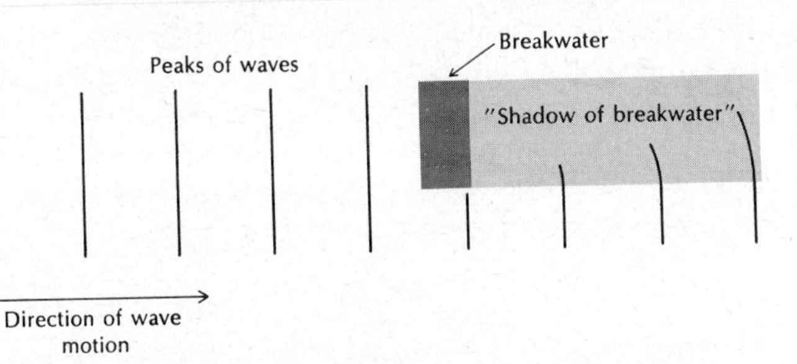

FIGURE 16.2 Ocean waves bend into the shadow region behind a breakwater. This is the phenomenon of diffraction.

traveled only in straight lines. Figure 16.4 shows the effect of changing the wavelength while keeping the size of the opening fixed. As the wavelength becomes shorter, the amount of bending due to diffraction becomes less and less. This result is important for the study of light, since it is the explanation for the fact that diffraction effects for light were not observed clearly in Newton's time.

In fact, the diffraction of light had already been recorded by Francisco Grimaldi in 1655. Grimaldi observed that, when sunlight was allowed to pass through a pinhole, the circle of light that resulted was slightly larger than should be expected if no light went into the shadow region. Since diffraction effects are most pronounced when the wavelength is comparable to the size of the barrier, Grimaldi's results indi-

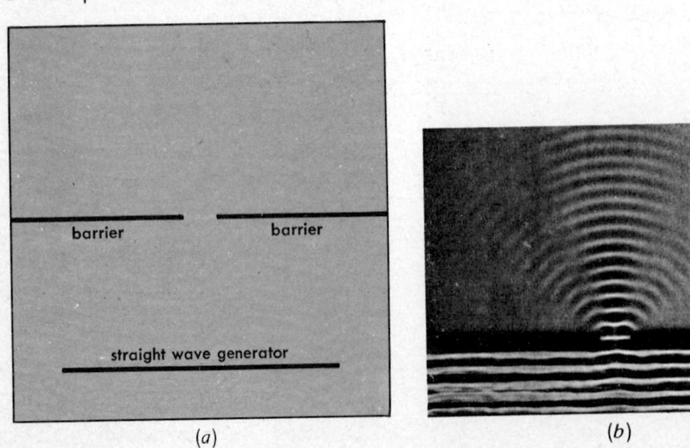

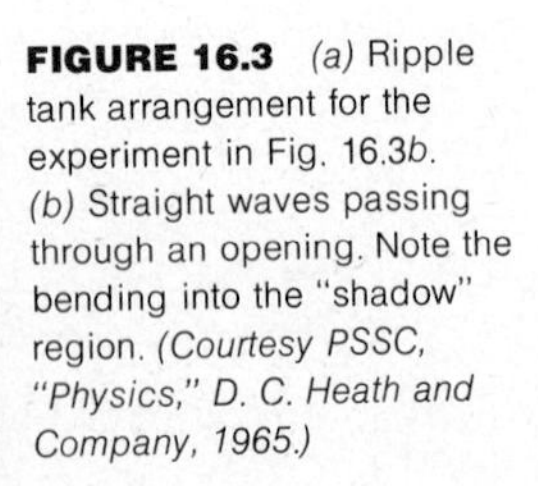

FIGURE 16.3 *(a)* Ripple tank arrangement for the experiment in Fig. 16.3*b*. *(b)* Straight waves passing through an opening. Note the bending into the "shadow" region. *(Courtesy PSSC, "Physics," D. C. Heath and Company, 1965.)*

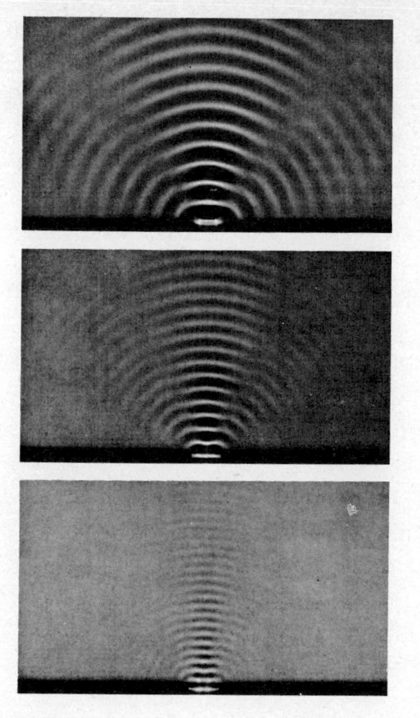

FIGURE 16.4 Three views of waves passing through the same opening. Note the decrease in bending at shorter wavelengths. *(Courtesy PSSC, "Physics," D. C. Heath and Company, 1965.)*

cate a very small wavelength for light. The effects were so small that Newton was unable to convince himself that these were true diffraction effects, and not some interaction of the light with the edges of the pinhole. He therefore concluded that light is a particle phenomenon, not a wave phenomenon.

16.2 YOUNG AND INTERFERENCE

The first clear-cut evidence for waves associated with light came with the work of Thomas Young in 1802, almost a century after Newton. His work apparently settled the problem, but we shall see in Chapter 18 that during the twentieth century additional modifications have had to be made. Young showed that light is a wave phenomenon by demonstrating that light exhibits the phenomenon of *interference*. To explain what we mean by interference we shall return to a discussion of water waves. Let us describe waves in water in terms of peaks and valleys, as shown in Fig. 16.1. Suppose we now have a situation in which water waves from two different sources arrive at the same point, such as A in

FIGURE 16.5 Diffraction of light, as observed by Grimaldi, was so small an effect that Newton was unconvinced of its reality.

Fig. 16.7. If they arrive so that the peaks and valleys of each coincide, the amplitudes of the waves are added together and the combined wave has larger peaks and larger valleys, as shown in Fig. 16.7. This is called *constructive interference*. But, if the situation is such that the peak of one wave arrives at the same time as the valley of the other, the result is cancelation, or *destructive interference*. The two kinds of interference are shown in Fig. 16.8. When the peaks and valleys arrive together, interference is constructive, and the resulting amplitude is the sum of the two sets of waves. When they arrive out of step, cancelation, or destructive interference, results. Note that, for destructive interference to occur, the two sets of waves must be displaced, or out of step, by one-half wavelength.

FIGURE 16.6 Thomas Young (1773–1829.) *(Courtesy University of Pennsylvania Library.)*

How can it happen that two sets of waves will be a half-wavelength out of step, so that when we bring them together we have peaks on valleys and valleys on peaks? Let us consider the analogy of two trains traveling on tracks side by side. Each train alternates boxcars and flatcars. In the beginning we have boxcars opposite boxcars and flatcars opposite flatcars and, if the two trains have the same velocity, boxcars will remain opposite boxcars and flatcars opposite flatcars, as in Fig. 16.9*a*. The boxcars and flatcars in each train are numbered so that we can keep track of them. Now suppose train *B*, *without changing its speed*, travels somewhat farther than *A*. That is, the track on which *B* is traveling takes some sort of detour. If train *B* travels exactly one car length farther on this detour than *A*, we shall have the situation shown in Fig. 16.9*b*. Now there are boxcars opposite flatcars and flatcars opposite boxcars. If at some later time train *B* travels an additional car length farther than *A*, along another detour in the track, boxcar *A*1 will be opposite boxcar *B*3, and *A*3 opposite *B*5, but again there will be boxcars opposite boxcars and flatcars opposite flatcars. This last situation corresponds to constructive interference, and that of Fig. 16.9*b* to destructive interference.

Now we can see how interference phenomena in light, or in any other wave phenomena, occur. We begin with light from two separate sources and arrange that the waves start out with peak opposite peak and valley opposite valley (that is, they start out *in phase*). But now, if the path

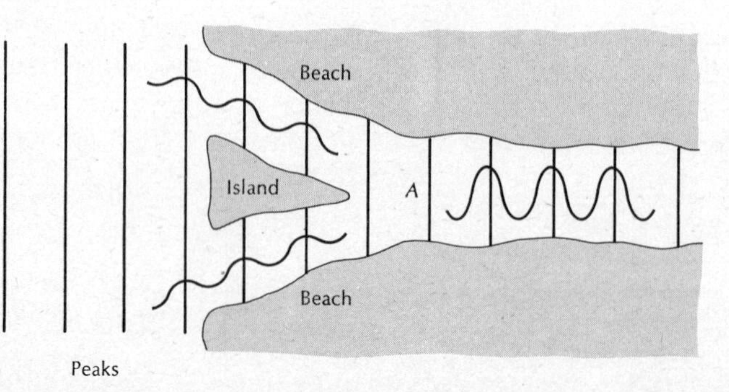

FIGURE 16.7 Two sets of waves from different sources arrive at point *A*. If the peaks of each arrive at the same time, the amplitudes are added together and the resulting wave is larger.

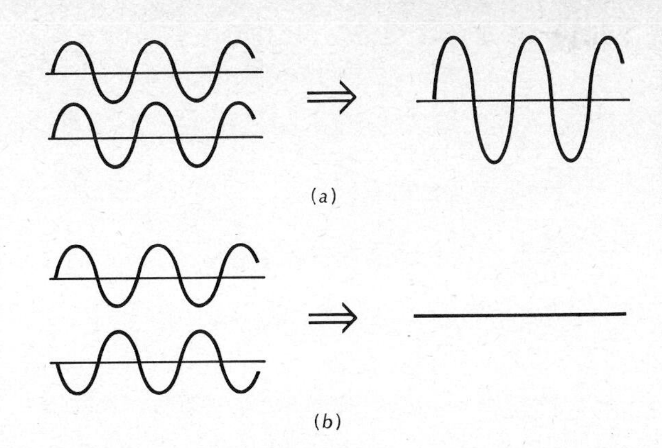

FIGURE 16.8 *(a)* Constructive interference occurs when waves arrive peak opposite peak and valley opposite valley. *(b)* Destructive interference occurs when waves arrive peak opposite valley and valley opposite peak.

length from these two sources to the point of measurement is different, when the light beams arrive at that point, no longer will peak be opposite peak and valley opposite valley in general. Only under the very special circumstance that the path difference is exactly the same as some whole multiple of the distance between peaks (boxcars) will the light arrive at the point of detection with peak opposite peak. That is, when one light beam travels farther than the other, if it travels farther by one, two, or any whole number of wavelengths, the beams will arrive peak opposite peak and reinforce.

A diagram of Young's experiment is shown in Fig. 16.10. Light from a distant source arrives at the two slits *A* and *B*. *A* and *B* then act as two sources of light waves. These will interfere constructively or destructively, depending on whether their paths differ such that they fall in or out of phase. At point 1 on the screen the paths are identical in length, and the waves interfere constructively. At point 2 the light wave from *B* travels a distance Δ farther than those from *A*. Whether constructive or destructive interference occurs at 2 will depend on whether Δ causes the waves to become $\frac{1}{2}$ wavelength, 1 wavelength, $1\frac{1}{2}$ wavelengths, or more out of step. If they are out of step by $\frac{1}{2}$, $1\frac{1}{2}$, etc., they will arrive with

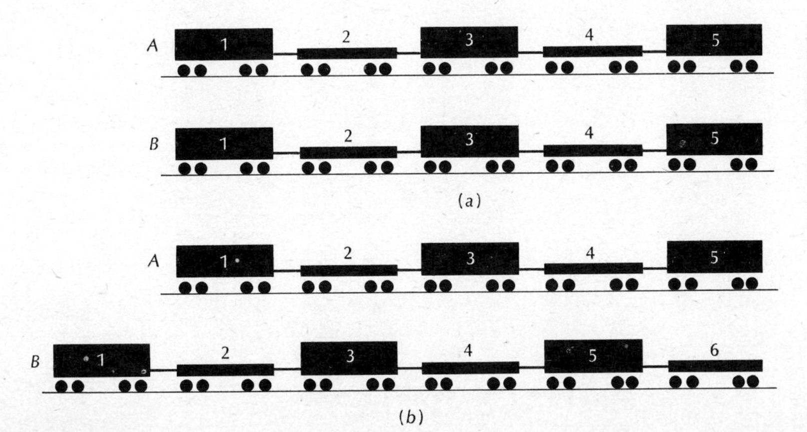

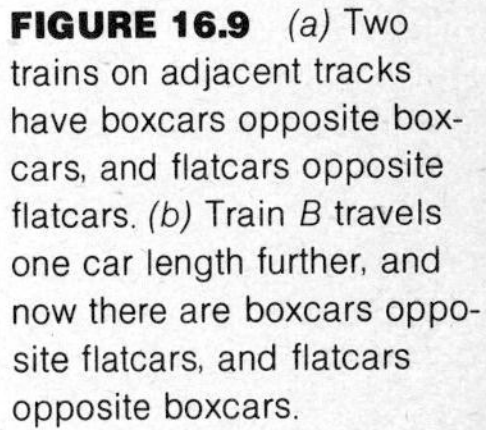

FIGURE 16.9 *(a)* Two trains on adjacent tracks have boxcars opposite boxcars, and flatcars opposite flatcars. *(b)* Train *B* travels one car length further, and now there are boxcars opposite flatcars, and flatcars opposite boxcars.

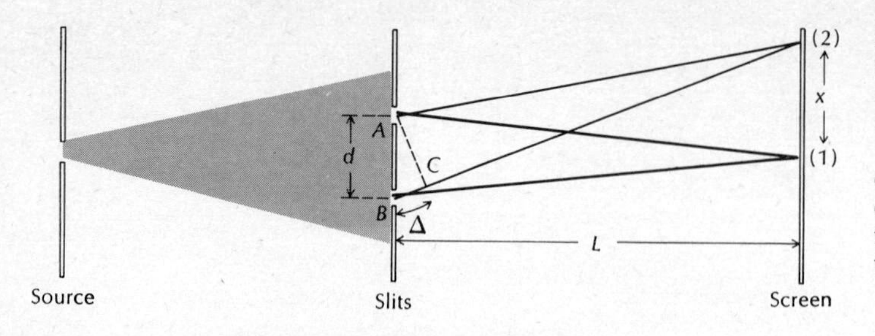

FIGURE 16.10 Young's interference experiment. Light from sources *A* and *B* interferes constructively at 1. Interference at 2 is either constructive or destructive, depending on the length of the path difference Δ and the wavelength λ.

peaks opposite valleys, and destructively interfere. If they are out of step by 1, 2, 3, or more whole wavelengths, they arrive peak opposite peak and constructively interfere.

Thus, without doing any mathematics, we can say that on the screen there will be regions of constructive interference and regions of destructive interference. Where they occur will depend on the distance between the slits *A* and *B*, and on the wavelength of light. Figure 16.11 shows the pattern of light intensity in such an experiment. These experiments allow a measurement of the wavelength of light. Because the separation between maxima using red light is found to be greater than that for blue, it is shown that red light has a greater wavelength than blue.

Approximate wavelengths found for light of various colors are:

$$\text{Red} = 6.5 \times 10^{-7} \text{ meter}$$
$$\text{Yellow} = 5.9 \times 10^{-7} \text{ meter}$$
$$\text{Green} = 5.0 \times 10^{-7} \text{ meter}$$
$$\text{Violet} = 4.0 \times 10^{-7} \text{ meter}$$

The range of wavelengths for visible light runs from 3.8×10^{-7} meter (deep violet) to 7.5×10^{-7} meter (deep red). Frequently they are expressed in a unit called the *angstrom* (Å). One angstrom is 1×10^{-10} meter. Red light has a wavelength of 6500 angstroms.

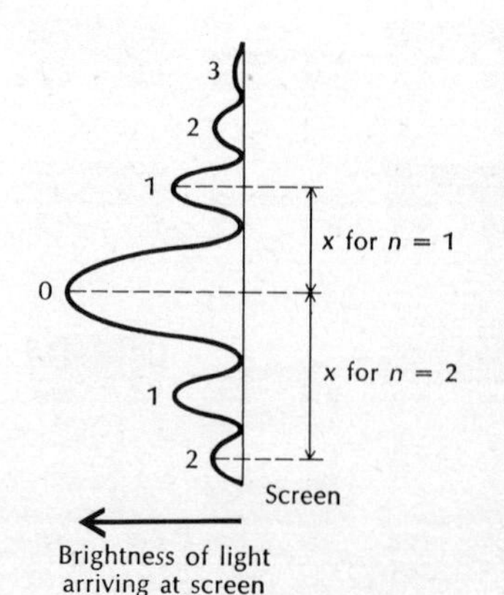

FIGURE 16.11 Intensity, or brightness, of light arriving at the screen in Young's experiment. The peak labeled 0 corresponds to no path difference for the two light beams; $n = 1$ corresponds to one wavelength path difference; $n = 2$, to two wavelengths path difference. The displacement x of the peaks depends on the wavelength, the separation of the two slits, and the distance to the screen.

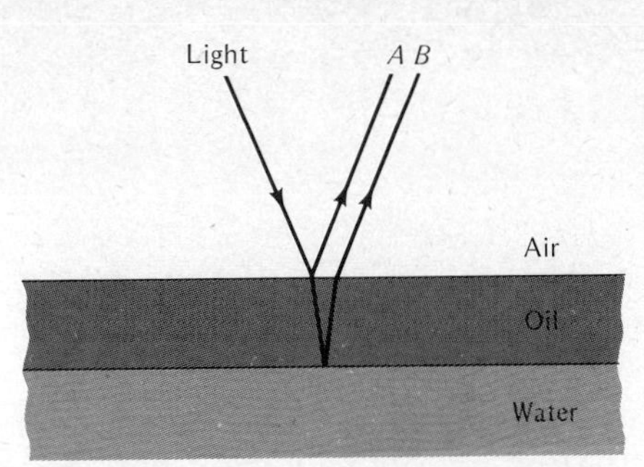

FIGURE 16.12 Beam *A* and beam *B* have traveled different distances. Therefore, when they combine, interference can occur. The most brilliant colors will occur when the film is about one-fourth the wavelength of light thick.

OIL FILMS

There are a number of examples of interference effects observable in nature. One of the commonest is the colored film observed when oil spreads out on water. Figure 16.12 illustrates the idea. Light from above is partly reflected from the upper surface, and partly reflected at the oil-water interface. These two reflected beams have traveled different distances. When they recombine there will be constructive interference for some colors, and destructive interference for others. When the oil film is observed in white light, some colors are removed from the white light by destructive interference. If violet is removed, for instance, the remaining light appears yellow or straw-colored. When yellow is removed, it appears blue-violet.

Thick oil films will meet the destructive interference condition for several colors. So, as the film becomes thicker, there will be a sequence of color variations depending on which is removed. This can easily be observed by placing a drop of light oil on a *clean* water surface and watching it spread, using white light reflected off the surface. Turpentine works also. After the turpentine spreads it evaporates, so one can watch the color changes as the film becomes thinner.

16.3 THE RELATIONSHIP BETWEEN WAVELENGTH, FREQUENCY, AND VELOCITY

We have demonstrated that light behaves like waves in that it shows diffraction and interference effects characteristic of all waves. As a wave it must have wavelength, frequency, and velocity. Figure 16.13 shows a wave moving with velocity v to the right. How many peaks per second pass a given point? If the velocity is 10 meters/second, every peak in a 10-meter length will pass any given point in 1 second. Therefore the number of peaks per second, which is called the frequency, is given by

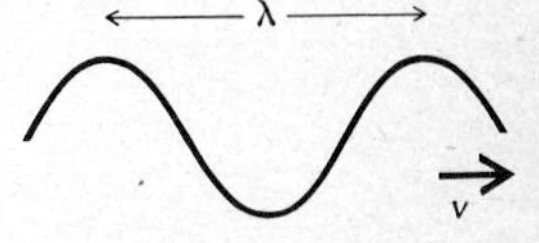

FIGURE 16.13 A wave moving to the right with velocity v and wavelength λ.

$$\text{Frequency (waves/second)} = \frac{\text{velocity (meters/second)}}{\text{wavelength (meters/wave)}}$$

or $$\text{Frequency} \times \text{wavelength} = \text{velocity} \tag{16.1}$$

We can see that this is correct by noting that $1/\lambda$ is the number of peaks per meter. For light, the velocity is customarily given the symbol c, so that Equation 16.1 is written

$$f\lambda = c \qquad (16.1)$$

Equation 16.1 allows a conversion from frequency to wavelength, or vice-versa, for light or any other wave phenomena.

The unit for frequency, waves per second, has been given the name *hertz* (hz) for H. Hertz whose work is discussed in Section 16.6. Thus 10 peaks per second is 10 hertz. Radio frequencies are frequently given in kilocycles (thousands of waves per second) or megacycles (millions of waves per second); 500 kilocycles is 500,000 hertz, and 10 megacycles is 10,000,000 hertz (10^7 hertz).

When light, or any electromagnetic wave passes from one media to another, the frequency remains constant. Since the velocity changes in going from one media to another, the wavelength must also change, if the frequency is to remain constant.

Example 16.1

What is the frequency of green light whose wavelength is 5×10^{-7} meter?

$$f = \frac{c}{\lambda}$$

$$= \frac{3 \times 10^8 \text{ meters/second}}{5 \times 10^{-7} \text{ meter}}$$

$$f = 0.6 \times 10^{15} \text{ hertz}$$

$$f = 6 \times 10^{14} \text{ hertz}$$

16.4 OPTICAL SPECTRA

One of the important areas of scientific study in the latter nineteenth century was the study of the colors of light emitted by different chemical elements under various conditions. If gases are studied in cathode-ray tubes, it is found that they emit various colors. Neon signs are an example of this, but only the bright orange-red signs contain neon. Other colors are produced by other gases, such as oxygen, xenon, hydrogen, etc. Each chemical element produces its own characteristic set of colors or wavelengths, which can function as a fingerprint for the particular element. To be useful one needs to be able to make accurate measurement of the wavelengths of light emitted. Gases in discharge tubes emit only a few very well-defined wavelengths. They are called *bright-line spectra.*

Young's two-slit experiment demonstrates the principles behind an optical device that can be used for very precise measurements of light

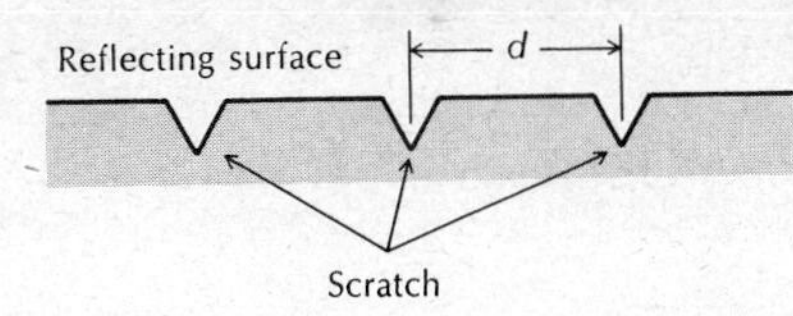

FIGURE 16.14 A large number of parallel scratches on a glass surface form a diffraction grating. Each reflecting surface acts as a source, so that instead of two sources, as in Young's experiment, there are many. This leads to a brighter and sharper interference pattern.

wavelengths. A *diffraction grating* is a piece of glass, ruled by a precision engine, with parallel scratches about 10^{-4} inch apart (Fig. 16.14). This produces a large number of parallel reflecting surfaces a distance *d* apart. If light shines on this grating from above, each of these reflecting surfaces acts as a source, like one of the slits in Young's experiment. This leads to an interference pattern similar to that for the two-slit experiment, except that, because there are so many slits, the bright lines are much sharper, and much more precise measurements of their position can be made. The extra precision possible with the diffraction grating means that more precise values for wavelengths result.

The specific wavelengths observed in the spectrum of hydrogen gas were later important in testing various theories of the structure of the atom. This will be discussed in Chapter 19. The spectrum of hydrogen and several other elements is shown in the color insert.

16.5 X-RAYS

In 1895 a major new discovery was made that in some ways can be said to have launched twentieth-century physics. Wilhelm Conrad Roentgen was experimenting with a cathode-ray tube. He found that a ray was being emitted by the anode of this tube that produced fluorescence in some nearby objects like that produced by ultraviolet light. But, in addition, these new so-called x-rays were able to penetrate where known forms of light could not. They could even penetrate human flesh and outline the bones of the body. X-rays found an immediate practical use. Within a few weeks of Roentgen's announcement of their properties, the broken arm of a boy in Hanover, New Hampshire, was set using x-rays to guide the doctor.

In 1912 Max von Laue showed that x-rays produced diffraction effects, like other forms of light, if the size of the grating was small enough.

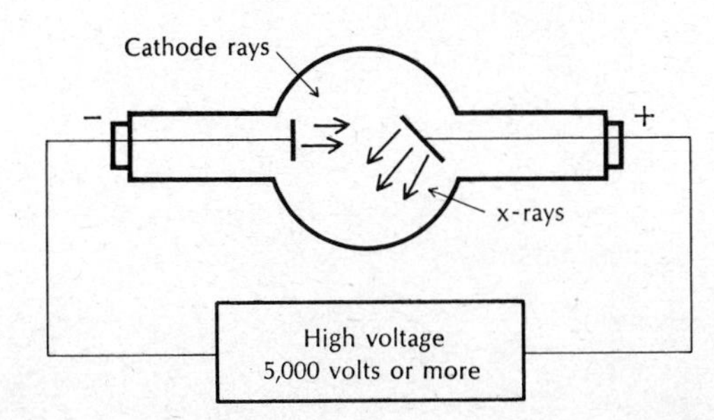

FIGURE 16.15 X-rays are generated by the impact of high-energy cathode rays (electrons) on the positive electrode on a cathode-ray tube.

Type of radiation	Wavelength
Radio waves	Miles to fractions of a meter
Microwaves	10^{-2} to 10^{-4} meter
Infrared	10^{-4} to 10^{-6} meter
Visible	7×10^{-7} to 3×10^{-7} meter
Ultraviolet	3×10^{-7} to 10^{-9} meter
X-rays	10^{-9} to 10^{-12} meter
Gamma-rays	10^{-10} to ? meter

TABLE 16.1 Kinds of electromagnetic radiation. The boundaries between the different forms are approximate. Each type merges smoothly with the next as the wavelength changes

A standard diffraction grating was much too coarse, but von Laue used a crystal, so that the grating spacing was that of the atoms in the crystal, about 10^{-10} meter. He showed that this is about the size of x-ray wavelengths and that therefore x-rays are a form of light, except that their wavelengths are much less than those of visible light.

16.6 MAXWELL AND ELECTROMAGNETIC RADIATION

We have been discussing light and its wavelength, and have shown that visible light has wavelengths between 3×10^{-7} and 7×10^{-7} meter. But we have also implied that there are wavelengths greater or smaller than these that we cannot see—ultraviolet light and x-rays, for example. When discussed in this general sense, light is called *electromagnetic radiation*. It is found with wavelengths from miles in size (radio waves) to less than the size of an atom. Table 16.1 is a survey of the kinds of electromagnetic radiation and their wavelengths.

The name electromagnetic radiation used above implies that there is a connection between light, electricity, and magnetism. Such a relationship was suspected for many years, but the development of this connection was made by James Clerk Maxwell (1831–1879). Maxwell developed a set of equations that summarized the laws of electricity and magnetism of Coulomb, Oersted, Ampère, and Faraday in a compact and elegant fashion. He added one idea to those previously existing. This idea can be stated: A changing electric field produces a magnetic field. Note the similarity of this to Faraday's law of electromagnetic induction. As a bonus in the construction of this set of equations describing electrical and magnetic phenomena, Maxwell was able to

FIGURE 16.16 James Clerk Maxwell (1831–1879). *(Courtesy American Institute of Physics.)*

E B E B E B
B E B E B E
Direction of motion of light →

FIGURE 16.17 Electromagnetic waves are waves of electric and magnetic fields related as shown. The magnetic field *B* is always at right angles to the electric field *E*.

manipulate the equations into a form which predicted a type of wave motion. The equations predicted the velocity of this wave in terms of constants which could be measured in the laboratory. When the velocity of the waves was calculated using these results, it agreed with the velocity of light, as well as it was then known. Maxwell thus predicted that his electromagnetic waves were in fact light. Since the equations place no restriction on the wavelength, electromagnetic waves of almost all conceivable wavelengths exist, as summarized in Table 16.1.

In 1887 Heinrich Hertz demonstrated the correctness of Maxwell's theory by generating and detecting electromagnetic waves of a wavelength of 10 meters in a manner predicted by Maxwell's equations. Figure 16.17 shows the waves of Maxwell. These waves consist of alternating electric fields and, at right angles to these, alternating magnetic fields. The entire structure moves to the right at the speed of light. The fact that waves of this form really exist can be shown experimentally in a number of ways.

SUMMARY

The wavelength λ of a wave is the distance from peak to peak or from valley to valley. The amplitude of a wave is the height of a peak above the undisturbed situation. The velocity of a wave is the velocity with which the peaks move. The frequency f is the number of peaks passing a given point per second.

Diffraction is the bending of a wave around an obstacle into the geometric shadow. It is characteristic of all wave phenomena.

Interference is the adding or canceling of two sets of waves arriving at the same point from separate sources. The addition of two waves is constructive interference. The cancelation of one wave by another is destructive interference. Young's demonstration of interference in light convinced most scientists for 100 years that light is a wave phenomenon.

Frequency, wavelength, and velocity for any wave are related by

$$\text{Frequency} = \frac{\text{velocity}}{\text{wavelength}}$$

A diffraction grating is used to produce a sharp interference pattern in light. Since the peaks for different colors occur in different directions, it can be used to separate light into its various colors.

Visible light ranges in wavelength from about 7.5×10^{-7} meter (deep red) to 3.8×10^{-7} meter (deep violet). Expressed in angstroms, this is 7500 to 3800 angstroms (1 angstrom $= 10^{-10}$ meter).

The optical spectrum or bright-line spectrum of a gas is the pattern of colors or particular wavelengths exhibited when the light from that gas in a discharge tube is passed through a prism or diffraction grating.

Light is one class of electromagnetic wave. Electromagnetic waves exist with all wavelengths from many miles to distances less than the size of an atom. X-rays are electromagnetic radiation whose wavelength is about 10^{-10} meter, roughly the size of an atom.

Maxwell developed the set of equations that describe electromagnetic waves.

SELECTED READING

Minnaert, M: "The Nature of Light and Color in the Open Air," Dover Publications, Inc., New York, 1954. A classic on light and color, covering a wide variety of phenomena that occur in nature. Included are light, color, rainbows, halos, light and color in the sky, and other natural phenomena involving light and color.

QUESTIONS

1 Some butterflies have beautifully iridescent colors on their wings. Can you explain this?
2 Sometimes concert halls are said to have "dead spots." What explanation can you propose for this?
3 Occasionally at the ocean shore a wave of abnormal height appears. Can you suggest one explanation for this?
4 Would Newton have accepted the wave model for light if he had had access to modern experiments?
5 How are photographs using infrared light used?
6 What fraction of neon signs actually contain neon?
7 How many examples of interference in light in everyday life can you think of?

SELF-TEST

____________ diffraction
____________ destructive interference
____________ constructive interference
____________ wavelength
____________ frequency
____________ amplitude
____________ spectrum
____________ electromagnetic wave
____________ x-rays

__________ infrared
__________ ultraviolet
__________ Maxwell
__________ Grimaldi
__________ Young

1 The height of a wave from the undisturbed situation
2 Light of wavelengths shorter than that of the visible region
3 The colors or wavelengths emitted from a gas which has been excited
4 Waves from different sources arrive out of step so as to cancel one another
5 The bending of light away from the straight-line path at a barrier
6 Very-short-wavelength radiation generated in a cathode-ray tube
7 The number of peaks of a wave passing per second
8 Light of wavelength longer than that of the visible region
9 First observed diffraction of light
10 Developed the equations of electromagnetic waves
11 Showed that light is a wave by doing interference experiments
12 Peak-to-peak distance of a wave
13 Waves arriving in step so as to add amplitudes
14 The waves predicted by Maxwell as a result of his equations

PROBLEMS

1 (*a*) Calculate the frequency of orange light whose wavelength is 4000 angstroms (4×10^{-7} m). (*b*) Calculate the wavelength of an FM radio station whose frequency is 88 megacycles (88×10^{6} hertz).
2 Calculate the frequency of green light of wavelength 5000×10^{-10} m (5000 angstroms).
3 (*a*) Calculate the wavelength in air of microwaves whose frequency is 15×10^{9} hertz. (*b*) Calculate the wavelength of these same microwaves in glass, where the velocity of light is 75 percent of that in air.
4 Calculate the wavelength of radio waves of frequency 5.7×10^{5} hertz (570 kilocycles).
5 Calculate the wavelength in crown glass and diamond of green light whose wavelength in air is 5×10^{-7} m.

THE TWENTIETH CENTURY

In the period just before 1895 the point of view existed that physics was essentially complete, except for cleaning up a few loose ends and refining the accuracy of existing measurements. In the period since that time, almost every aspect of our view of physics has required modification or complete revision.

The first topic we will discuss is Einstein's theory of special relativity, which led to a modification of our ideas of matter and energy. In Chapter 18 we will study the revision of our view of the nature of light and the revival of a particle theory of light, as required by Einstein's work on the photoelectric effect. In Chapter 19 we will look at the revisions in ideas about the structure of the atom pointed to by the work of Rutherford and Bohr. Then, in Chapter 20 we will find that electrons and all matter must be conceived of as having wavelike, as well as particlelike, properties. In Chapter 21 the discovery and development of radioactivity and the understanding of the structure of the nucleus are covered. In recent years a bewildering variety of new particles has been discovered. Physics is just beginning to bring some order out of this chaos. These developments are the subject of Chapters 22 and 23.

The theory of special relativity was first proposed by Albert Einstein in 1905. Its basic ideas have since been abundantly verified experimentally, so that it is now accepted as one of the important statements about our world and our universe. Relativity theory has acquired an entirely undeserved reputation for difficulty. Mathematically, we can understand the basic ideas of special relativity using only simple algebra and a little geometry. Conceptually, relativity theory has produced many curious paradoxes. These require careful analysis to understand, because they tend to violate our common sense. Einstein once defined common sense as "that body of prejudices that we acquire up to age 18."

RELATIVITY

17.1 RELATIVITY

Just what does the term *relativity* mean? As you sit reading this ask yourself how fast you are moving. Your first answer may be "Not at all." But that is, in fact, not true. The earth rotates on its axis once a day; which means that at the latitude of New York City you are moving about 750 miles/hour eastward. In addition, the earth moves in its orbit around the sun at a speed of about 10^5 miles/hour. Also, the sun, with the entire solar system in tow, is moving toward the star Vega at a speed of about 10^6 miles/hour. So—how can you say you are not moving?

Now you will say that you have been cheated. What you meant to say was that you could not sense or measure any motion. Stated more precisely, what you meant was that you were not moving *relative* to the earth. That is, the distance between you and nearby points on the earth was not changing.

So in order to talk precisely about velocity, we must always specify velocity measured with respect to something. Your velocity may be measured with respect to the center of the earth, the center of the sun, or the center of our galaxy. When we speak of your velocity with respect to an object which is also moving, we are speaking of the relative velocity between you and the other object. In the mechanics of Newton, only relative velocities appear. Until now we have not emphasized this, but have always used velocities relative to the earth's surface, or velocities relative to the center of the earth. Nowhere in Newton's mechanics does the idea appear that perhaps there is some point in the universe which is at rest and from which all velocities could be measured. Nevertheless, this idea was assumed by all scientists until quite recently. Until Einstein began his work, no one had clearly pointed out that the idea of a point at absolute rest is not a necessary part of Newton's mechanics.

During the nineteenth century, as the properties of light were studied, the question of absolute velocities came up again. By absolute velocity we mean velocity as measured with respect to some point at rest with respect to the universe as a whole. You know that, if two cars are traveling alongside each other, their relative velocity can be zero, even though their velocity with respect to the road is 50 miles/hour. The question then arises, does light behave in a similar way? Einstein wondered what a light beam would look like if he could travel alongside at the velocity of light.

Perhaps we can see part of the difficulty by considering a simple experiment. Since we have shown that light is a wave, this experiment will involve waves. Stand on a bridge over a slow-moving stream and drop a pebble. A ring of waves will spread out from the point where the pebble entered the water, as seen in Fig. 17.1. We want to look at the velocity of these waves. Presumably we could set up some method to measure these velocities. The simplest might be to take a series of photographs a known time interval apart. If you have imagined the experiment well, or actually performed it, you know that two things will happen. First, the waves spread in a circle from the point where the pebble struck the water. Second, since the stream is moving, the entire circle of waves moves downstream. The velocity of the waves can be measured relative to the moving water, or relative to the bridge. Since the waves spread in a circle, they must be moving with the same velocity in all directions *with respect to the water*. Let us call that velocity v. But if the stream is moving with velocity V, the whole circle of waves moves downstream with velocity V. The velocity of the waves with respect to the bridge will depend on which part of the circle we study. This is indicated in Fig. 17.2.

The waves on the downstream side of the circle will move at a velocity $V + v$ relative to the bridge. The waves on the upstream side will move with velocity $V - v$ relative to the bridge. Waves on other parts of the circle will have more complicated velocities. The basic point here is that the velocity of the wave relative to the water v must be added

FIGURE 17.1 A stone is dropped in a moving stream.

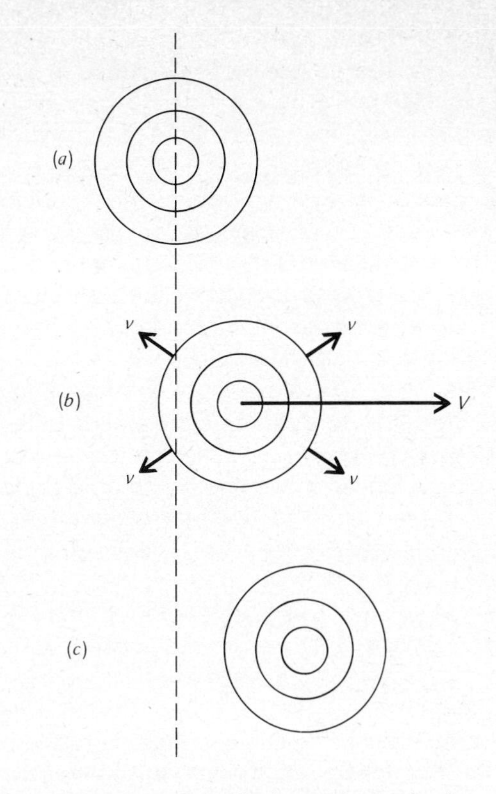

FIGURE 17.2 Circular waves in a moving stream. The entire pattern moves downstream with velocity V. The individual waves move outward from the center with velocity v.

to or subtracted from the velocity of the stream V to find out what velocity will be measured relative to the bridge for that part of the wave.

To conclude, velocities are *always* measured between two objects, or points in space. They are always relative. We can have one car moving at 50 miles/hour relative to the earth, and another moving in the same direction at 70 miles/hour relative to the earth. The second car will then have a velocity of 20 miles/hour relative to the first. You may walk up the aisle of a jetliner at a speed of 1 mile/hour, relative to the airplane. If its speed relative to the ground is 600 miles/hour, yours will be 601 miles/hour relative to the ground.

We frequently find it convenient to use the earth as a reference frame. That is, we find it convenient to measure velocities relative to the earth and to imagine that the earth is fixed in space. But this is a convenience—not a necessity. Because we do not sense the earth's motion through space, we do not include this velocity in calculations regarding what goes on on its surface.

17.2 THE MICHELSON-MORLEY EXPERIMENT

The experiment with the pebble and the bridge involves waves in water. Notice that the velocity of the water relative to the bridge comes directly into any calculation of the velocity of the waves relative to the bridge.

Water is the *medium* in which these waves move. After light was shown to be a wave, the question was asked: What is the medium in which light moves? All waves then known move in some medium: sound waves in air, water waves in water, etc. It seemed reasonable to suppose that light waves also move in something. Science in the 1800s was not able to conceive of a wave that moved without a medium. The medium for light was given the name *luminiferous ether* (not to be confused with the medical anesthetic). The idea that light moves in something suggests an experiment. Why not measure the motion of the earth through the ether? Since the earth moves in different directions at different times of the year, no matter what direction the ether is moving, there should be some time of year when the earth is moving with respect to it. In our original experiment we were able to determine the motion of the stream with respect to the bridge by studying its effect on the velocity of the waves relative to the bridge. Similarly, we should be able to determine the motion of the ether by measuring the effect of its motion on the velocity of light waves with respect to the earth. The assumption buried in all of this is that the ether might be at rest with respect to the universe. Velocities measured relative to the ether would then be unique, because they would be absolute velocities, in the sense that they were measured with respect to a coordinate system at rest.

The experimental test of these ideas involves the following idea. The earth moves in its orbit around the sun at a speed of 30 kilometers/second. No matter what the motion of the ether, the earth changes its velocity with respect to the ether every 6 months as it moves around the sun (unless the ether moves with the earth, which sets up the earth as the unique place in the universe, an idea that had long since been abandoned; also, experimentally, there are tests to show that this does not happen).

The experiment involves measuring any changes in the speed of light relative to the earth caused by the different motions of the earth with respect to the ether at different times of the year. Since the speed of the earth in its orbit is 2×10^4 meters/second and the velocity of light is 3×10^8 meters/second, this experiment requires measuring changes of the velocity of light of about 1 part in 10^4. If v is the velocity of the earth in its orbit, and c is the velocity of light, we are talking about an experiment that measures an effect of the order of v/c. In fact, practical means for measuring the effect are sensitive not to effects of the order of v/c, but to effects of the order v^2/c^2. An experiment sensitive to changes in the speed of light of 1 part in 10^8 therefore had to be devised.

A. A. Michelson developed such an experiment, based on the phenomenon of *interference*. This experiment was first performed in 1881, and later repeated with greater sensitivity by Michelson and E. W. Morley in 1887. Consider green light whose wavelength is 5×10^{-7} meter. For this wavelength there are 2×10^6 waves/meter, or 2×10^7 waves/10 meters. Ten meters was the effective length of Michelson's apparatus.

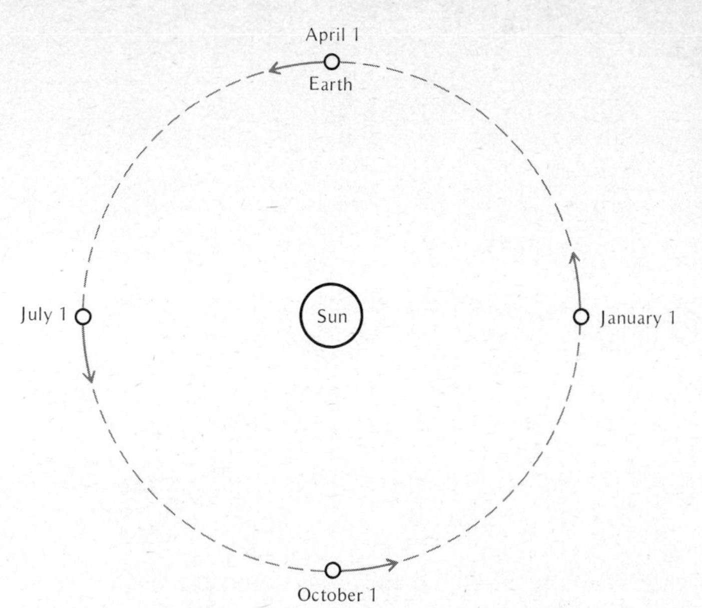

FIGURE 17.3 At different times of the year, the earth's velocity about the sun is directed in different directions in space. No matter what the direction of the ether velocity, the earth's velocity with respect to it will change during the year.

To measure to 1 part in 10^8 meant measuring effects of about one-fifth of a wavelength of light in 10 meters. This can be done quite easily using interference, which we discussed in Chapter 16.

The effective idea of Michelson's experiment was to measure the difference in the velocity of light parallel to and perpendicular to the ether wind. (Ether wind is the term used to describe the effect of the earth moving through the ether.) If the ether hypothesis is correct, there should be a detectable difference.

A schematic top view of the interferometer invented by Michelson is shown in Fig. 17.4. A beam of light enters from the left and hits the partly silvered mirror. About one-half of the light goes through the mirror and continues on path *A*. The other half is reflected by the mirror and travels along path *B*. Both beams are reflected by mirrors back along their original paths. At the partly silvered mirror a fraction of each beam continues toward the observer. Since this is a mixture of light beams that have traveled two different paths, interference will result. The combination of the two beams produces what are called interference fringes. These are regions of constructive and destructive interference that can be observed as light and dark bands in the apparatus. A slight change in the velocity of light in either arm of the apparatus will produce a measurable shift in the position of these fringes.

In the Michelson-Morley experiment the entire apparatus was mounted on a large stone block, to make it mechanically rigid and vibration-free. This block was floated in mercury, which made it easy to rotate. The experiment was performed by setting up and adjusting the apparatus and then rotating the stone block 90°. The rotation of the apparatus interchanges the two arms *A* and *B*. If the ether wind had been

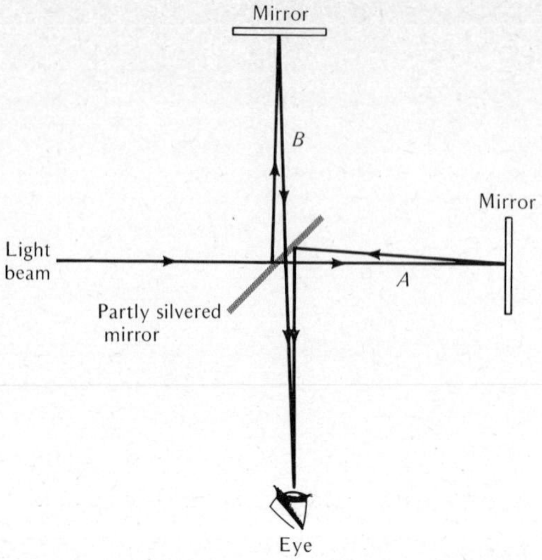

FIGURE 17.4 The geometry of the Michelson-Morley experiment.

blowing along arm *A*, for example, it would now be along arm *B*. If the ether wind had any effect on the velocity of light, a shift of the interference fringes in the apparatus would be observed. No such effect was ever seen, even though the experiment was repeated many times at various times of the year.

This null result of the Michelson-Morley experiment was very hard to explain. There were many ingenious attempts, including one by H. A. Lorentz, who suggested that perhaps the apparatus shrank in its dimensions parallel to the direction of motion through the ether. Lorentz was able to calculate how much the apparatus would have to shrink in order to account for the absence of any effect in the Michelson-Morley experiment. Later we will find this Lorentz contraction turning up in Einstein's relativity theory.

17.3 EINSTEIN'S HYPOTHESIS

One way of stating the result of the Michelson-Morley experiment is to say that it is impossible to measure any motion through the ether or, in other words, that ether does not exist. (Does it make any sense to suggest that something exists if it has no measurable consequences?) Another way—although it is not obvious—is to say that the velocity of light will be measured to be the same by any two observers, even if they are in relative motion. This last statement violates our common sense. It means that two observers, one on a rocket and one on earth, will measure the same velocity for the same beam of light, despite their relative motion. How can such a peculiar result be true? (Figure 17.5.)

The question at the end of the last paragraph is a cheat. The proper way to describe the problem is: "We have here an experiment result (the Michelson-Morley experiment). If the experimental result is true,

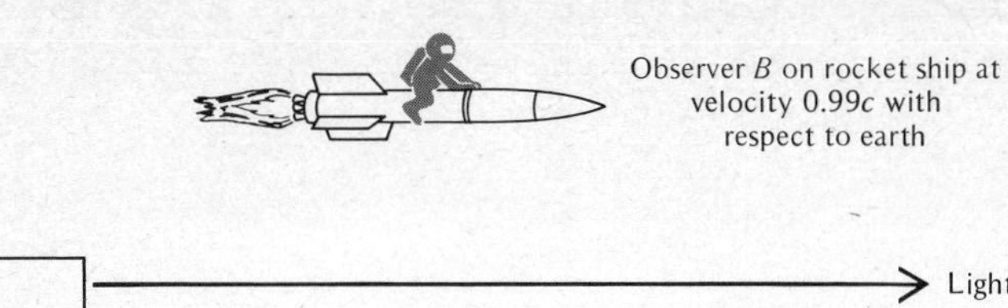

FIGURE 17.5 Both observer *A* and observer *B* will measure the same value *c* for the velocity of the light beam, even though *B* is moving at 0.99*c* with respect to *A*.

we should try to understand it. What is wrong with our ideas about time, distance, and velocity that this result appears to be a paradox?" Einstein approached the problem in this manner and discovered that a significant modification in what is meant by time and distance is required in order to understand the Michelson-Morley experiment. Why distance and time? Velocity is distance divided by time, so any difficulties with velocity must go back to difficulties with distance and time.

What can be wrong with our idea of distance or length? We originally defined length essentially by defining a unit, the meter, and then comparing the length of an unknown object with a meter stick. But suppose we want to measure the length of a moving object. Can this be done? What about comparing the ends of the object with a meter stick, as the object moves by. But how is this to be done? By looking—that is, by using light. But suppose the speed of the moving object is not small compared to the speed of light? Perhaps this will give us trouble in determining its length. We will find that, when objects move at speeds approaching the speed of light, there are difficulties with both length and time measurements. To understand these we must first carefully state Einstein's starting point.

THE POSTULATES OF RELATIVITY

Einstein began his analysis with two postulates. A *postulate* is an assumption used as a starting point. Postulates are assumed to be true based on experimental or other types of reasons. The fundamental test of the value of any postulate is whether or not it leads to correct conclusions (conclusions that agree with experiment, even though they may seem to violate common sense). Newton's three laws are really postulates which are accepted because the conclusions they lead to are found to agree with experiment, at least for objects whose velocity is much less than the velocity of light.

Einstein's first postulate is based on the Michelson-Morley experiment. It says:

The velocity of light will be measured the same for any observer, no matter what his state of motion. The velocity of light will be measured the same for any observer no matter what the velocity of the source of the light may be.

This postulate, if true, is sufficient to explain the results of the Michelson-Morley experiment, because it says that the motion of the apparatus will give no measurable change in the velocity of light. It leads to some paradoxical conclusions, such as that shown in Fig. 17.5. The basic idea is to start from this point and rethink our ideas about distance and time.

Einstein's second postulate is sometimes called the principle of relativity:

It is impossible to do any experiment which will determine an observer's state of absolute motion. The only thing that can be determined is the observer's motion relative to some other object.

This postulate specifically denies the possibility that there is a point which is at rest in some absolute sense. Objects can only be at rest with respect to other objects. This principle was not new with Einstein. It is also part of Newton's physics, although this was not clearly recognized until Einstein's work. In fact, as we have seen, there was much interest during the nineteenth century in using experiments with light to find the state of absolute rest. The success of any such experiment, such as the Michelson-Morley experiment, would violate the principle of relativity.

We now use these two postulates to examine the measurement of space and time when objects are moving at speeds approaching the speed of light.

17.4 TIME DILATION

Let us first look at the question of comparing two clocks—one at rest and one moving with some large velocity v. We will first discuss a rather simple clock, and then argue that all clocks must give the same result. Our simple clock will consist of two mirrors a distance L apart, with a beam of light bouncing back and forth. Your first reaction may be that this is not a clock at all. But back in Chapter 2 we pointed out that any repetitive physical phenomenon could be used to define a unit of time. In the clock just proposed, the light beam bouncing back and forth is the repetitive phenomenon. (In Chapter 18 we discuss lasers briefly. A laser behaves very much like our simple clock. In fact it has been proposed to make a fundamental standard of time using a laser.)

We want to compare two of these clocks, constructed as identically as possible. One is at rest, and the other moving with some velocity v

with respect to the first. We will orient the moving clock as shown in Fig. 17.6 for reasons that will be clearer later. The question to be answered is, Do the two clocks agree? Are time intervals as measured by one the same as time intervals as measured by the other?

The key idea is this: As measured by *any* observer, the velocity of light in the two clocks will be the same, by Einstein's first postulate. So for an observer at rest with respect to one clock the light beam will make a round trip in a time $2L/c$. (Distance divided by velocity gives time.) Let $2L/c = t_0$. Then t_0 is the time interval as measured by the observer at rest with respect to the first clock. The path of the light beam of the moving clock as seen by the same observer is shown in Fig. 17.6*b*. Clearly the light beam travels further and therefore takes longer. The exact result for the time to make this longer trip is

$$t = \frac{2L}{c}\frac{1}{\sqrt{1-v^2/c^2}} = \frac{t_0}{\sqrt{1-v^2/c^2}} \tag{17.1}$$

Here v/c is a number between 0 and 1, and therefore the quantity $\sqrt{1-v^2/c^2}$ is always 1 or smaller. Note that we have assumed in this statement that our clock cannot move faster than the speed of light. We will later show why this is so.

We therefore find that t, the round-trip time for the light beam in the moving clock, is larger than t_0. A clock in which the time between ticks is long will run slow. To our observer at rest the moving clock appears to run slow. But only if v/c is very nearly 1 will the difference be very large.

It is useful at this point to show the value of the quantities $\sqrt{1-v^2/c^2}$

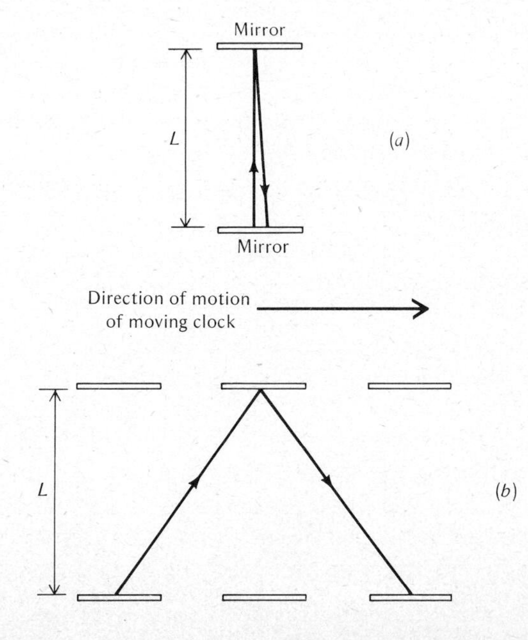

FIGURE 17.6 *(a)* Path of a beam of light in our "light clock" at rest. *(b)* Path of light beam as seen when the clock is in motion to the right with respect to the observer.

and $1/\sqrt{1 - v^2/c^2}$ for various values of v/c. Note that only for v/c quite near the velocity of light is $\sqrt{1 - v^2/c^2}$ very different from 1. This is shown in Table 17.1. The quantity $\sqrt{1 - v^2/c^2}$ occurs throughout the discussion of relativity.

Let us call our two clocks *A* and *B*. *A* is at rest and *B* is moving. If we examine the situation as seen by an observer riding with *B*, he sees clock *A* running slow because it is moving with respect to him. But an observer at rest with respect to *A* sees clock *B* running slow. That is, two observers in relative motion will disagree as to whose clock is running slow. Each believes that it is the clock which is moving (with respect to him) that is running slow. This may seem to be an utterly impossible result. The result violates your common sense. But your common sense is based on ideas of the meaning of time which Einstein showed to be false. If the velocity of light is accepted as constant to all observers, this apparently strange result appears as a natural consequence.

You may claim that it is impossible for each observer to see the other's clock run slow. What is really happening? The key to that question is in the word really. It implies that there is some fundamental reality that is more than just the measurements a scientist may make. Modern physics denies this. It assigns meaning only to questions capable of experimental measurement or verification. What we have shown is that each observer will find that the other's clock runs slow *when he makes measurements*, if we assume the constancy of the velocity of light.

You may claim that the clocks we have used are very special, and that a *real* clock, of course, would not behave this way. But if that were true, a measurement of absolute velocity would be possible, as follows. In only one coordinate system would *real* clocks and our light clock agree with one another. This coordinate system would then be unique. Velocities with respect to this unique coordinate system would be absolute velocities. One of the postulates of relativity is that no unique coordinate system, and thus no absolute velocities, can exist. Therefore *all* clocks must behave as our light clock does.

Is there any experimental evidence that moving clocks run slow? The answer is yes. An experiment has been performed using muons which come from cosmic rays. Muons are discussed in more detail in Chapters 22 and 23. For the moment let us just use the following facts. Muons are particles produced in the upper atmosphere by cosmic ray collisions. They move down toward the surface of the earth at a speed that is about 0.99 of the velocity of light. From other experiments muons are known to have a lifetime of only 2.2×10^{-6} second before undergoing radioactive decay. This means they should travel about 600 meters before decaying (since the speed of light is 3×10^8 meters/second). On the average, they are observed to travel more than 6,000 meters. This is evidence that the moving clock (the muon decay) runs slowly. While the moving clock measures 2.2×10^{-6} second, we (at rest) measure about 10 times as long a time interval, because the muon

v/c	v^2/c^2	$\sqrt{1 - v^2/c^2}$	$1/\sqrt{1 - v^2/c^2}$
0.1	0.01	0.995	1.005
0.2	0.04	0.980	1.021
0.3	0.09	0.954	1.048
0.4	0.16	0.916	1.091
0.5	0.25	0.866	1.155
0.6	0.36	0.800	1.250
0.7	0.49	0.714	1.400
0.8	0.64	0.600	1.667
0.9	0.81	0.436	2.294
0.95	0.902	0.312	3.202
0.99	0.980	0.141	7.092
0.999	0.998	0.0447	22.366
0.9999	0.9998	0.0141	70.712
0.99999	0.99998	0.0045	223.6

TABLE 17.1 Values for the quantities appearing in relativity for various values of v/c.

travels 10 times as far as we expect that it should during its lifetime. From Table 17.1 we can see that this data implies a speed of greater than $0.99c$ for the muons.

We will discuss another of the paradoxes associated with these ideas, the "twin paradox," in Section 17.8.

Example 17.1

A radioactive atom has an average lifetime of 1×10^{-3} second. What would the measured average lifetime be for atoms of this kind moving (*a*) at a speed $v = 0.6c$ and (*b*) at a speed $v = 0.99c$?

The measured time is equal to the time measured at rest t_0 divided by $\sqrt{1 - v^2/c^2}$. (Use Table 17.1 for $1/\sqrt{1 - v^2/c^2}$.)

$$t = \frac{t_0}{\sqrt{1 - v^2/c^2}}$$

(*a*) $t = (1 \times 10^{-3} \text{ second}) \times (1.250)$

$t = 1.25 \times 10^{-3}$ second

(*b*) $t = (1 \times 10^{-3} \text{ second}) \times (7.092)$

$t = 7.1 \times 10^{-3}$ second

17.5 THE LORENTZ CONTRACTION

We now turn to the question of measuring length. Suppose we take our light clock and set the mirrors exactly 1 meter apart. We can measure this distance by measuring the time it takes a light pulse to go back and forth. Now let the mirrors move parallel to the line between them, not perpendicular as before. (See Fig. 17.7.) Now we ask: What is the apparent distance between the moving mirrors? An equivalent question is: How long does a moving meter stick appear to be if it is moving parallel to its length? We use the word appear to emphasize the fact that we are measuring the length of a moving object in terms of meter sticks and clocks that are *not* moving. To an observer at rest with respect to the moving object, that is, to a moving observer, the length is still 1 meter.

By going through an analysis similar to that used for time dilation, we find that a moving meter stick appears shortened. If L_0 is its length when not moving, the length of the meter stick moving with velocity v appears to be

$$L = L_0 \sqrt{1 - v^2/c^2} \tag{17.2}$$

This is called the *Lorentz contraction*. H. A. Lorentz proposed that a contraction of this magnitude for motion through the ether would account for the result of the Michelson-Morley experiment. We constructed our light clock, with the light running perpendicular to the direction of motion to avoid difficulties with the Lorentz contraction. The Lorentz contraction is found to occur *only* for dimensions parallel to the direction of motion. Dimensions perpendicular to the motion are not affected. Please note that the Lorentz contraction is a statement about *measurements* of moving objects. The question, "Is it *really* shorter?" is not relevant, and leads to false difficulties.

So we find that, if we observe moving objects, time intervals appear longer and meter sticks appear shorter. Both of these results are absolutely required if we believe the postulates of relativity. Since the conclusions drawn from these postulates have been verified in a number of different ways, essentially everyone now believes that they are true.

Example 7.2

What is the apparent length of a meter stick moving parallel to its length at a speed of $0.9999c$?

Use Equation 17.2 and Table 17.1:

$$L = L_0 \sqrt{1 - v^2/c^2}$$

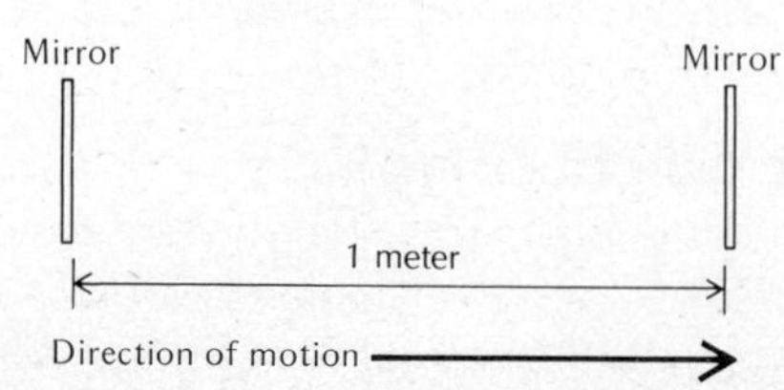

FIGURE 17.7 A meter stick moving parallel to its length.

$L = (1\text{ meter}) \times (0.0141)$
$L = 0.0141\text{ meter}$
$L = 1.4\text{ centimeters}$

17.6 COMPARISON OF VELOCITIES

We have seen that distance and time must be handled very carefully when measurements are made on systems moving with velocities approaching the velocity of light. We might suspect that the measurement of velocities themselves would also have some unusual aspects. This turns out to be true, and can be illustrated as in Fig. 17.8.

In this figure we see two rocket ships moving in opposite directions. As measured by an observer on earth, each has a velocity of $0.90c$. What velocity does an observer on rocket A measure for rocket B? According to all our experience with cars and the like, we would say $0.90c + 0.90c = 1.80c$. But the difficulties with time and distance produce a different result. When the proper equations are worked out and solved, it turns out that the velocity of B as measured by an observer on A is $0.99c$! This is true even though the observer on A still measures the relative velocity between A and the earth as $0.90c$.

In fact, the relativistic relationships for velocities lead to the conclusion that no observer can measure a speed larger than c for any object. This means that the velocity of light is the limiting velocity in our universe. We will shortly find another reason for believing this.

17.7 ENERGY AND MOMENTUM

Energy and momentum are two of the most important quantities in physics. Since each involves velocities, and velocities must be handled carefully, these two quantities may also have to be modified.

The point of view taken is that momentum conservation and energy conservation must not be abandoned, since these laws are too valuable and apparently universal. Rather, we will examine the properties of objects moving at high velocities, using the ideas already presented,

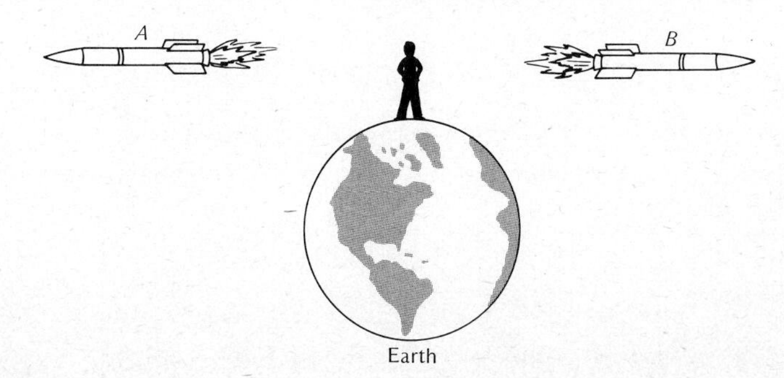

FIGURE 17.8 The observer on earth measures a velocity of $0.9c$ for each rocket. An observer on A measures $0.9c$ for the relative velocity of earth and A, and $0.99c$ for the relative velocity of A and B.

and find those quantities that behave like we believe energy and momentum should behave. These we will identify as the correct forms for energy and momentum. Obviously these forms must reduce to the ones we previously used when velocities were small. Then we will test the predictions of these new forms in experiments.

We will not go into the mathematical details, but it is possible to study collisions between high-speed atomic particles and learn what the proper forms must be for the momentum and energy. For a particle such as an electron, moving at velocity v, momentum turns out to be

$$\text{Momentum} = p = \frac{m_0 v}{\sqrt{1 - v^2/c^2}} \tag{17.3}$$

In this equation m_0 is the mass of the electron measured when it is at rest (how can the mass of something moving rapidly be measured?), and v is the velocity of the electron. Notice that, if v is much smaller than c, v/c is a small number, and v^2/c^2 is even smaller. Therefore, if v is small, $\sqrt{1 - v^2/c^2}$ is just about equal to 1 (see Table 17.1). Then Equation 17.3 is identical to the result for momentum we discussed previously in Chapter 5.

Equation 17.3 is frequently written

$$p = mv \tag{17.4}$$

Here m is called the *relativistic mass*, and is given by

$$m = \frac{m_0}{\sqrt{1 - v^2/c^2}} \tag{17.5}$$

Since $\sqrt{1 - v^2/c^2}$ is always 1 or smaller, m is equal to or larger than m_0. This result is frequently described as an increase in mass as the velocity of a particle (or anything else) becomes large. This is a useful and convenient point of view. However, it is not the only way to look at the result. Another point of view is that, for fast-moving particles, momentum and velocity are not simply related through a quantity called the mass, but in a more complex way. The point to this remark is simply that the term relativistic mass increase is useful only if one defines mass as that quantity which is multiplied by velocity to obtain momentum.

ENERGY

As you might expect by now, energy also has a somewhat different form when large velocities are considered than the form you have previously learned. The form found necessary is

$$E = \frac{m_0 c^2}{\sqrt{1 - v^2/c^2}} \tag{17.6}$$

It appears that this equation does not resemble at all the result $E = \frac{1}{2}mv^2$

you have previously used. But we will see that indeed it does, almost.

Physicists frequently find it useful to find nearly correct, or approximate, values for complicated mathematical expressions. If the quantity v/c is small compared to 1, the quantity $1/\sqrt{1 - v^2/c^2}$ is approximately equal to

$$\frac{1}{\sqrt{1 - v^2/c^2}} \cong 1 + \frac{1}{2}\frac{v^2}{c^2} + \frac{3}{8}\frac{v^4}{c^4} \cdots \tag{17.7}$$

The dots indicate that there are other smaller terms that we have not written down. Using Equation 17.7 we find that

$$E \cong m_0c^2 \left(1 + \frac{1}{2}\frac{v^2}{c^2} + \frac{3}{8}\frac{v^4}{c^4} \cdots\right) \tag{17.8}$$

$$\cong m_0c^2 + \frac{1}{2} m_0v^2 + \frac{3}{8} m_0 \frac{v^4}{c^2} \cdots$$

The second term in this equation is our old friend kinetic energy. The third term is smaller than the second. But what is the first term? It seems to imply that even at zero velocity an object of mass m_0 has an amount of energy m_0c^2. This is called the *rest energy*. The implication is that, in some way, mass and energy are related. An object at rest has an amount of energy determined by its rest mass.

Because c is such a large number, the amount of rest energy is very large. Using $c = 3 \times 10^8$ meters/second, the energy equivalent of 1 kilogram of matter is given by

$$E = m_0c^2$$

$$= (1 \text{ kilogram}) \times (3 \times 10^8 \text{ meters/second}) \times (3 \times 10^8 \text{ meters/second})$$

$$E = 9 \times 10^{16} \text{ joules}$$

This amount of energy would raise roughly 2×10^{11} kilograms of water (about 10^8 tons) from its freezing point to its boiling point!

The question is: Is this rest energy ever observable? And if so, how? We have learned previously about energy conservation—that energy is never created or destroyed. It is merely rearranged from one form to another. If mass is a form of energy, then possibly it can be rearranged to other forms of energy. The conversion of mass to other forms of energy has been found experimentally. It occurs in the processes that occur in the nuclei of atoms, called *radioactivity*. We will study these processes in more detail in Chapter 21. For now, let us just say that in a radioactive process the products of the reaction have less mass than the starting material. This difference shows up as kinetic energy of the products. In fact, the same is true of ordinary chemical processes, but the amount of energy released is about 1 million times smaller than in radioactivity. The mass changes in chemical processes are too small for our measuring instruments.

If we look at Equation 17.6 for energy, we see that, as the velocity of an object approaches the velocity of light ($v/c = 1$), the energy becomes very large because the term $\sqrt{1 - v^2/c^2}$ becomes very small (see Table 17.1).

$$E = \frac{m_0c^2}{\sqrt{1 - v^2/c^2}} \tag{17.6}$$

In fact, if $v/c = 1$, the energy becomes infinitely large. Since we still believe that work goes to increase energy, this means that it would require infinite work to bring any object whose rest mass was not zero to a velocity equal to c. This is another reason to believe that c is the limiting velocity in the universe.

17.8 THE "TWIN PARADOX"

One of the most frequently discussed paradoxes produced by relativity is the "twin paradox." Imagine a pair of twins, one of whom remains on earth while the other takes a trip in a rocket at very high speed. When the rocket returns, which twin will be older?

We must remind ourselves that the biological processes of the body also represent a clock. We have showed that all clocks must behave identically, so whatever we said about the light clock must also apply to bodily processes.

The twin on earth sees his brother's rocket moving with very high speed, so he concludes that his brother's biological clock will run slow. As far as he is concerned, the traveling twin will be the younger. But the twin on the rocket observes that the velocity of earth relative to the rocket is very large. Therefore he concludes that the clock on earth will run slow and that his brother who stayed on earth will be younger. Obviously both brothers cannot be right? So what is going on?

FIGURE 17.9 Which is the correct description for the twin paradox, *(b)* or *(c)*?

In order to actually compare ages, it is clear that the traveling brother must return to earth. So the problem loses the symmetry it had in the first simple statement. The brother in the rocket must turn around and return. So it is possible to tell which brother took the trip, which a simple statement about relative velocities will not allow. When a careful analysis is carried out, it becomes clear that the traveling brother is the younger by precisely the amount predicted by our time dilation formula.†

17.9 GENERAL RELATIVITY AND THE EQUIVALENCE PRINCIPLE

The special theory of relativity deals only with velocities, not accelerations. After the publication of the special theory, Einstein turned his attention to what is known as the general theory of relativity.

It might be assumed that an observer can always tell when he is being accelerated; that is, accelerations are absolute and not relative. If we accelerate our car, we feel forces, and objects may slide off the dashboard. But consider and compare two situations. First we look at an astronaut sitting in his rocket on the launch pad on earth. He feels a force due to gravity, and objects dropped in the rocket fall with an acceleration given by g. Now let the same astronaut be far out in space, away from the earth and the sun. He fires his rockets so that the acceleration of the rocket is exactly g. He will feel forces, just as he did on earth, and objects dropped in the rocket will fall with acceleration g. Einstein pointed out that, if there are no windows in the rocket, there is no experiment the astronaut can do that will tell him whether he is sitting still on earth or in outer space being accelerated. The two situations are equivalent.

Einstein used these ideas to state what is called the *equivalence principle*. This principle can be stated:

> *No experiment is possible that will distinguish between the effect of a uniform gravitational field and the effect of a constant acceleration.*

There are several immediate consequences of this principle, some reasonably obvious and some not so much so. We have previously written down two laws involving masses. One is Newton's second law:

$$a = \frac{F}{m} \tag{17.10}$$

The other is Newton's law of gravity:

†See, for example, R. Resnick: "Introduction to Special Relativity," John Wiley and Sons, New York, 1968, p. 201.

$$F = G \frac{m_1 m_2}{R^2} \quad (17.11)$$

Notice that these are two totally different uses of the term mass. It has appeared up to now that accidentally these two uses of mass are identical. The equivalence principle requires that this be true. The terms used for the two quantities are *gravitational mass* (Equation 17.11) and *inertial mass* (Equation 17.10). The equivalence of gravitational mass and inertial mass has been tested experimentally to an accuracy of 1 part in 10^{11}, most recently by Robert Dicke at Princeton University.

Another consequence of the equivalence principle is that light beams must be bent in a gravitational field. Light, although it has no mass, must be affected by and "fall" in a gravitational field. The experimental test of this proposition involves measuring the path of light from stars that has passed very close to the sun during an eclipse. Six months later the same stars are photographed, with the sun gone, and the two photographs compared. The results of these experiments also agree with the equivalence principle. This is illustrated in Fig. 17.10. (The exact numerical result is somewhat different from that calculated using the equivalence principle.)

A further consequence of the equivalence principle is the so-called *gravitational red shift*. Light emitted at the surface of the earth will lose energy as it "climbs" up the earth's gravitational field. We will see in Chapter 18 that the energy of light is related to its frequency, and thus to its color. Light that loses energy is shifted in color toward the red end of the spectrum. For small distances at the surface of the earth, this is a very small effect. For a vertical distance of about 100 feet it is only a little more than 1 part in 10^{15} in wavelength or frequency. Nevertheless, using an effect called the *Mössbauer effect*, two Harvard scientists were able to make this measurement and demonstrate the existence of the gravitational red shift.

Einstein also created a theory of gravity within the framework of general relativity. In this theory the effects of gravity propagate with the speed of light (not infinite speed, as implicit in Newton's theory of gravity). One of the major areas of physics and astronomy today involves studying the detailed predictions of this theory in a variety of ways. Delicate experiments in earth satellites, telescopic study of distant galaxies, and many new theoretical predictions are all involved in this work.

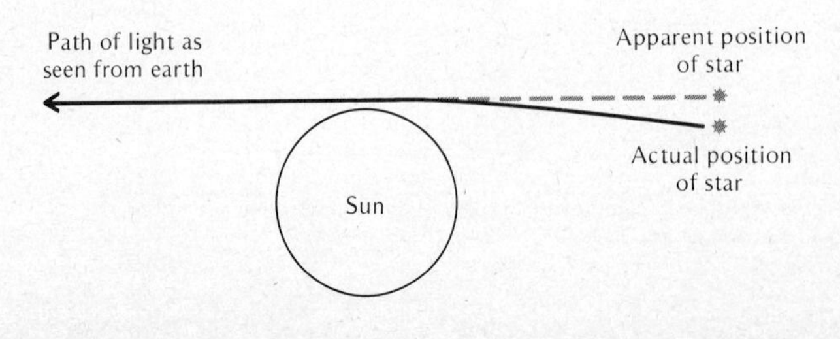

FIGURE 17.10 Light passing near the sun is bent through a small angle. This shifts the apparent position of the star as shown.

SUMMARY

In physics, we always measure velocities relative to some reference frame such as the earth, the sun, the center of the galaxy, etc. We cannot measure absolute velocity.

The Michelson-Morley experiment showed that the velocity of light is independent of the motion of the observer through the ether. Therefore it was suggested that the ether does not exist.

Einstein's postulates of relativity are:

1 The velocity of light will be measured the same for any observer, no matter what his state of motion. The velocity of light will be measured the same for any observer, no matter what the velocity of the source of light may be.

2 It is impossible to do any experiment which will determine an observer's state of absolute motion. The only thing that can be determined is the observer's motion relative to some other object.

Moving clocks appear to run slow, compared to clocks at rest with respect to the observer. The magnitude of this effect is given by the time-dilation expression

$$t = \frac{t_0}{\sqrt{1 - v^2/c^2}}$$

Moving objects appear, to an observer at rest, shortened in the direction of their motion. The magnitude of this Lorentz contraction is given by

$$L = L_0\sqrt{1 - v^2/c^2}$$

Consideration of energy and momentum lead to the result that the mass of an object appears to increase as its velocity increases. This is given by

$$m = \frac{m_0}{\sqrt{1 - v^2/c^2}}$$

The energy of an object moving very fast can be written as

$$E \cong m_0c^2 + \frac{1}{2}m_0v^2 + \frac{3}{8}m_0\frac{v^4}{c^2} \cdots$$

The term m_0c^2 is called the rest energy. Because c^2 is large, the rest energy of any object is very large.

The equivalence principle of Einstein is:

No experiment is possible that will distinguish between the effect of a uniform gravitational field and the effect of constant acceleration.

QUESTIONS

1. Astronauts in orbit around the earth travel at about 8,000 m/second. By how much do their clocks run slow as observed from earth?
2. If you stand in a truck moving at 30 miles/hour and throw a baseball forward at 30 miles/hour, the speed of the baseball with respect to the truck is 60 miles/hour. If we imagine shining a light forward from the truck, does the velocity of the truck add to the velocity of the light when the velocity of light is measured with respect to the road? Why?
3. How is it possible for each of two observers to conclude that the other's clock runs slow?
4. Why do we not observe relativistic effects in everyday life?
5. On page 310 there is a discussion of what science means by reality. In what way does this differ from your own ideas of reality?

SELECTED READING

Mermin, N. David: "Space and Time in Special Relativity," McGraw-Hill Book Company, New York, 1968. A very detailed discussion of the basic principles and results of relativity, with a minimum of mathematics.

SELF-TEST

__________ Michelson-Morley experiment
__________ ether
__________ time dilation
__________ Lorentz contraction
__________ rest energy
__________ equivalence principle
__________ gravitational mass
__________ inertial mass

1. Moving clocks appear to run slow
2. Energy associated with the mass of an object
3. The property of mass which resists acceleration
4. Medium in which light waves were assumed to travel
5. Showed that the ether does not exist
6. The property of mass that leads to gravitational attraction between masses

7 Meter sticks moving in the direction of their length are measured to be shorter than 1 m

8 The effect of a constant acceleration cannot be distinguished experimentally from the effect of a uniform gravitational field

PROBLEMS (Use Table 17.1 if needed)

1 A high-energy particle is moving at speed $v = 0.9c$. If its lifetime when at rest is 1.0×10^{-6} second, how long will the lifetime of the moving particle appear to be?

2 A rocket ship 200 m long (at rest) is moving with a velocity of $0.8c$. How long does it appear to be to an observer at rest?

3 An astronaut travels in his spaceship at a speed of $0.9c$. If his rest mass is 80 kg, what is his apparent mass when he travels at this speed?

4 An observer notes that his watch changes its reading by 20 minutes, while a moving clock changes its reading by only 12 minutes during the same interval. What is the speed of the moving clock with respect to the observer?

5 The rest mass of an electron is about 9×10^{-31} kg. What is its apparent mass if its speed is (*a*) $0.8c$? (*b*) $0.9c$? (*c*) $0.99c$?

6 Two pi mesons (pions) are created, one at rest in the laboratory, and the other moving at speed $v = 4c/5$ with respect to the laboratory. Each decays in 2.5×10^{-8} second in its own rest frame. (*a*) Find the lifetime of the moving pion as measured in the laboratory. (*b*) Find the distance the moving pion travels in the laboratory before it decays. (*c*) What is the lifetime of the pion at rest in the laboratory as viewed by an observer perched on the moving pion?

7 A spaceman looks out the window of his rocket ship and sees a rod of length 1.5 m (measured by him, of course, as it moves by) flash by him at a velocity of $4c/5$. (*a*) How long would the spaceman claim it took the rod to pass a point on his ship? (*b*) An observer is sitting on top of the rod. What length does he claim the rod to be? (*c*) How long would the observer on the rod claim it took his rod to pass the point on the spaceship?

8 The disk of the Milky Way galaxy is about 10^5 light years in diameter. A cosmic ray proton enters the galactic plane with speed $v = 0.99c$. (*a*) How long does it take the proton to cross the galaxy from our viewpoint? (*b*) How long does the proton think it takes? (*c*) How wide is the galaxy to the proton (in its direction of motion)?

9 A physicist makes observations of the lifetime of rapidly moving particles in his laboratory. He finds that the particles move a distance of 10 m in a time of 5×10^{-7} second, before they decay into other particles. (*a*) What is the velocity of these particles relative to the physicist? (*b*) What distance would an observer moving with these particles claim he had traveled? (*c*) What is the lifetime of these particles as measured by an observer at rest with respect to them?

10 An observer measures the length of a rod moving at a speed of $3c/5$ in the direction of its length. The rod is 2 m long when measured at rest. (*a*) What is the length of the rod as measured by the observer? (*b*) How long does the rod take to pass him? (*c*) If the rod were perpendicular to its direction of motion, what would the observer measure as its length?

11 In the fusion of four protons to form a helium nucleus in the center of the sun about 5×10^{-29} kg of mass is converted into energy. (*a*) How much energy is liberated in the process? (*b*) One kilogram of hydrogen contains about 6×10^{26} protons. How much energy would be liberated by fusion if all the protons in 1 kg of hydrogen reacted? (*c*) How many 100-watt light bulbs could be powered for 1 year by this energy?

THE PHOTO-ELECTRIC EFFECT

The next topic of twentieth-century physics to be discussed is an effect which revived the particle view of the nature of light, apparently put to rest 100 years before by Young's experiments on interference. This phenomenon is called the *photoelectric effect*, and its first successful explanation was given by Albert Einstein (1879–1955) in 1905.

FIGURE 18.1 Albert Einstein (1879–1955.) *(Courtesy American Institute of Physics.)*

18.1 THE EXPERIMENTAL RESULTS

The photoelectric effect was first observed by H. Hertz in 1887 in the course of experiments which demonstrated that Maxwell's equations for electromagnetic radiation were valid. Hertz observed that light, particularly ultraviolet light, striking a metallic surface changes the electrical properties of the air surrounding the surface. He found that air became a conductor and allowed charge to leak away from his apparatus. This effect can also be demonstrated using a charged electroscope, as shown in Fig. 18.2. Light shining on the ball of a negatively charged electroscope discharges it. The effect is observed to be more pronounced, the more active, in a chemical sense, the metal of the electroscope on which the light shines. Only negative charge is emitted from the metal, since only the negatively charged electroscope is discharged.

At the time of its discovery the photoelectric effect was an experimental nuisance, for it interfered with the proper functioning of Hertz' apparatus. Once again, as on so many occasions, a new effect was discovered as an adjunct to a totally different experiment. Fortunately, the investigator realized that something new, different, and in need of explanation was happening. The ability to sort such wheat from the chaff of other experimental nuisances that are not significant is one mark of a truly great experimental scientist.

Hertz' announcement of the photoelectric effect produced a flurry of investigation, and certain facts were very quickly established. The air becomes conducting because negative electric charges are emitted from the illuminated metal surface, as shown in Fig. 18.3. Each metal behaves somewhat differently. Sodium, potassium, and other alkali metals emit negative charges even in visible light. Less chemically reactive metals show the effect only when illuminated by ultraviolet light.

The apparatus which evolved in the hands of a number of observers for the detailed study of the photoelectric effect resembles in many ways that used in the study of cathode rays, except that provision is made for light to strike the electrode whose photoelectric effect is being studied. Figure 18.3 is a schematic drawing of this apparatus. Its essential features are the following. (1) A glass bulb containing the electrodes, which can be evacuated to a high vacuum if desired. (2) One electrode, called the cathode, which can be illuminated by light and emits the photoelectric carriers. Note that in other situations the term cathode is always used for the negative electrode. Here we shall use it

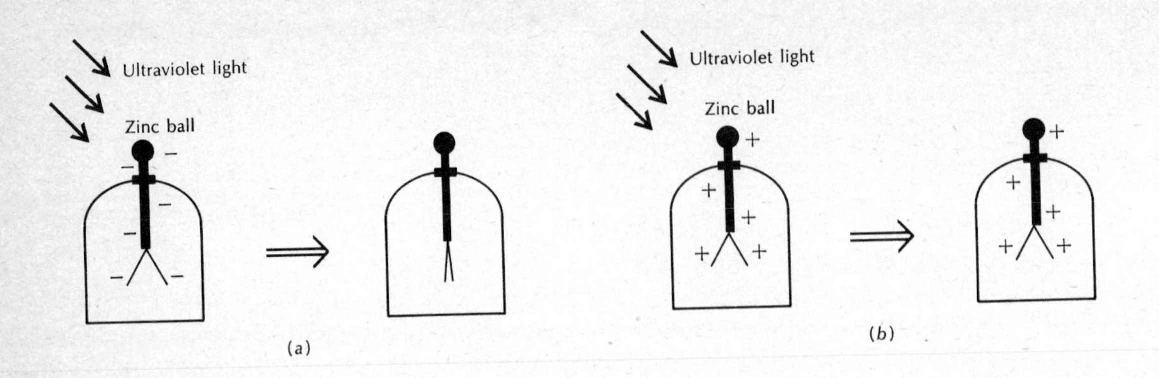

FIGURE 18.2 (a) Ultraviolet light on a negatively charged electroscope causes it to lose charge by the emission of electrons. (b) Little or no change is observed when ultraviolet light shines on a positively charged electroscope. Any electrons emitted are attracted back to the electroscope by its positive charge.

for the electrode which *emits* the negative carrier of electricity. When a battery is connected between the two electrodes, the cathode can be made either positive or negative. (3) A second electrode, the anode. (4) An arrangement whereby the electric potential, or voltage, between the two electrodes can be varied. This has been indicated in the diagram as a battery. Normal potentials for these experiments are small, not more than a few volts. (5) A galvanometer or other current-measuring device for measuring the small electric currents. Current is the total charge passing a given point per second. Each photocarrier, which is an electric charge emitted from the cathode because of the action of light, carries a certain charge from cathode to anode. Therefore the current is proportional to the number of photoelectric carriers emitted from the cathode and collected at the anode per second.

The first problem to be settled was the nature of the charge carriers in the photoelectric effect. Two immediate possibilities existed. Gas molecules in the region between the electrodes could be acting as charge carriers, or atoms from the cathode could be broken loose in some way and carried to the anode. The first possibility was eliminated very early by the discovery that the photoelectric effect was not modified by pumping air out of the bulb. This seemed to eliminate the possibility that gas atoms, or ions, were involved in the process. By constructing a cathode of a sodium amalgam (sodium metal dissolved in mercury),

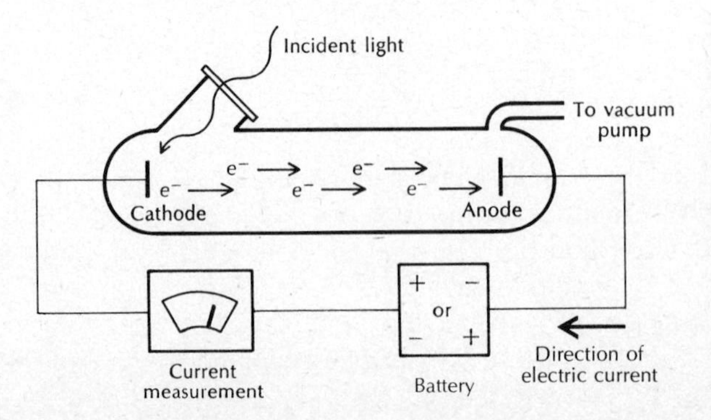

FIGURE 18.3 Schematic drawing of apparatus for studying the photoelectric effect. Electrons emitted from the cathode are collected at the anode, and counted by measuring the current. The battery can be inserted so that either the cathode or the anode can be made the positive terminal.

Philipp Lenard showed in 1900 that the materials of the cathode were not involved in the process. He allowed the photoelectric current to flow for a sufficiently long time so that, if the sodium atoms were the charge carrier, measurable amounts of sodium would be found deposited on the anode, as in an electrolysis experiment. No sodium was found. Thus both of the first ideas regarding the nature of the charge carriers were eliminated.

PHOTOELECTRONS

In 1897, Thomson measured the ratio of charge to mass for the electron in cathode rays. It was by then natural to suspect the possibility that the photoelectric charge carriers were electrons, since other likely possibilities had been eliminated. Several different scientists studied this possibility. The results showed that the charge-to-mass ratio of the photoelectric carriers and that of cathode rays are identical within experimental error. Although this did not eliminate the possibility that a particle with the same ratio of charge to mass as the electron, but of different charge and mass, was involved, it made a strong case for the idea that the photocarriers are electrons. No other particle was then known that had the value of charge to mass known for the electron. The photocarriers thus are called *photoelectrons.*

THE CURRENT-INTENSITY RELATIONSHIP

The most interesting experiments involving photoelectrons were quantitative studies of the relationship between the intensity and wavelength of the incident light and the magnitude of the photoelectric current and energy of the photoelectrons. The intensity of light is a measure of the amount of energy in the light beam. Intensity can be defined in terms of the energy arriving at a unit area of surface per second. The intensity of a beam of light can be varied by moving the source farther away, so that a smaller fraction of its total output strikes the surface, or by inserting appropriate filters in the light beam.

It was found that over a range of light intensities of 50 million to 1, the photoelectric current, and therefore the *number* of photoelectrons, is directly proportional to the intensity of the incident light. Therefore the number of photoelectrons ejected from the cathode is directly proportional to the rate at which light energy arrives at the metal surface.

KINETIC ENERGY OF PHOTOELECTRONS

The surprise came when the kinetic energy of photoelectrons was studied. When electrons leave the metal surface, they must have some kinetic energy for, if they left the surface with zero velocity, they would not move away from the metal of the cathode. The energy of photoelectrons was determined by studying the photocurrent while varying the voltage between the anode and the cathode.

In order to understand this technique, it is necessary to recall some of the results in Chapter 10. We defined the voltage V between points A and B in terms of the work done in moving a unit charge between A and B. If an electron of charge e is moved through a potential V, a certain amount of work eV must be done. This work increases the kinetic energy of the electron if the electric field produces a force in the direction of its motion, or decreases the kinetic energy if the force is in the opposite direction, as seen in Fig. 18.4*a* and *b*. In the photoelectric experiments the direction of the electric force is determined by whether the anode is made positive or negative with respect to the emitting plate, the cathode. The battery placed in the circuit of Fig. 18.3 creates a voltage between the anode and the cathode. Since the terminals of the battery can be connected in either direction, the voltage can be either positive or negative. The work done on an electron moving from the cathode to the anode is then given by

$$W = eV \tag{18.1}$$

If the potential difference is positive, as in Fig. 18.4*a* (anode positive with respect to cathode), the work done on the electron is eV and *increases* the kinetic energy. If the potential difference is negative, as in Fig. 18.4*b* (anode negative with respect to cathode), the work done on the electron *decreases* the kinetic energy and slows its motion. In particular, there must be a value of V such that eV is exactly equal to the kinetic energy of an electron as it leaves the emitting plate. This value of the voltage reduces the kinetic energy of an electron to zero if the anode is made negative with respect to the cathode. The velocity of the electron is then zero, and the positive charge of the cathode pulls it back. Such an electron will not reach the anode and will not contribute to the measured current. Therefore, if the voltage between the electrodes is V, the minimum kinetic energy required for an electron leaving the cathode, in order to reach the anode, is

$$\tfrac{1}{2}mv^2 = eV \tag{18.2}$$

If a photoelectron does not reach the *receiving plate*, it will not be part of the current measured in the external circuit.

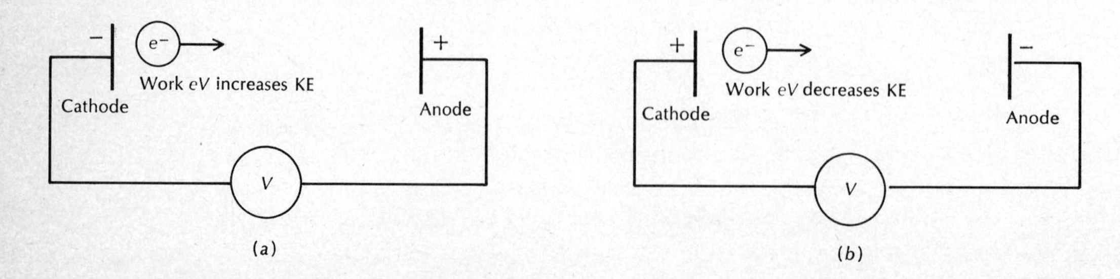

FIGURE 18.4 (*a*) The electric work, eV acts to increase the kinetic energy of the photoelectrons when the anode is made positive with respect to the cathode. (*b*) The electric work eV acts to decrease the kinetic energy of the photoelectrons when the anode is made negative with respect to the cathode.

We have seen that the units of kinetic energy are joules. If e is in coulombs and V is in volts, volts times coulombs is a unit of energy. Frequently, the *electron-volt* (eV) is a unit of energy used in problems of this kind. This is the energy acquired by one electron accelerated by a potential of one volt. Since the charge on the electron is 1.6×10^{-19} coulomb (Chapter 14), 1 electron-volt must equal 1.6×10^{-19} joule.

The student should note the similarity between the photoelectric experiment and a stone thrown in the air. The stone is thrown upward with some velocity, and thus some kinetic energy. The force on the stone due to gravity acts to slow it. When the work mgh done by gravity on the stone exactly equals the initial kinetic energy, the kinetic energy and velocity will be zero and gravity will start it in motion back toward the earth. If, however, the stone is initially thrown downward, the work done by gravity adds to the initial kinetic energy given to the stone. The only difference in the electrical problem is that we can vary the electric potential V, and thus the force, whereas we cannot easily vary the force of gravity on a stone.

Experiments on the energy of photoelectrons are carried out by changing the voltage between the cathode and the anode of Fig. 18.3. As the anode is given a larger and larger negative voltage with respect to the cathode, fewer and fewer photoelectrons are able to reach it. This is interpreted to mean that the photoelectrons have a variety of energies and that only those with energies greater than eV reach the anode. Finally, when V is made sufficiently large, no photoelectrons at all are observed to reach the anode. This value of V, which we designate V_0, measures the maximum energy of the photoelectrons under the given conditions. Using this principle, we can thus study the maximum kinetic energy of photoelectrons at varying conditions of intensity and frequency of incident light. A plot of photocurrent versus V is given in Fig. 18.5 to illustrate the foregoing discussion. This also shows the increase in photocurrent, or number of photoelectrons, as the intensity of light is increased from curve A to B to C. The maximum kinetic energy of the

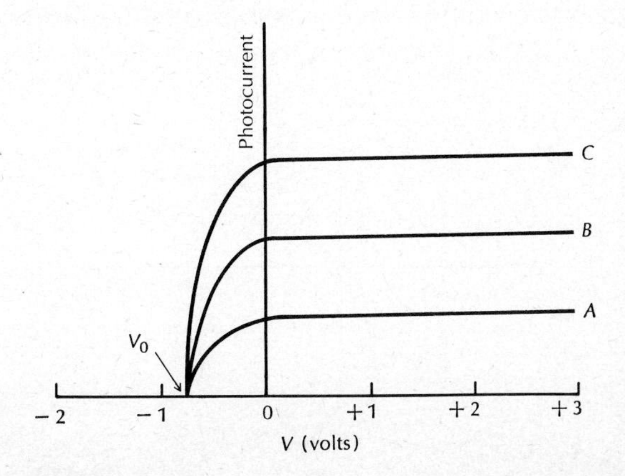

FIGURE 18.5 The photocurrent, and therefore the *number* of photoelectrons, increases with the intensity of light. In this graph, curve C corresponds to the highest intensity of light, curve B the next, and curve A the least intensity. Note that V_0, the voltage required to stop all photoelectrons, does not change with intensity. This means that the maximum kinetic energy of the photoelectrons does not change with the intensity of the light.

photoelectrons is given by $\frac{1}{2}mv^2 = eV_0$ and does not depend on the intensity of the light, as seen in Fig. 18.5.

In terms of the wave theory of light discussed in Chapters 15 and 16, the expectation is that the maximum energy of photoelectrons should be proportional to the intensity of the light. Remember, the intensity of a beam of light is directly proportional to the energy in the beam. The observed maximum energy does not depend at all on the intensity of the light, as can be seen in Fig. 18.5. However, it is found to depend on the wavelength of the light. The shorter the wavelength of the incident light, the greater the maximum energy observed for the emitted photoelectrons. Ultraviolet light produces more energetic photoelectrons than does violet light, which produces more energetic ones than blue light. For wavelengths longer than some critical wavelength, found to depend *only* on the cathode material, *no* photoelectrons are observed, no matter how intense the light. This is shown in Fig. 18.6, where the photocurrent for a metal such as sodium or potassium is shown for several colors of light. For this example, no photoelectrons are observed for wavelengths longer than 6.6×10^{-7} meter (red).

TIME DELAY

There is one further experimental fact of great importance. According to the wave theory of light, the energy in the light wave is uniformly distributed over the wavefront. It seems reasonable to assume that, in order to eject a photoelectron of a given energy, that amount of energy must in some manner be accumulated, or concentrated, in the vicinity of one electron. Presumably, the region would be about the size of one atom, approximately 10^{-10} meter in diameter. It is possible to calculate how long this should take, using the minimum light intensities for which the photoelectric effect had been observed. The calculated results indicated that it should take from 1 minute to 100 days before enough energy is received in an area the size of an atom to eject a photoelectron with the observed energies. The differences between these values depend on certain assumptions of the calculation, but the fundamental conclusion is that, in light of low intensity, appreciable delays should

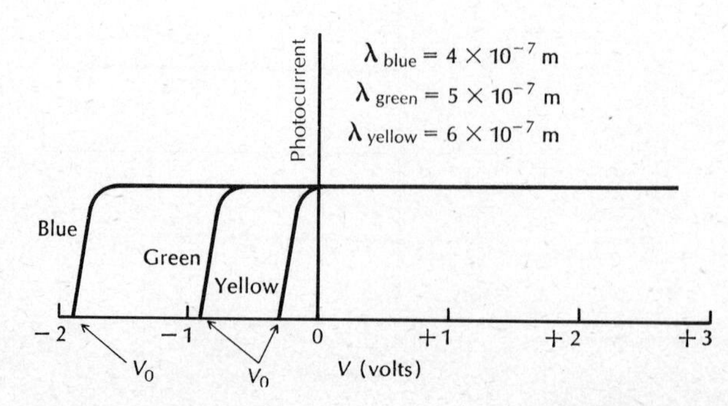

FIGURE 18.6 This graph shows the change in maximum kinetic energy of the photoelectrons with the wavelength of light. eV_0 is the maximum kinetic energy observed for the photoelectrons. This increases as the wavelength of light becomes shorter in going from yellow to green to blue. The curves shown are approximately correct for the metal sodium.

occur before any photoelectrons are observed. Very precise experiments indicate that, if any delay occurs, it is less than 3×10^{-9} second. This vanishingly small time delay is very difficult to account for using the wave model for light.

These three facts, (1) that the photocurrent, and therefore the number of photoelectrons, is proportional to the intensity of the light, (2) that the maximum kinetic energy increases as the wavelengths of light become shorter, or their frequencies larger, and (3) that no delay occurs when weak illumination is used, are the keys to understanding the photoelectric effect. These are experimental facts, and any model for the nature of light must explain them. Particularly the last two are completely at odds with the expectations of the wave model of light. According to the wave model, the energy in a light beam is proportional to the intensity, not the frequency, and at low light intensities significant delays would occur in the emission of photoelectrons.

18.2 THE EINSTEIN HYPOTHESIS

An explanation of the photoelectric effect was suggested by Albert Einstein in 1905. His idea was, essentially, to revive a particle theory for the nature of light. Young, nearly 100 years earlier, had presumably killed this theory. Einstein proposed that light consists of massless particles, or photons, and that the energy of each photon is proportional to the frequency of the light.

This model accounts for the observation that the maximum energies of photoelectrons are proportional to the frequency of the light. It can also explain the lack of any appreciable delay since, if light is a shower of particles, as soon as even one particle of sufficient energy strikes the metal surface, it can eject a photoelectron. If each photon has the same probability of ejecting a photoelectron, Einstein's hypothesis also shows why the photocurrent is proportional to the intensity of the light. One merely assumes that the intensity of a light beam is a measure of the number of photons in the beam.

Einstein's hypothesis is an extension of ideas proposed by Max Planck 4 years earlier to account for the distribution of wavelengths in radiant energy from a hot object. Planck showed that this distribution could be accounted for if the energy associated with light was proportional to its frequency. Einstein used the same proportionality constant between energy and frequency as Planck did, with the result that the energy of a photon E is related to the frequency of the light f by

$$\text{Photon energy} = \text{frequency} \times \text{a constant}$$

or

$$E = hf \tag{18.3}$$

The constant h is called *Planck's constant.*

Einstein suggested that each photon gives its total energy hf to *one* electron in a metal. If a certain amount of work ϕ must be done to remove this electron from the metal, the maximum kinetic energy that the emitted electron could have would be $hf - \phi$. Some electrons might dissipate an additional portion of their energy as heat before leaving the metal surface. Einstein therefore proposed an equation relating the *maximum* observed kinetic energy of photoelectrons to the frequency of the light and the work ϕ necessary to remove an electron from the metal. This equation is

Maximum electron kinetic energy = photon energy − work to remove electron

$$\left(\tfrac{1}{2}mv^2\right)_{max} = hf - \phi \qquad (18.4)$$

This can also be expressed in terms of the wavelength of the light.

$$\left(\tfrac{1}{2}mv^2\right)_{max} = h\left(\frac{c}{\lambda}\right) - \phi \qquad (18.5)$$

Here ϕ is characteristic of each cathode material. The physical content of Equation 18.4 is as follows. The maximum kinetic energy $\left(\tfrac{1}{2}mv^2\right)_{max}$ is given by the energy of one photon of light hf minus whatever energy is needed to remove the electron from the metal ϕ. If we plot on a graph the maximum observed energy of photoelectrons vertically and the frequency of the incident light horizontally, Equation 18.4 should yield a straight line whose slope is h (see Section A.5 in the appendix). At the time Einstein proposed this result the available data were not good enough to establish its accuracy. The American physicist Robert A. Millikan, in 1916, performed a careful series of experiments which verified Einstein's equation and provided for a number of years the best experimental value for h. A plot of some data of this kind is given in Fig. 18.7 for sodium, aluminum, and gold. The slope of each line is h. The currently accepted value for h is 6.63×10^{-34} joule-second. In 1921 Einstein was awarded the Nobel Prize in physics for his work on the photoelectric effect.

What conclusions can be drawn from the photoelectric effect and Einstein's interpretation of it? Young's experiment on the interference of light clearly showed that light behaves just like any other wave phenomenon. Only the view of light as waves seems adequate to account for these experiments. On the other hand, Einstein's theory of the photoelectric effect is based on a model which says that the energy in a beam of light is concentrated in bundles of energy, or photons. His model is confirmed by experiment, in the sense that the relationship between energy and frequency predicted by the model agrees with experimental observation. Thus there appear to be two utterly irreconcilable views as to the nature of light. The present view is that light is both wavelike

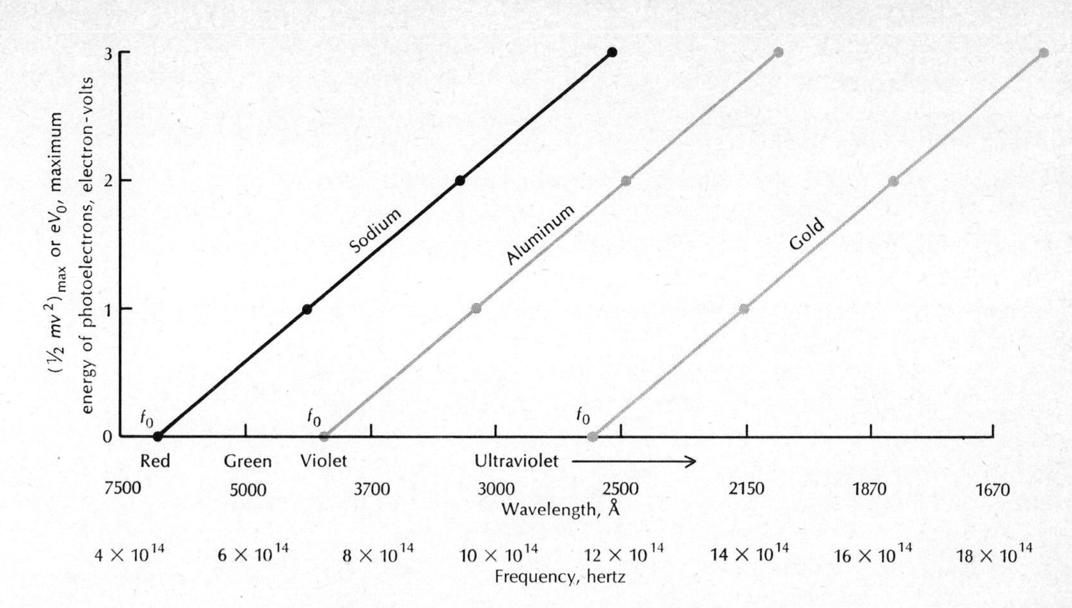

FIGURE 18.7 A plot of maximum kinetic energy of photoelectrons as a function of wavelength or frequency of light for the metals sodium, aluminum, and gold. The slope of each line depends only on Planck's constant, and therefore all are the same. The frequency f_0, below which no photoelectrons are observed, is lowest for sodium and highest for gold because the work required to remove an electron is smallest for sodium and largest for gold.

and particlelike in nature. Whether one deals with light as a wave or a particle depends on the particular experimental situation. In general, experiments at high frequencies or very low intensities are conducive to the observation of particlelike effects, and experiments at low frequencies and high intensities to the observation of wavelike effects. Either type of description, when properly formulated, can be used, however, to account for any type of experiment.

Figure 18.8 is an attempt to indicate pictorially how light can be at one and the same time a wave and a particle. It shows a group of waves, called a *wave packet*. Such a group can be observed when a stone is thrown in a pond. A wave packet can be considered a particle, in the sense that it is localized in a region of space. But it will show wavelike characteristics because of its wave structure. In Chapter 20 we shall discuss wave-particle relationships further.

18.3 THE PRODUCTION OF X-RAYS

Another piece of evidence that light consists of photons whose energy is related to their frequency through *Einstein's frequency condition* $E = hf$ can be found in the production of x-rays. In Chapter 16 it was pointed out that x-rays are generated at the anode of a cathode-ray tube and that they have very short wavelengths. An important relation-

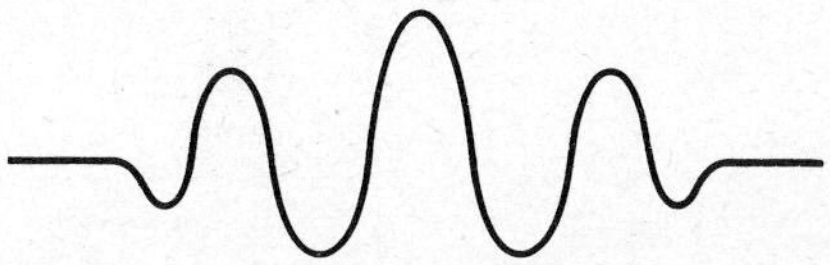

FIGURE 18.8 A wave packet or group of waves. The group is localized in space, and to this extent is an entity which can be called a particle, but the group will still show wavelike behavior.

ship exists between the voltage across the x-ray tube and the wavelengths of the x-rays found.

Figure 18.9 shows an x-ray tube. The target anode is shaped so as to emit x-rays in the most useful direction. Electrons are accelerated from the cathode to the anode by the high voltage across the tube, usually 5,000 volts or more. When they strike the anode, x-rays are emitted. The production of x-rays can be looked on as a photoelectric effect in reverse. In the photoelectric effect, photons of light give their energy to electrons; in the production of x-rays, electrons give their energy to photons.

If electrons give their energy to create x-ray photons, there should be a relationship between the energy of the x-ray photons and the electrons in the tube. Just such a relationship is found.

Electrons accelerated through a potential V acquire an energy $E = eV$. The maximum energy of the x-ray photons, as measured by their wavelength, is found to be precisely this value. That is,

$$E_{\text{photon(max)}} = hf = \frac{hc}{\lambda} = eV \tag{18.6}$$

This relationship has been verified experimentally.

The significance of Equation 18.6 is that it indicates that each electron striking the anode can at best give to an x-ray photon all its energy. The fact that the energy of the photons is shown to be proportional to their frequency and inverse to their wavelength bears out the photon hypothesis and Einstein's frequency condition.

18.4 OTHER EVIDENCE FOR THE PARTICLE NATURE OF LIGHT

A great deal of space has been devoted to the photoelectric effect and the production of x-rays in order to demonstrate the particle nature of light. Before leaving this subject, it should be pointed out that there are other experiments which indicate the validity of this point of view. One of these, called the *Compton effect*, is indicated schematically in Fig. 18.10. In this experiment x-rays incident on matter are observed to be deflected in direction with a *change* in their energy. The results of

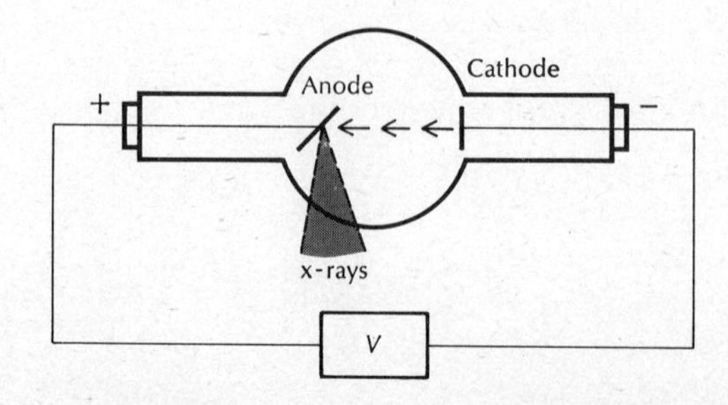

FIGURE 18.9 X-ray photons have an energy determined by the energy of the electrons striking the anode.

FIGURE 18.10 A photon of energy *hf* collides with an electron at rest. The result is that some of the energy of the photon goes into kinetic energy of the electron, and the photon has a reduced energy *hf′*. The analysis of this effect by Compton deals with photons as distinct particles which an undergo collisions with electrons like billiard balls.

this experiment can be completely accounted for by assuming an elastic collision between a photon of light and an electron, as shown. The ability to deal with collisions between light and electrons in the same manner in which one deals with collisions between billiard balls is another piece of evidence for the particle character of light.

It is also possible to build detectors for x-rays which respond to the effect of single photons. This result is very hard to understand if the photon model of light is wrong.

18.5 THE LASER

One of the more fascinating developments of modern science is the *laser*. (Laser stands for *l*ight *a*mplification by *s*timulated *e*mission of *r*adiation.)

To use a laser one arranges a bunch of atoms so that they are just about to emit a specific wavelength of light. In a helium-neon laser this wavelength is 6328 angstroms, and comes from the neon atoms. Various means are used to excite atoms so that they will then emit at the desired wavelength. An atom which is excited so that is ready to emit light can be looked at as a loaded mousetrap. If a photon of exactly the same frequency comes by, it springs the trap, and the excited atom emits its photon. This is called *stimulated emission*. The two photons are exactly in phase. (By in phase we mean peak on peak, and valley on valley.) They continue to stimulate the emission of more photons. All the photons are exactly in phase and are traveling exactly in the same direction.

It is the phase coherence, directionality, and purity of color that give a laser its unique properties. Lasers can be focused to very small points, so that all the energy in their beam can be concentrated in a very small region. In this way they can be used to burn holes in razor blades, etc. But no laser emits more energy than is put into it. In fact, the most efficient lasers put between 1 and 5% of the energy input into the coherent beam. So to vaporize large objects using lasers, very large amounts of energy must be put into them. Also, the technological problems of handling such large amounts of energy must be solved. There has been much nonsense written about lasers being used to vaporize tanks, and the like, without taking such questions into account.

SUMMARY

Study of the photoelectric effect yielded the following experimental facts:

1. The photocurrent, or number of photoelectrons, is proportional to the intensity of the light.
2. The maximum kinetic energy of the electrons is inversely proportional to the wavelength of the light and directly proportional to the frequency of the light.
3. No time delay occurs.
4. There is a minimum frequency of light which will eject photoelectrons. This frequency is different for different metals and is lowest for chemically active metals.

The first three of these facts are difficult to explain on the basis of the wave model for light. Einstein proposed a solution based on the idea that light comes in massless packets, called photons. He related the energy of a photon to the frequency, f, of the light by

$$E = hf$$

where h is Planck's constant ($h = 6.6 \times 10^{-34}$ joule-second).

The maximum kinetic energy of photoelectrons is given by Einstein's equation

$$\left(\tfrac{1}{2}mv^2\right)_{max} = hf - \phi$$

The Compton effect and the generation of x-rays can also be used to verify the photon model for light.

SELECTED READING

Gamow, George: "Thirty Years That Shook Physics: The Story of Quantum Theory," Science Study Series S45, Doubleday and Company, Garden City, N.Y., 1966. The quantum theory is one of the major revolutions of physics in the twentieth century. One of its beginnings is in the photoelectric effect. This book is a sprightly history of the whole period, 1895–1925.

QUESTIONS

1. How can one reconcile the wave and particle models of light?
2. Is it possible to obtain more energy from a laser than one puts in when exciting the atoms?

3 Which model of light is easier to use when discussing radio waves?
4 Which model of light is easier to use when discussing x-rays?

SELF-TEST

__________ photoelectron
__________ cathode
__________ H. Hertz
__________ A. Einstein
__________ photon
__________ Planck's constant
__________ x-rays
__________ Compton effect
__________ laser

1 The "particle" of light
2 Very-short-wavelength electromagnetic radiation
3 Electrode that emits photoelectrons
4 X-rays collide with electrons with the result that their energy is changed
5 Electron emitted through the action of light
6 First discovered the photoelectric effect
7 Explained the photoelectric effect
8 Used in Einstein's work to relate energy and frequency of photons
9 Emits very monochromatic light

PROBLEMS

1 Calculate the energy of a photon of red light whose wavelength is 6500 angstroms (6.5×10^{-7} m).
2 Calculate the energy of a photon of green light whose wavelength is 5000 angstroms.
3 Calculate the energy of a γ-ray photon whose frequency is 3×10^{18} hertz.
4 Calculate the energy of a radio-wave photon whose frequency is 1×10^{6} hertz (1,000 kilocycles).
5 Calculate the energy of an x-ray photon whose wavelength is 1 angstrom (10^{-10} m).

6 Assume the following data for the maximum energy of photoelectrons in the photoelectric effect. Find the value of Planck's constant from these data. (The value you will obtain is not the true value.)

$(\frac{1}{2}mv^2)_{max}$, joules	Frequency of light, hertz
1×10^{-17}	1×10^{15}
4×10^{-17}	2×10^{15}
7×10^{-17}	3×10^{15}

7 Blue light of wavelength 4500 angstroms is incident on a metal surface whose work function ϕ is 1.5 electron-volts. Calculate the maximum energy, in joules and electron-volts, of the resulting photoelectrons.

8 Ultraviolet light of wavelength 3000 angstroms causes photoelectrons with a maximum energy of 2 electron-volts to be emitted from a metal surface. Find the energy necessary to remove these electrons from the metal surface.

By about 1870, it was clear that matter consisted of atoms, but nothing could be said concerning atomic structure except that atoms behaved much like billiard balls in collisions, and that they entered into chemical reactions by a mechanism as yet obscure, although Faraday's work on electrolysis had indicated that it has an electrical character.

THE ATOM HAS A STRUCTURE

19.1 CANAL RAYS

In 1886 E. Goldstein discovered another kind of ray in a cathode-ray tube. When the cathode of a tube was perforated and a screen of fluorescent material placed behind it, a bright spot of light appeared on the screen directly behind the "canals" in the cathode (Fig. 19.1). The rays that produced the spot became known as *canal rays*. Remember, cathode rays are generated at the cathode and move away from it. These new rays seemed to move toward the cathode and through the holes.

After Thomson demonstrated that cathode rays carry a negative charge and have very little mass, it was possible to make the logical surmise that canal rays probably were positively charged particles, since they moved in the opposite direction in the electric field between the anode and the cathode. It was found that canal rays could be deflected, but only by much stronger electric or magnetic fields than necessary to bend a stream of electrons. Canal rays are positively charged and have a charge-to-mass ratio q/m of about 3×10^6 coulombs/kilogram. (Thomson's value for the charge-to-mass ratio of the electron is $e/m = 1.76 \times 10^{11}$ coulombs/kilogram.) Therefore, if one assumes the charge on cathode and canal rays to be similar, the mass of the canal ray must be about 10^5 times that of the electron. Cathode-ray experiments and several other lines of evidence had already suggested that electrons exist within atoms. But most of the mass of the atom was still to be accounted for. This strongly suggested the idea that canal rays are atoms which have had one or more electrons knocked out by bombardment in the cathode-ray beams. This, in fact, is the case.

19.2 THE THOMSON ATOM

If there are electrons in an atom, and atoms are observed to be electrically neutral, there must also be a positively charged part. A compari-

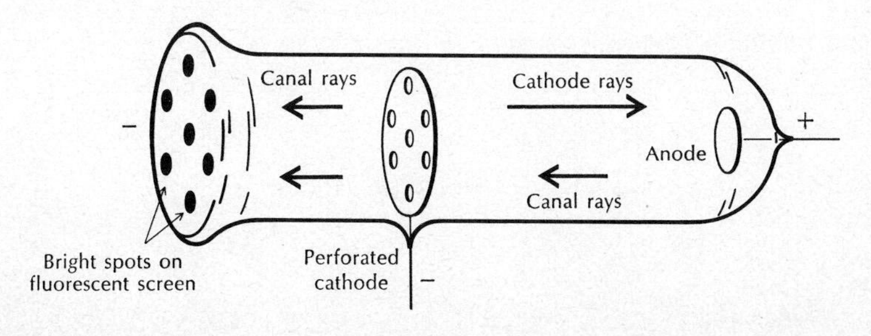

FIGURE 19.1 Canal rays travel in the opposite direction from cathode rays. They move toward the cathode, and can be detected by drilling holes in the cathode so that the canal rays come through.

son of the charge-to-mass ratio for canal rays and for electrons implies that most of the mass of the atom is contained in the positively charged portion. Using this information, Thomson in 1903 proposed a model of the atom that accounted for the experimental facts as they were then known. This model has become known as the *"plum pudding" model* of the atom. The size of atoms was known to be about 10^{-10} meter by that date, and Thomson proposed that the atom consists of a uniformly positively charged sphere with mass equal to the mass of the atom, as in Fig. 19.2. Embedded in this material are electrons distributed much like raisins or plums in a pudding. Thomson then tried to calculate the equilibrium positions various numbers of electrons would occupy in such a positively charged sphere. He was able to show that certain patterns for the most stable arrangement were repeated as more and more electrons were added, indicating a possible explanation for the periodic table. He was unable, however, to account for the observed colors or wavelengths of spectral lines on the basis of this model.

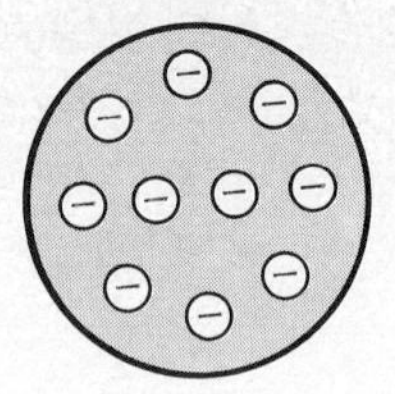

FIGURE 19.2 The "plum pudding" model of the atom. The shaded sphere is uniformly positively charged, and the negative electrons are embedded in it like plums in a pudding.

19.3 RUTHERFORD SCATTERING AND THE NUCLEAR MODEL OF THE ATOM

Ernest Rutherford (1871–1937), a pioneer in the study of radioactivity, in 1909 was using a product of radioactive atoms to probe the structure of the atom. This product was a beam of α particles. All Rutherford knew about them at the time, and all we need to know to understand these experiments, is that an α particle is small (about the size of an atom or smaller), that it is positively charged, and that it has a ratio of charge to mass about that of a charged hydrogen atom. The α particles emitted by the sources Rutherford used had very high velocities (about 10^7 meters/second, or nearly one-tenth the speed of light). When a beam of α particles was allowed to strike a thin film of matter, such as a gold foil, they penetrated it and could be detected on the other side (Fig. 19.4). Since α particles are positively charged, they were expected to interact with the positive and negative charges of an atom in a way which could be predicted from models of the atom. The positively charged α particle should be attracted by the negatively charged portion of the atom and repelled by the positive part. Calculations based on Thomson's "plum pudding" model predicted that none of the α particles would be deflected by very much from their original path. The Thomson model assumed that, whatever material an atom was com-

FIGURE 19.3 Ernest Rutherford (1871–1937). *(Courtesy University of Pennsylvania Library.)*

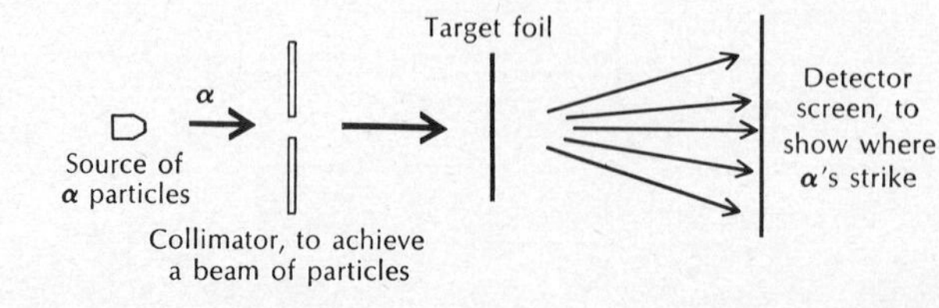

FIGURE 19.4 The α-particle scattering experiment. The beam of α's strike the target and are scattered in various directions.

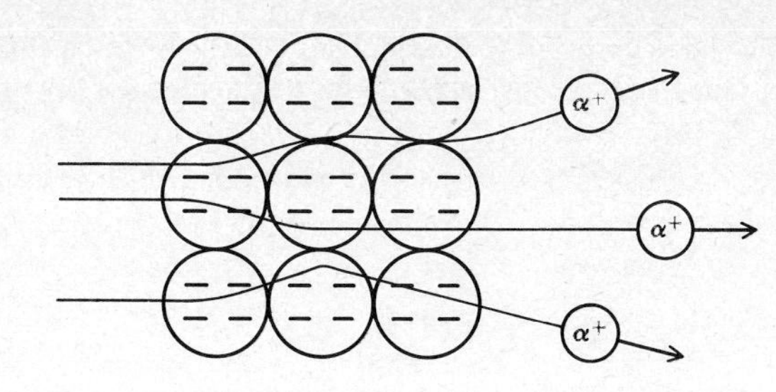

FIGURE 19.5 α-Particle scattering using the "plum pudding" model for the atom. The α particles are scattered many times through small angles.

posed of, α particles could pass right through it. The α particles might be scattered many times in even a thin film by interaction with positive and negative charges, but the random nature of this process would lead to a deflection of more than a few degrees in only a vanishingly small number of cases. Figure 19.5 indicates how this was proposed.

Two of Rutherford's students, Hans Geiger and Ernest Marsden, began a study of the number of α particles scattered in various directions. They found, to everyone's surprise, that a significant number of them were scattered backward, as in Fig. 19.6. In Rutherford's words, "It was almost as incredible as if you had fired a fifteen-inch shell at a piece of tissue paper and it came back and hit you." There was no way, on the basis of the Thomson model of the atom, to account for this occurrence.

In Chapters 5 and 6 we studied the conservation of momentum and energy as they apply to collision processes. When these results are applied to α-particle scattering, it is found that the only way an α particle can be scattered nearly backward is in a *single* elastic collision with a body *more* massive than itself. This type of collision implies that there exists some massive part of the atom that the α particle cannot penetrate. The Thomson model of the atom had its positive charge and mass distributed evenly throughout the size of the atom. On this basis, an α particle would penetrate only a few atomic thicknesses into a foil before being scattered, since atoms fill up the whole space of the foil. But experimentally, they are observed to penetrate much thicker foils than this.

Geiger and Marsden used foils about 1,000 atoms thick, and yet most of the α particles passed through with only small deviations. Only a few were scattered in the backward direction. Therefore Rutherford was led to the idea that most of the mass and positive charge of the

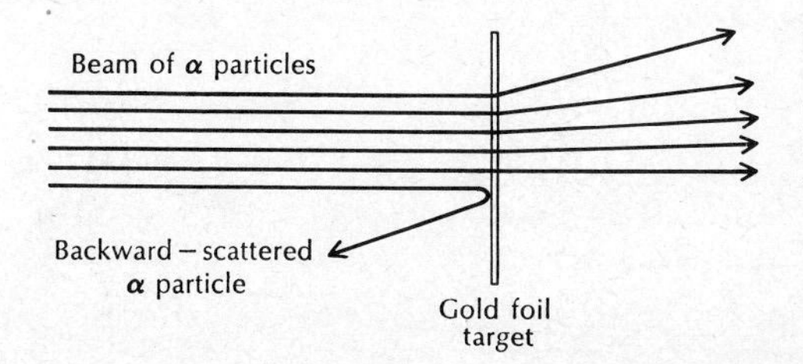

FIGURE 19.6 In the experiment of Geiger and Marsden, a few α particles were found to be scattered in the backward direction.

foil atoms must be associated with a very small body, called the *nucleus*, somewhere in the atom. The rest of the atom must be mostly empty space. On this assumption the probability of a head-on collision would be very small, and the α particles could penetrate films many atomic diameters thick, with only a small number undergoing head-on collisions. But when a head-on collision did occur, the α particle would be scattered directly backward. Since the α particle was much heavier than an electron, collisions with electrons would not lead to backward scattering. Figure 19.7 shows α-particle scattering in the Rutherford model.

Rutherford calculated the probability that an α particle would be scattered in a given direction on the assumption that it made a single elastic collision with a massive body of charge Q. The repulsive force between the positive charge of this body and the positive charge of the α particle was responsible for deflecting the α particle. His results showed that the number scattered in a given direction was proportional to the thickness of the film and to the square of the charge on the nucleus Q. It should be inversely proportional to the fourth power of the velocity of the α particle. He also predicted the dependence of the number of α particles on the direction of scattering. This search for the dependence of an effect on the known or variable parameters of an experiment is characteristic of good experimental physics. It is frequently much easier to show that the phenomenon depends in the predicted way on some variable parameter than it is to show that the absolute magnitude of the result is correct.

Geiger and Marsden were able to verify that the dependence on thickness, velocity, and angle was as predicted by Rutherford. They could not verify the predicted dependence on the charge Q, since the value of Q was not known. So they turned the experiment around and used the α-particle scattering to measure the charge on the nucleus, assuming that the theory was correct. This technique occurs frequently. Once major portions of a theory can be shown to be correct, the theory can be used to interpret experiments to obtain numbers otherwise inaccessible, in this case the charge of the nucleus of the target atom. If the numbers obtained in this way are consistent with other data, the theory is given additional verification.

The magnitude of the nuclear charges determined for targets of several elements by Geiger and Marsden can be expressed as $Q = Ze$, where Z is a whole number and e is the magnitude of the charge of the

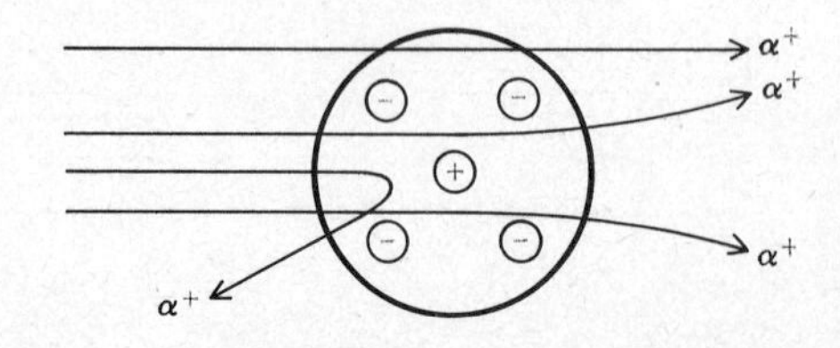

FIGURE 19.7 α-Particle scattering in the Rutherford model. When the α particle collides with the small, massive, positively charged nucleus, it can be scattered through a large angle.

electron. Their results showed that Z was approximately one-half the atomic mass of the element in question. More importantly, Z was found to be equal to the atomic number of the element in the periodic table. In the original construction of the periodic table, the atomic number was given no physical significance. It was simply the numerical position of the element in the table. Rutherford's work thus shows the physical significance of this number.

Shortly afterward, in 1913–1914, Henry Mosely found an even better way to measure the nuclear charge of atoms. This came out of a study of the wavelengths of x-rays emitted from various elements. He was able to study more elements than Rutherford's group, and to show that the nuclear charge and atomic number are the same. His method showed clearly that nuclear charge and not atomic mass should be used to construct the periodic chart. He was able to show that vacancies still existed in the chart at hafnium (number 72) and technetium (number 43).

Rutherford's experiments also allowed an estimate of the size of the nucleus. Rutherford reasoned that the distance of closest approach of an α particle and the massive charged part of the atom in a head-on collision could be no less than the size of this charged portion. He was able to show that this distance, the diameter of the nucleus, was about 10^{-14} or 10^{-15} meter, one ten-thousandth that of the atom. Not long after Geiger's and Marsden's experiments, α particles themselves were shown to be nuclei of helium atoms.

19.4 THE NUCLEAR ATOM

The following, then, is a description of the model of the atom that resulted from Rutherford's work. Almost all the mass of the atom is concentrated in a tiny (10^{-14} to 10^{-15} meter) nucleus which has a positive charge equal to the atomic number Z times the charge of the electron. Somewhere outside this nucleus, presumably at distances corresponding to the size of the atom, are Z electrons, the number necessary to balance electrically the positive charge of the nucleus and to yield an atom which is electrically neutral. Electrons were presumed to travel in orbits around the nucleus, like planets around the sun. As a result, this model is sometimes called the *planetary model* of the atom.

The concept of atomic number now has been given a physical meaning. Previously, it denoted the position of an element in the periodic table. The atomic mass was considered by Mendeleev to be the important property of a substance. But now we see that the periodic table is a listing of elements arranged according to nuclear charge and, consequently, number of electrons. In most cases this order is the same as an arrangement by atomic mass; but in a few cases in which the two arrangements differ, ordering by atomic number agrees more satisfactorily with chemical properties. For example, the elements argon and potassium are out of order in terms of their masses when they are put in the positions indicated by their chemical properties. (See the periodic chart, Table 12.3.) Ordering by atomic mass would place potassium (atomic mass 39.1) under neon, and argon (atomic mass 40.0) under sodium; but in terms of their chemical properties argon is an inert gas like neon, and potassium is a light, silvery metal like sodium. Precise measurements of nuclear charge show that Z is 18 for argon and 19 for potassium. Therefore the atomic number Z, and not the atomic mass, is the determining factor for the position of elements in the periodic table.

There were a number of deficiencies in the model of the atom proposed by Rutherford, and these were well known to him. His model proposed that a large amount of positive charge be concentrated in the very small volume of the nucleus. The mutual repulsion of these charges should act to drive them apart. No theory existed at that time to explain why this did not happen. Also, according to Maxwell's theory of electricity and magnetism, the electrons in the atom ought to be attracted by the positive nucleus and fall into the nucleus, losing their energy as light. This was not observed to happen.

Why, then, should Rutherford propose a model with these defects? He was driven to it as the only explanation for the α-particle-scattering experiments. The experiments convinced him that the model was correct. Therefore there must exist some explanation for its obvious defects, even though it was not apparent to him what that explanation could be. Again we see that the final recourse of the physicist is to experimental results. That is, the test of any theory is always its relevance to the real world, in the sense of correctly accounting for the results of experiments. The balance of this chapter will be a discussion of the way out of one part of Rutherford's difficulties. Chapter 21 is a discussion of the structure of the nucleus, and the "glue" that holds it together, to solve the second part of Rutherford's problem.

In 1911 a young Danish physicist who had just received his Ph.D., Niels Bohr, arrived in Rutherford's laboratory. He set about trying to understand how the electrons in Rutherford's atom stay in orbit.

FIGURE 19.8 Niels Bohr (1885–1962). (*Courtesy University of Pennsylvania Library.*)

19.5 STATEMENT OF BOHR'S PROBLEM

The planetary model of the atom pictures it as a sort of miniature solar system, with electrons in the role of planets in orbit around the nucleus. Before such a model can be accepted, two aspects of electron behavior must be understood. We have already discussed the particular colors emitted from gases of elements, their bright-line spectra. Any model of the atom must account for the particular colors, or wavelengths, emitted. Furthermore, when light is emitted from an atom, it loses energy. It seems therefore that, as energy leaves the atom, the electron should slow down and fall into the nucleus.

The theories of electricity as they existed in 1910 made very specific predictions about the light emitted from atoms. These predictions are based on the fact that *any* accelerated electric charge was believed to emit energy in the form of light. An electron in a circular orbit is constantly accelerated toward the nucleus, and therefore should emit light. The frequency of this light should be the frequency of revolution of the electron about the nucleus. But the frequency of revolution depends on the distance of the electron from the nucleus. Therefore a wide range of frequencies of light was predicted for any atom, in contrast to the few very specific colors observed. Furthermore, the existing theory led to the conclusion that the electron should continuously emit light and energy and spiral into the nucleus in a very short time. This would amount to the collapse of all atoms, a suggestion not borne out by the facts. Bohr's problem was to account for the specific colors observed in the spectra of atoms and for the fact that atoms do not collapse.

19.6 BOHR'S PROPOSALS FOR HYDROGEN

Because hydrogen is the simplest of all atoms, having only one electron and a single positive charge on the nucleus, Bohr chose it as the subject of his study. His first step was to say, essentially, "Classical physics won't work." That is, he considered the possibility that the classical laws of physics, of Newton and Maxwell, created for and tested in experiments involving large objects, were simply inadequate to describe events that occur in systems as small as atoms. After all, none of these laws had been tested in atoms or in any other object of that size. Bohr was convinced, however, that if the mechanics of atoms were not the same as the mechanics of Newton, the two were not totally unrelated. He believed that one might be a special case of the other, and that they would merge slowly one into the other as one went from systems of atomic size to larger and larger systems. Bohr called this the *correspondence principle*. It was an important tool in the development of his model of the atom.

Einstein had shown in 1905, in his analysis of the photoelectric effect, that a beam of light consisted of particles, or photons, and that the energy of each photon is related to the frequency of the light by the equation

$$\begin{aligned} \text{Energy} &= h \times \text{frequency} \\ E &= hf \end{aligned} \tag{19.1}$$

This implies that *each* photon leaving an atom must carry away an amount of energy given by this relationship. The existence of certain discrete frequencies, or wavelengths, and therefore colors, in the light emitted from atoms implies therefore that only photons of certain energies are emitted. To explain this, Bohr proposed that electrons in an atom could exist only in certain well-defined energy states. He further assumed that, when an electron is in one of these special states, it is stable and does not radiate light, but that the emission of light occurs only when the electron "jumps" from one such state to another. The frequency of the light emitted is then given by $hf = E_2 - E_1$, where E_2 is the energy of the electron before it jumps, and E_1 is its energy after it jumps. That is, he accepted and extended Einstein's relationship between energy and frequency for light.

This process of making assumptions, or postulates, in the creation of a physical theory is a common one. Bohr's postulates were not derived or proved in the mathematical sense. Essentially the process is one of saying, "Let's see what happens if we assume such-and-such to be the case." If the conclusions obtained from this assumption agree with experiments, then we feel they have some validity. If they do not, the assumptions will be discarded, and others tried. In this case these two assumptions accounted for the existence of discrete frequencies in the spectra of atoms, and satisfied the conservation-of-energy principle; but they did not give a method for determining what the energies of the electron states in the atom were.

We need at this point to present a picture of what is meant by the energy states of an electron in an atom. Suppose we first consider a very simple example—a child's marble rolling down the stairs. If each step has a height d, the marble loses an amount of potential energy given by mgd each step it falls. Where does this potential energy go? If the marble does not bounce (the collision with the step is inelastic), the change in potential energy turns up as heat in the marble and in the step. Now assume a different situation. Suppose that, while the marble is falling from A to B, it emits a photon of light, and that the photon has an energy $E = mgd$; the photon carries off an amount of energy exactly equal to the loss in potential energy. That is, instead of being converted into heat, the lost potential energy is carried away by the photon. Here we have a situation quite analogous to that postulated by Bohr for the atom. The marble starts with a well-defined potential energy, it falls to another state with a well-defined energy, and emits a photon that carries away the difference in energy so that energy is conserved. This is illustrated in Fig. 19.9.

19.7 THE HYDROGEN ATOM

Now let us see how this is to be applied to the simplest of atoms, hydrogen. If we accept Rutherford's model for the atom and note that hydrogen has atomic number $Z = 1$, the model we use for the hydrogen atom is a nucleus containing most of the mass of the atom, which has a positive charge equal in magnitude to the charge of the electron. Wandering about the nucleus in some kind of orbit, which resembles the orbit of a planet around the sun, is one electron. Since the nucleus and the electron have opposite charges, they must attract each other, according to

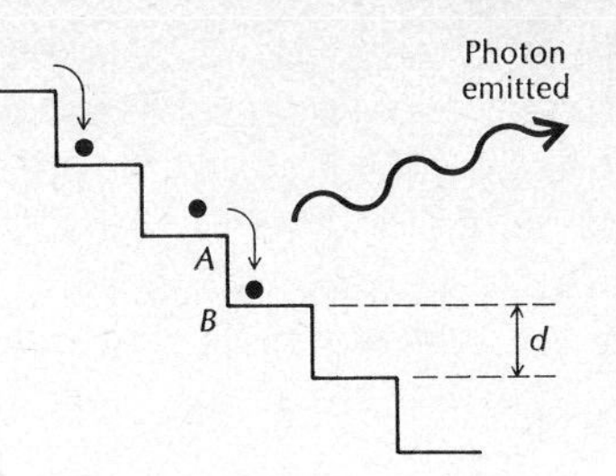

FIGURE 19.9 As the marble falls, it loses an amount of potential energy given by *mgd* for each step. If this lost energy were emitted as a photon, this situation would be analogous to the electron's changing energy levels in the atom.

Coulomb's law. Therefore, to move an electron farther away from the nucleus requires a certain amount of work to be done against the coulomb force of attraction. This work must turn up as a change in energy. Therefore the energy states we are looking for correspond to states which have the electron in an orbit at different distances from the nucleus. This means that, if an electron can exist in certain energy states, as Bohr postulated, it can be found only at certain special distances from the nucleus, corresponding to these energies. We now want to try to find a formula for these special distances.

The force between the electron and a nucleus of charge $+e$ is given by Coulomb's law:

$$F=\frac{Ke^2}{r^2} \tag{19.2}$$

Here r is the distance between the nucleus and the electron. But we have already shown that, for a planet or an electron in a circular orbit, the acceleration toward the center times the mass must exactly equal the applied force. In fact, it is the applied force that provides this acceleration. Therefore

$$\frac{mv^2}{r}=\frac{Ke^2}{r^2} \tag{19.3}$$

We have used this same result for a planet in orbit around the sun, but there the attractive force was the gravitational attraction between the two bodies. Equation 19.3 contains both v and r, neither of which we know. So now we shall use another relationship to remove v and leave an equation involving only r. This other relationship is the key to the whole development, since it specifies the condition for the special distances, or allowed energy states. Bohr originally obtained this result from his correspondence principle, but we shall propose it as a postulate.

When we studied the behavior of the planets, we discovered that Kepler's second law showed that the quantity mvr (mass times velocity times distance to the sun) is constant for the motion of a planet in a circular orbit. This quantity, which we discussed in Chapter 5, is called angular momentum. For planets, any value of angular momentum is allowed but, for any given planet, the angular momentum is constant. For the planetary electron orbits in the hydrogen atom Bohr postulated that *not* all possible values for angular momentum were permitted to the electron and that this restriction defined the allowed energy states. The allowed amounts of angular momentum he proposed were one, two, three, or any whole-number multiple of a basic quantity. This basic unit of angular momentum is given by Planck's constant h divided by 2π. This proposal is called *Bohr's quantum hypothesis*. From it we shall be able to calculate a value for the allowed energy states. But before this postulate of Bohr's can be said to be true, we must see whether the conclusions drawn from it agree with experiment.

It is well to stop at this point to summarize Bohr's proposals. Two of these ideas involved brand new ideas in physics. Bohr's postulates are:

1. *Electrons in atoms exist only in special, or allowed, states of energy.*
2. *Electrons radiate light only when jumping from one of these special states to another.*

These two postulates taken together are a drastic modification of the way accelerated charges are expected to emit light.

3. *The special, or allowed, states are determined by Bohr's quantum condition for angular momentum,* $mvr = n(h/2\pi)$, *for* $n = 1, 2, 3. \ldots$

Postulate 3 is brand new in that it proposes that angular momentum comes only in distinct quantities; it is *quantized*. We have already seen that matter comes in distinct units, atoms and electrons, and that electric charge comes in a distinct unit, the charge of the electron. The energy in a beam of light also comes in distinct units, photons. So this proposal of Bohr's is one more example of this same type of development.

4. *The force between electron and nucleus is the coulomb force, and the relationships for kinetic and potential energy are those which have already been used.*

This statement simply points out those aspects of the Bohr theory in which no change is proposed in the basic physics of the situation.

19.8 BOHR'S RESULTS

We will not reproduce the algebra needed to obtain Bohr's results. Rather, we will describe the answers he found. One reason for this is that Bohr's ideas have been superceded by the ideas to be described in Chapter 20. The basic notion of quantization, that quantities like angu-

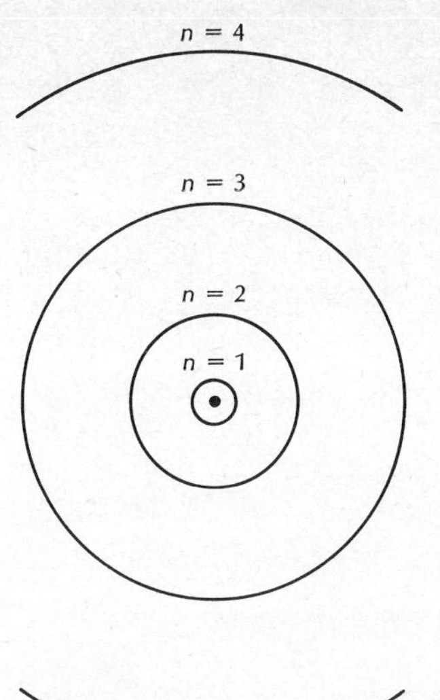

FIGURE 19.10 Diagram, to scale, of the size of the first four Bohr orbits for hydrogen.

lar momentum and energy come in discrete bundles, is maintained, however, in the newer theories.

Bohr's theory leads to a numerical prediction of the size of the allowed, or special orbits for the electron in a hydrogen atom. The smallest of these corresponds closely to measured values for the size of the hydrogen atom. His result for the allowed sizes is

$$r = 0.5 \times 10^{-10} n^2 \qquad \text{in meters} \tag{19.4}$$

Here n is 1, 2, 3, 4, or any other whole number. $n = 1$ corresponds to the smallest orbit. The value of n in this equation gives the number of units of angular momentum in the orbit. Figure 19.10 shows the sizes of the first four Bohr orbits.

It is not possible for us to measure the radius of the orbit of the electron in a hydrogen atom easily; so we cannot use this result directly to check the theory. But Bohr had also proposed that the light emitted from an atom would have a frequency f related to the change in energy of the atom. He proposed that

$$hf = E_2 - E_1 \tag{19.5}$$

where E_2 and E_1 are the energies of two of the allowed states of the atom. That is, E_2 is the energy corresponding to a particular choice of n in Equation 19.4, and E_1 is the energy corresponding to a different choice of n. Remember that all possible values of n are allowed, as long as they are positive whole numbers. To see whether the theory agrees with experiment, we must find the energy of an electron a distance r from the nucleus. Then we can use the values for r in Equation 19.4 to determine the allowed energies.

When Bohr made this calculation, he found that he was able to predict exactly the colors (wavelengths) of light emitted by hydrogen gas. Therefore his unique proposals were taken seriously, even though many people (including Rutherford) had a great deal of difficulty understanding how they could possibly be true.

Bohr's theory thus permitted a calculation of the wavelengths of the observed light emitted from hydrogen gas, on the basis of a very few assumptions. The crucial assumptions of the model are the two nonclassical assumptions:

1. An electron does not radiate away its energy when it is in one of the special allowed energy states.
2. The allowed values of the angular momentum are given by

$$mvr = n \frac{h}{2\pi} \tag{19.6}$$

Neither of these assumptions has any basis in classical physics. Each must be looked at as an arbitrary assumption introduced to understand the observations. But the fact that these assumptions do allow a calculation that produces the observed colors of light implies that there is something to them. It implies that Bohr's assumptions have hit on something essential to the description of events which occur in the atom. The two assumptions mentioned above—the existence of stationary, or "quantum" states, and the restriction on the allowed values of angular momentum—have proved to be very basic ingredients in the construction of any theory describing events in the realm of atomic dimensions.

The Bohr theory has its limitations, however, and has proved incapable of dealing with atoms more complicated than hydrogen. The model gives no particularly compelling reason why the quantum hypothesis should hold, except that it seems to work. Its main strength is in the pictorial model for the structure of the atom which it yields, even though this picture is at times somewhat misleading. In the next chapter, we will discuss the further developments that have been carried on from the base Bohr provided.

SUMMARY

The scattering of α particles by thin metal foils showed that the atom has a small, massive, positively charged nucleus and much empty space. The positive charge on the nucleus is given by $Q = +Ze$, where e is the magnitude of the charge on the electron, and Z is the atomic number. The atomic number gives the amount of positive charge on the nucleus and the number of electrons around it.

Bohr's theory of the atom included two new assumptions:

1 An electron in an atom does not radiate away its energy when it is in an allowed energy state. It radiates a photon when going from one allowed state to another.
2 The allowed values of angular momentum are given by Bohr's quantum condition,

$$mvr = n\frac{h}{2\pi}$$

Using these ideas, Bohr showed how to calculate the frequencies of the spectral lines of hydrogen that agreed with the values found by experiment.

SELECTED READING

Andrade, E. N. da C.: "Rutherford and the Nature of the Atom," Science Study Series S35, Doubleday and Company, Garden City, N.Y., 1964. A biography of the man who, more than any other, is associated with the atomic nucleus, its discovery, and the early study of its structure and properties.

QUESTIONS

1 What are the differences between the Rutherford and Thomson models of the atom?
2 Which of the objections to Rutherford's model of the atom does Bohr's theory attempt to answer?

SELF-TEST

__________ canal rays
__________ "plum pudding" model of the atom
__________ α particle
__________ atomic number
__________ planetary model of the atom
__________ correspondence principle
__________ nucleus
__________ Bohr's quantum condition
__________ Rutherford
__________ Bohr
__________ Geiger and Marsden

1 The model of the atom proposed by Rutherford
2 Particle from radioactive atoms
3 Small, massive, positively charged part of the atom
4 Relates classical physics and Bohr's results
5 Thomson's model of the atom
6 Positive particles in a cathode-ray tube
7 Angular momentum comes in certain allowed amounts
8 Measured α-particle scattering
9 Number of positive charges on the nucleus
10 Proposed the nuclear atom
11 Proposed the quantization of angular momentum

PROBLEMS

1 (*a*) Calculate the radius of the allowed Bohr orbit for an electron in the hydrogen atom with $n = 3$. (*b*) Calculate the angular momentum in this state.
2 Calculate the radius of an electron in the Bohr orbit with $n = 2$.
3 Calculate the angular momentum of an electron in a Bohr orbit for hydrogen with $n = 4$.

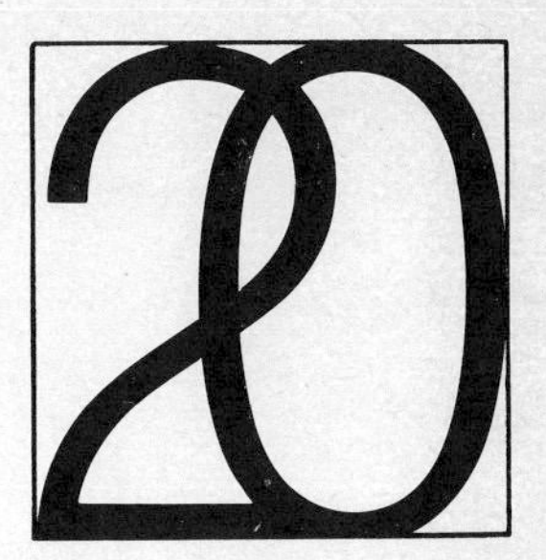

THE WAVE NATURE OF MATTER

We come now to one of the most startling developments of twentieth-century physics. We have seen that, historically, there were two views as to the nature of light. Young's experiments on interference and others on diffraction swung the scale completely to the view that light is a wave phenomenon. Maxwell's development of a description of light waves on the basis of his electromagnetic theory reinforced this idea. On the other hand, the theory of the photoelectric effect proposed by Einstein and verified experimentally by Millikan demanded that light be considered as particles as well, and that both wave and particle viewpoints be considered together in an adequate description of light. Until 1924, however, no thought had been given to a model for the behavior of electrons other than that they were particles of charge e and mass m, which followed Newton's law of mechanics, except for the modifications proposed by Bohr.

In 1924 Louis de Broglie proposed an amazing idea. He said, essentially, "Light behaves as though it is both waves and particles; perhaps nature is symmetric and electrons behave as particles and also as waves." De Broglie deduced a relationship between the "wavelength" an electron might have and its momentum, based on this proposal. Rather than follow his reasoning, we shall discuss an experiment first performed in 1927, which demonstrates the point we are discussing.

FIGURE 20.1 Louis de Broglie (1892–). (*Courtesy American Institute of Physics.*)

20.1 THE DAVISSON-GERMER EXPERIMENT

In 1927 American physicists, C. J. Davisson and L. H. Germer, reported experimental results that confirmed de Broglie's electron-wave theory. They performed an experiment much like the one described in Chapter 16 that verified the wave nature of x-rays by demonstrating x-ray diffraction and interference. In both cases the regular arrangement of atoms in a crystal served as a diffraction grating (see Chapter 16) in which the separation of reflecting surfaces is about the same as the wavelength of the incident "light." Figure 20.2 shows the structure of a single crystal. The lines indicate some of the many different rows of atoms that act like the scratches on a grating. Davisson and Germer directed a beam of electrons toward a single crystal of nickel. The velocity of the electrons was as nearly uniform as their apparatus permitted, so that each electron reached the crystal with the same kinetic energy and momentum. A movable detector measured the number of electrons reflected from the face of the crystal in every direction (Fig. 20.3). Electrons were found to be reflected in all directions. But at specific values of the velocity of the electrons, and therefore of their momentum, strong peaks were observed in the number of reflected electrons in certain specific directions. These peaks were interpreted as interference effects, and thus as indicating that electrons have a wavelike character, since they show interference and diffraction effects.

FIGURE 20.2 The orderly rows of atoms in a crystal function in a manner similar to the scratches in a diffraction grating, and lead to interference effects in waves of the proper wavelength when reflected from the crystal.

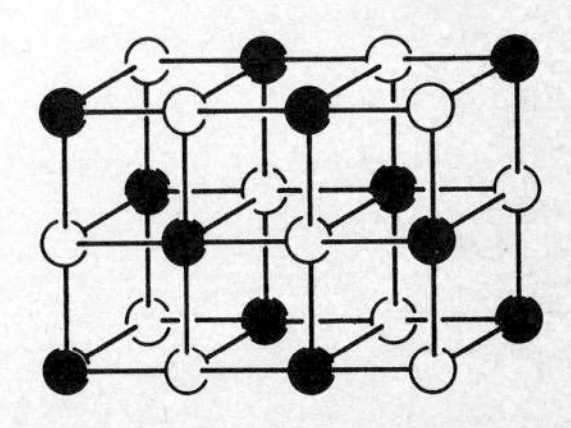

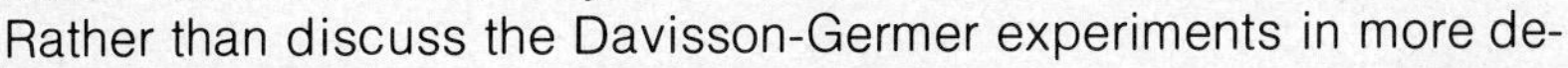

Rather than discuss the Davisson-Germer experiments in more de-

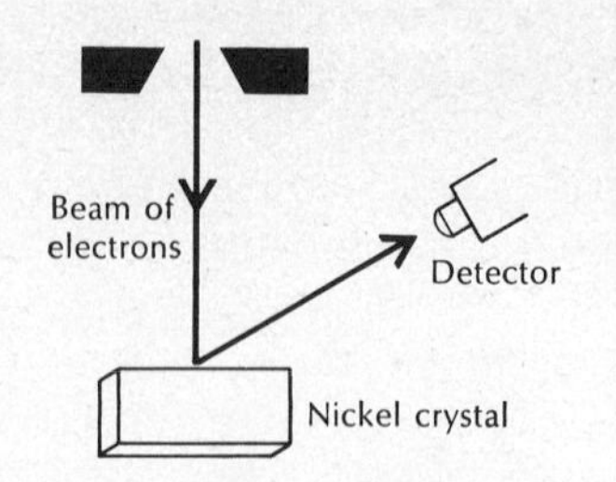

FIGURE 20.3 The Davisson-Germer experiment. Electrons reflected from the nickel crystal show strong reflections in certain directions. If these are interpreted as interference effects, the "wavelength" of the electron can be found.

tail, let us compare two experiments, one performed with x-rays, which are electromagnetic radiation and known to show wave effects, and the second using a beam of electrons and essentially the same experimental arrangement. This experiment is shown in Fig. 20.4. The target is a crystal thin enough so that either the electron beam or the x-rays can pass through. When x-rays are used, the result is the pattern seen in Fig. 20.5. Von Laue showed, by taking such a photograph, that x-rays are a wavelike phenomenon. They are now called Laue photographs, or patterns. Figure 20.6 shows the results of using a beam of electrons. The similarity between the two patterns is evident. The bright rings are regions of high intensity of x-rays or electrons.

20.2 THE DE BROGLIE RELATIONSHIP

Davisson and Germer's data showed that the wavelength which must be assigned to the electron to account for the position of the interference maxima varies with the velocity of the electron. The higher the velocity of the electron, and therefore the higher its momentum, the smaller its wavelength must be, according to these experiments. This result had been predicted by de Broglie, although Davisson and Germer were unaware of it when they undertook the experiment. The relationship between wavelength λ and momentum mv for an electron, as given by both de Broglie's theory and Davisson and Germer's experiment, is

$$\lambda = \frac{h}{mv} \tag{20.1}$$

Here h is again Planck's constant.

Once again, experiments force us to accept a view of nature which seems to violate common sense. We are driven to assign some kind of wave property to the electron, which had previously been considered purely and simply a particle. Further experiments show that this result

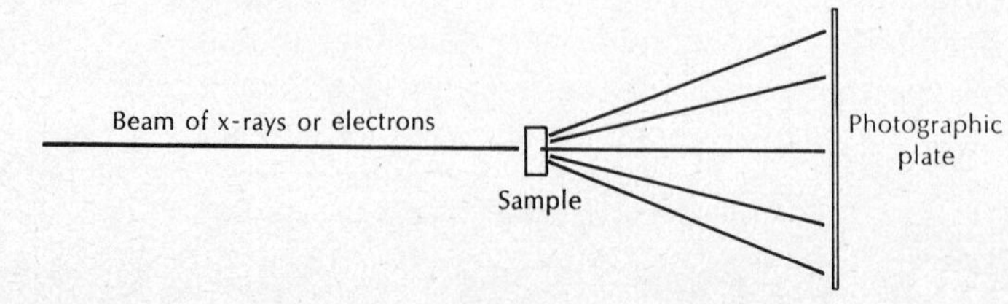

FIGURE 20.4 The experimental arrangement used to obtain the photographs, Figs. 20.5 and 20.6.

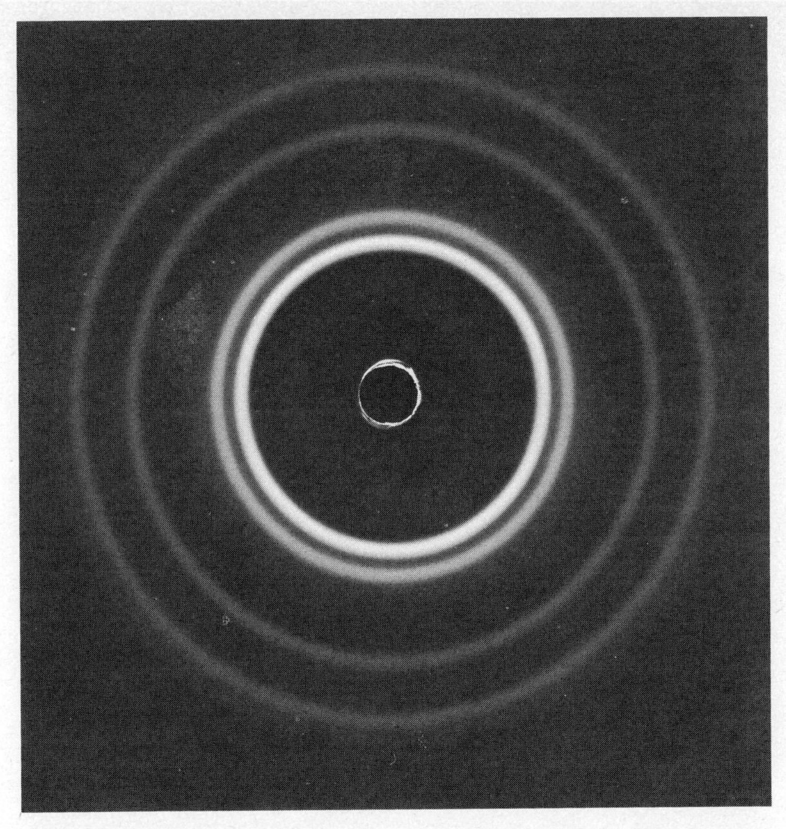

FIGURE 20.5 This photograph results from the passage of a beam of x-rays through an aluminum foil. (*Courtesy Educational Development Center.*)

is not limited to electrons. In experiments with helium atoms, similar effects can be observed if a velocity is chosen for the helium atoms which makes the apparent "wavelength" approximately the same as the spacing of atoms in a crystal. Let us calculate the "wavelength" of a helium atom whose velocity is 10^3 meters/second and whose mass is about 6.7×10^{-27} kilogram:

$$\lambda = \frac{h}{mv} = \frac{6.6 \times 10^{-34} \text{ joule-second}}{6.7 \times 10^{-27} \text{ kilogram} \times 10^3 \text{ meters/second}}$$

$$\lambda \approx 1 \times 10^{-10} \text{ meter} = 1 \text{ angstrom}$$

It seems, therefore, that *all matter* must behave in a wavelike manner. Why are wavelike effects not seen in ordinary living? This can quickly be answered if we calculate the wavelength of a 0.1-kilogram ball thrown at a speed of 10 meters/second:

$$\lambda = \frac{h}{mv} = \frac{6.6 \times 10^{-34} \text{ joule-second}}{0.1 \text{ kilogram} \times 10 \text{ meter/second}}$$

$$\lambda = 6.6 \times 10^{-34} \text{ meter}$$

$$\lambda = 6.6 \times 10^{-24} \text{ angstrom}$$

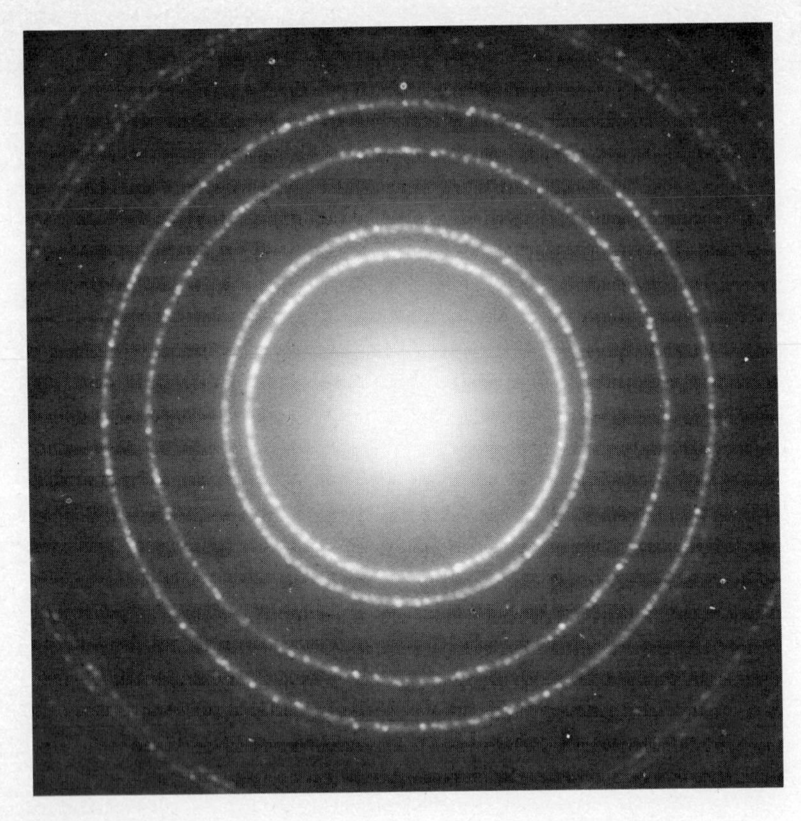

FIGURE 20.6 This photograph is a result of passing a beam of electrons through an aluminum foil. The similarity of this to Fig. 20.5 leaves little doubt that x-rays and electrons both have wavelike properties. (*Courtesy Educational Development Center.*)

The wavelength for such a massive object is incredibly small, about 10^{23} times smaller than the size of an atom. Therefore, although wave properties may exist for such a ball, they are hardly likely to be observed.

De Broglie's suggestion, even before the direct experimental confirmation by Davisson and Germer, was incorporated into a new theory of the behavior of electrons in atoms, and of matter in general. This theory has been called both wave mechanics, because the wave character of the electron plays such a central role, and quantum mechanics, because quantities such as energy and momentum are quantized, or come in discrete bundles. We have already seen how this occurs in the photoelectric effect and in the Bohr theory of the hydrogen atom. The development of quantum mechanics is associated with the names of Erwin Schrödinger and Werner Heisenberg. Quantum mechanics is sufficiently mathematical that we cannot discuss it in detail, but we can try to show how it works in a simple situation.

Let us take a look at the hydrogen atom, using the wave properties of the electron. We have seen that two waves of any sort will interfere constructively if they are in phase (peaks on peaks) and destructively if they are out of phase (peaks on valleys). Suppose we now bend a wave around in a circle, and suppose that this represents an electron in the vicinity of the hydrogen nucleus. This is shown in Fig. 20.7. We see that, in general, the wave will destructively interfere with itself, as in Fig. 20.7*a*. Only if the distance around is exactly a whole number of wavelengths (as in Fig. 20.7*b*), shall we find peaks on peaks and valleys on valleys, or constructive interference. Let us apply this to Bohr's idea that only certain states are allowed for the electron. Logically, we would expect these to be the states in which the electron wave interferes constructively with itself, that is, the states corresponding to a whole number of wavelengths in one circuit of the nucleus. Let us look at a state in which the distance around the nucleus is exactly one electron wavelength. For this we have

$$\text{Distance around the orbit} = 2\pi r = \lambda \qquad (20.2)$$

But, from the de Broglie relationship,

$$\lambda = \frac{h}{mv} \qquad (20.1)$$

Putting the two together, we have

$$2\pi r = \frac{h}{mv} \qquad (20.3)$$

or

$$mvr = \frac{h}{2\pi} \qquad (20.4)$$

This is exactly the relationship Bohr postulated for the quantum condition on angular momentum. The angular momentum is $h/2\pi$ or $2h/2\pi$ or $3h/2\pi$ or $nh/2\pi$, where n is any whole number. We obtained $mvr = h/2\pi$ because we chose the particular state in which there is only one full wavelength around the nucleus. Two whole wavelengths around will give the state in which $mvr = 2h/2\pi$. Although this is the same result as Bohr's postulate, Bohr was forced to obtain the quantum condition in a way which did not tell very much about the physics of the situation. In the wave model, once we have allowed electrons to have wavelike properties, we obtain the quantum condition simply by saying that the electron wave in the hydrogen atom must interfere constructively with

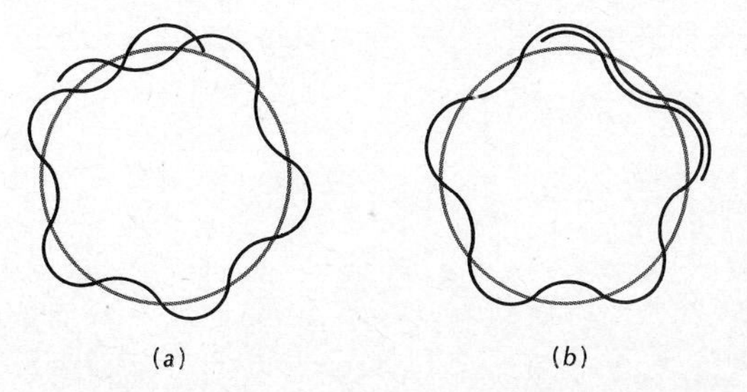

FIGURE 20.7 (*a*) In general, a wave bent around in a circle will destructively interfere with itself. (*b*) If the wavelength is just right, constructive interference will occur. This is exactly the same as the Bohr condition on angular momentum for allowed orbits.

itself. The student should be aware that this is a crude picture, and that the aim is to "get the idea" rather than to represent precisely what is going on.

20.3 THE WAVE-PARTICLE DUALITY

Although the concept that an electron behaves like waves allows us to obtain Bohr's quantum condition and, in fact, to achieve much more in the way of explaining the nature of the atom than Bohr's theory did, it leaves some very severe conceptual difficulties. How can we reconcile all the previous work, which indicates that electrons behave like particles, with the notion that electrons behave like waves? How can we reconcile the idea of a particle, which implies that the electron is at a very distinct point in space, with that of a wave, which leaves us a little fuzzy as to just where the electron really is? The fact of the matter seems to be that we are stuck with both points of view for electrons, just as we are for light. This dilemma has been called the wave-particle duality, and is a central feature of the current description of both light and electrons.

We can obtain an idea of how these ideas go together by considering a stone thrown into a lake. Outward from the point where the stone lands spread circular waves, as shown in Fig. 20.8. Some time after the stone has landed, we can look at the cross section of these waves along the line *AB*. It will look somewhat as shown on the right in Fig. 20.8. This is a small *group* of waves which move together as a unit outward from the center. We can consider the electron as such a small group of waves, moving as a unit. But if we do, where is the electron? It is somewhere in the region where the waves have a reasonable amplitude. We assume that the wave which describes an electron is big where the electron is, and is small or zero everywhere else. But we can make various groups of waves, or wave packets as they are called, to represent electrons. Some possibilities are shown in Fig. 20.9. These waves all have a different extent in space; so how are we to say where the electron is or which is the proper representation? None of these groups of waves is excluded by anything we have yet said. In fact, none of them is excluded as a valid picture of what the electron is like. If we look at these various wave packets, one interesting observation emerges. In *D* we are very unclear as to where the electron is, but we can measure rather precisely

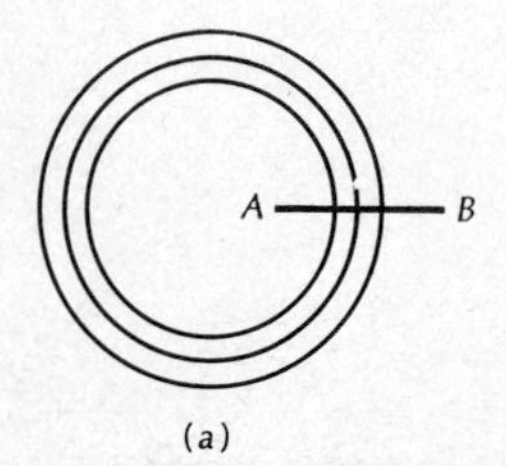

A B

(b)

FIGURE 20.8 (*a*) Waves in a lake spread out from the point where a stone lands. (*b*) Cross section along the line *AB* of the wave heights in (*a*) This is a wave packet.

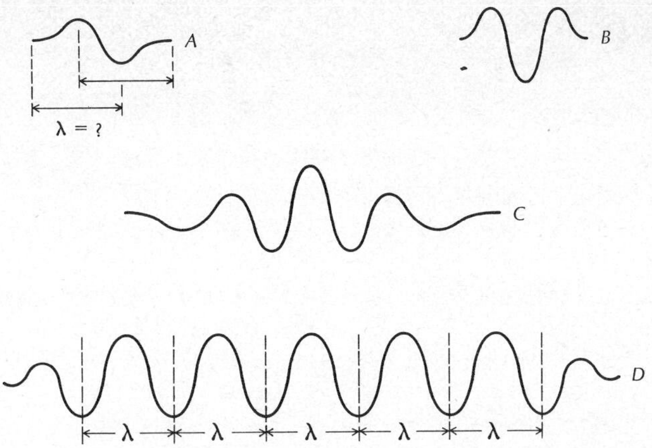

FIGURE 20.9 Several types of wave packets. The extent in space of *A* is small, but the wavelength is very uncertain. In *D* the wavelength is well defined, at the cost of much uncertainty in spatial position.

the wavelength, and therefore obtain the momentum from the de Broglie relationship. In *A* we can say rather precisely where the electron is, but only very inaccurately what its wavelength or momentum is. This observation is the basis of the *Heisenberg uncertainty principle*, proposed by the German physicist Werner Heisenberg in 1927. The uncertainty principle says that we cannot know simultaneously and precisely both the position and the momentum of an electron. If we wish to know precisely the momentum, and thus the wavelength, we must sacrifice some knowledge of position; and if we wish to know position precisely, we must sacrifice some knowledge of momentum.

Note that the uncertainty principle is a direct consequence of the wave-particle view of the electron. In order to construct a group of waves to represent the electron, we lose some information about its momentum, and if we make a very long string of waves so that the mo-

FIGURE 20.10 Werner Heisenberg (1901–). (*Courtesy American Institute of Physics.*)

mentum is accurately known, the position becomes unclear. Therefore, if we must accept the idea that electrons have the properties of both waves and particles, we must also accept the uncertainty principle.

Another way of looking at this principle was proposed by Heisenberg. Imagine trying to observe the position of an electron using a microscope, as in Fig. 20.11. The shorter the wavelength of light, the higher the resolving power of a microscope. Therefore, to find an electron, we should use very short-wavelength light, or x-rays. But bouncing x-rays off electrons causes a change in the electrons' position through the Compton effect. Therefore, the process of observing the electron interferes with our knowledge of its position. This means that, in general, the process of observing atom-sized systems will be disruptive to those systems. It will cause uncertainties in our knowledge about them.

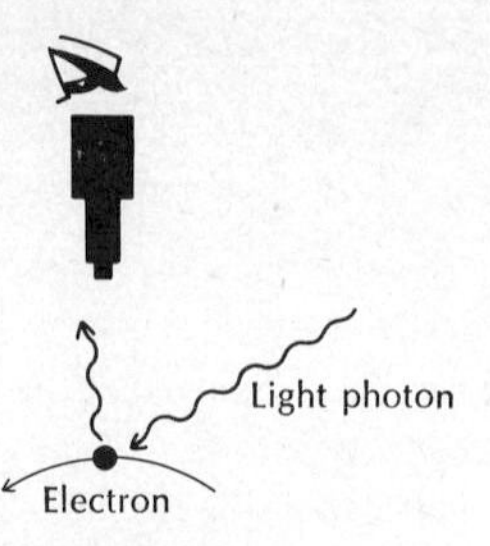

FIGURE 20.11 The collision of a photon with an electron as a result of Compton effect. This introduces an uncertainty in the measured position of the electron.

The uncertainty principle and the wave properties of matter have led to an interesting philosophical question. Before quantum mechanics, newtonian physics was, in principle, capable of calculating at any time in the past or future the exact position of every atom and electron in the universe. This meant that the future was completely determined by the present arrangement of matter and energy in the universe. Clearly, this leads to a deterministic universe, with no opportunity to change the course of events.

Quantum mechanics, on the other hand, tells us that all we can say about the electron are statements about the *probability* of finding it in a given region of space. The uncertainty principle says that we cannot know precisely its position and momentum, the exact information needed in newtonian physics to calculate the future course of events. So therefore quantum mechanics seems to have relieved us of determinism. But this point is by no means settled in either direction. Albert Einstein, for one, was completely opposed to this probability interpretation of quantum mechanics. He said, "God doesn't throw dice." But no other interpretation of quantum mechanics and the wave nature of the electron has yet come forth.

20.4 THE ATOM

The view of the atom which the wave picture of the electron leads to is quite different than that based on the Bohr model. For the hydrogen atom it gives numerical results identical to those based on the Bohr theory. For other atoms, with more than one electron, it gives much better numerical results than the Bohr model.

The basic principle used is the idea of probability. We do not say that an electron is here or there. We say that we can calculate the probability of finding it in a given region of space. (Of course, the probability of finding it somewhere is set equal to one.) The probability of finding the electron is related to the magnitude of the wave describing it. In fact, the probability is proportional to the square of the amplitude of the wave.

The net result of this is that, when we talk about an atom, we find a high probability of finding electrons in a region around the atom roughly equal to the value we measure in other experiments. The probability is low everywhere else. These electron clouds are frequently represented as shown in Fig. 20.12. Here the darker regions represent regions of high electron density, or probability. No information can be obtained about where the electron is at any given instant.

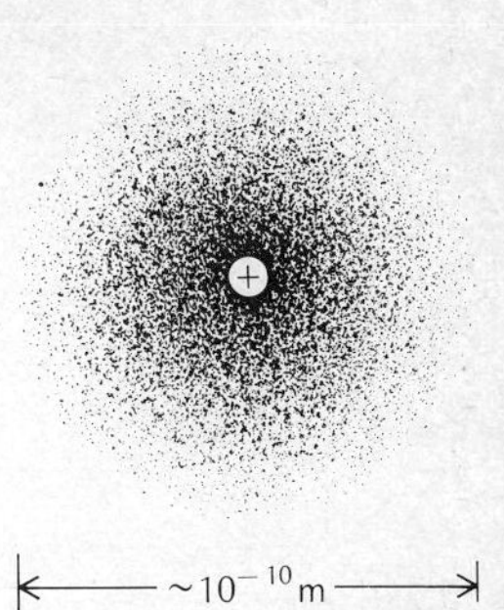

FIGURE 20.12 Electron density of a hydrogen atom. The probability of finding the electron is greatest where the cloud is shown densest.

SUMMARY

Davisson and Germer found experimentally that electrons must be treated as having wavelike properties. The relationship between wavelength and momentum, found by them and proposed theoretically by de Broglie, is

$$\lambda = \frac{h}{mv}$$

Electrons must be treated as having both wavelike and particlelike properties.

The Heisenberg uncertainty principle says that the position and momentum of an electron cannot be precisely known at the same time. A complete knowledge of one implies a sacrifice of some knowledge of the other.

QUESTIONS

1. Why can one not observe the wave properties of a baseball?
2. Why do we not observe the effects of the uncertainty principle in everyday life?
3. How can an electron behave both as a particle and a wave at the same time?

SELF-TEST

__________ de Broglie
__________ Davisson and Germer
__________ wavelength of an electron
__________ quantum mechanics
__________ wave-particle duality
__________ uncertainty principle

1. Discovered experimentally the wavelength relationship for electrons
2. Predicted theoretically the wavelength relationship for electrons
3. General theory that takes into account the wave properties of matter
4. The necessity for accepting simultaneously both wave and particle properties for electrons
5. Planck's constant divided by momentum

PROBLEMS

1. Calculate the de Broglie wavelength for an electron whose velocity is $v = 10^7$ m/second. The electron mass is 9.1×10^{-31} kg.
2. The de Broglie wavelength for a proton is found to be 10^{-10} m. The mass of a proton is 1.6×10^{-27} kg. What is its velocity?
3. Calculate the de Broglie wavelength for an α particle whose velocity is 10^7 m/second. The mass of an α particle is approximately 6.4×10^{-27} kg.
4. Calculate the de Broglie wavelength for an electron of velocity 10^6 m/second. The electron mass is 9.1×10^{-31} kg.

RADIOACTIVITY AND THE STRUCTURE OF THE NUCLEUS

The discovery of radioactivity was one of a series of major developments in physics which launched the twentieth century. It occurred a few months after the first of those events, the discovery of x-rays.

21.1 BECQUEREL AND THE DISCOVERY OF RADIOACTIVITY

Henri Becquerel, and his father before him, had for some time been interested in the phenomena of fluorescence and phosphorescence. *Fluorescence* is the property some materials have of shining visibly when irradiated by ultraviolet light. *Phosphorescence* is the property of glowing in the dark after the source of excitation has been removed. The two phenomena are intimately related. The difference is that fluorescence ceases immediately and phosphorescence persists for some time after the exciting light has been removed. One of the clues to the discovery of x-rays by Roentgen in 1895 was the fluorescence they induced in nearby objects.

Becquerel, in studying the phosphorescence and fluorescence of various substances when irradiated with light or x-rays, questioned whether x-rays might be emitted by phosphorescent substances. It was known that x-rays were capable of exposing a photographic plate wrapped in opaque paper to exclude all light. Therefore Becquerel wrapped photographic plates and placed various phosphorescent substances that had been fully excited by exposure to bright sunlight next to the wrapped plates. He observed no exposure of the photographic plate until he tried some crystals of uranium sulfate. These produced a clear image on the plate through the opaque paper. Becquerel therefore assumed that, in addition to visible light, these particular crystals emitted x-rays after exposure to bright sunlight.

Then there occurred one of the fortuitous events in the history of science. For several days the weather was cloudy, and Becquerel stored his wrapped plates and crystals away in a drawer, awaiting a sunny day. When no sun appeared for several days, he developed some of the plates, expecting to find a very faint image, since the crystals had not been exposed to sunlight and would not be expected to emit any x-rays. To his surprise the image on the photographic plate was as dense as before. He then stored uranium sulfate crystals in the dark for weeks until all the phosphorescence was gone, and discovered that the effect on the photographic plates was not diminished at all. Figure 21.1 shows the effect of a uranium-bearing mineral on a photographic plate.

This experiment established that uranium salts emit radiation, which, in its ability to penetrate black paper and affect a photographic plate, is very similar to x-rays. Once again, in the hands of a skilled observer, a chance observation had been turned into a major discovery. The phenomenon is called *radioactivity*.

By early 1896, Becquerel had shown that the radiation from uranium

(a)

(b)

FIGURE 21.1 (*a*) Photograph of a uranium-bearing mineral. (*b*) The effect of the same mineral on a photographic plate. The black regions of the mineral in (*a*) are radioactive. They show light in (*b*) because these are the regions where the radioactivity has affected the photographic plate. (*Courtesy Ward's Natural Science Establishment.*)

possessed another property similar to that of x-rays. It could discharge an electroscope, indicating that it made the air through which the radiation passed electrically conducting. This is called ionization. The rate at which the electroscope discharged was proportional to the amount of radiation, which thus provided a quantitative way to measure the intensity of radioactivity.

The primary excitement in physics at that time was the discovery and properties of x-rays (physics has its fads). But Becquerel's discovery attracted the attention of Marie Curie, a colleague of his in Paris. She quickly established that, of the known elements, only uranium and thorium showed the Becquerel rays. She also discovered that some of the ores of uranium, particularly pitchblende, showed more radioactivity than the amount of uranium in the ore accounted for. She and her husband, Pierre Curie, then embarked on a chemical search for the active substance. In 1898 they announced the discovery of a substance similar to the element bismuth in its chemical properties whose radioactivity was 400 times greater than that of uranium. They believed that this was a new element, to which they gave the name *polonium*. This element has the atomic number 84 in the modern periodic table (Table 12.3).

Shortly thereafter, the Curies announced the discovery of another radioactive element, this one chemically similar to the element barium and even more intensely radioactive than polonium. This element they called *radium*. It is number 88 in the periodic table. The process of obtaining pure radium, as the chemical compound radium chloride, from the ore was extremely tedious. First a chemical separation of barium was carried out. The radium separated with the barium because of the similarity of their chemical properties. Then it was necessary to use the fact that radium chloride dissolves very slightly less than barium chloride in a mixture of alcohol and water. The mixture of radium chloride and barium chloride was dissolved in the mixture of alcohol and water, and crystals were allowed to form. These crystals were slightly richer in radium chloride than the original mixture. This process had to be repeated many times, in order to separate pure radium chloride from barium chloride. Eventually, a few milligrams of pure radium chloride were obtained from several tons of ore. This separation process is called *fractional crystallization*. At each step in the process, the discharge of the electroscope was used to detect and follow the radioactivity through the chemical processes.

At this point there were two main lines of investigation to be pursued in order to understand radioactivity. One was the search for other radioactive species which would presumably be other new and unknown elements. The other was the study of the properties and the nature of the rays emitted from radioactive substances.

21.2 THE NATURE OF BECQUEREL RAYS

Two years after the discovery of radioactivity, in 1898, Ernest Rutherford isolated some radioactive material in such a way that he could measure the amount of ionization it produced in air, and thus the intensity of the rays it emitted. He then placed layers of aluminum foil between the radioactive source and his detector. He noted that a significant fraction of the ionization disappeared when the first few layers of foil were added. The remainder of the radiation required much more foil to block it. He thus determined that there are two kinds of rays whose ability to penetrate aluminum foil differs markedly. He called these α (Greek alpha) and β (Greek beta) rays, with the β ray the more penetrating. Later a third and even more penetrating radiation was found and designated γ (Greek gamma) rays.

Almost immediately these rays were tested by electric and magnetic fields in much the same type of apparatus as the one Thomson had used to measure the ratio of charge to mass of the electron e/m. The results showed that β rays had a ratio of charge to mass which was the same as that of cathode rays, while it appeared that α and γ rays were not at all deflected by a magnetic field. Therefore Rutherford concluded that β rays are electrons. When he had constructed a much larger magnet, it was found that α particles were also deflected in a magnetic field. They turned out to have a ratio of charge to mass about 1,000 times smaller than that of the electron, and a charge of the opposite sign. Also, they had a velocity nearly one-tenth the velocity of light!

Since by this time it was believed that the electron was one of the constituents of atoms, and yet had little of the atom's mass, it was easy to calculate the charge-to-mass ratio for an atom which was missing one electron, using the known atomic masses. The ratio found for α rays fitted what one would expect for a hydrogen molecule singly ionized (missing one electron) or a helium atom doubly ionized (missing both electrons). Rutherford's intuition led him to propose helium as the correct choice. If α rays were doubly ionized helium atoms, it should be possible to capture some and demonstrate the presence of helium gas. This was accomplished in 1903 by taking advantage of the fact that one of the radioactive materials formed in the decay of radium is a gas, radon, which itself emits α particles. Since α particles can penetrate a small amount of matter, radon was trapped on one side of a thin glass tube and the other side evacuated. This arrangement is shown schematically in Fig. 21.2. The radon gas could not penetrate the glass, but the much smaller α particles did. After about a week, the region which had been evacuated was analyzed spectroscopically for the presence of helium gas. It was found. Therefore one concludes that α particles are helium atoms with two electrons missing. (The student might wonder where this helium regains its electrons. In fact, it steals them from some part of the apparatus, leaving one area of the apparatus electron-poor and another electron-rich. Since no material is a perfect electric insulator, when this happens, current will flow to correct the imbalance.)

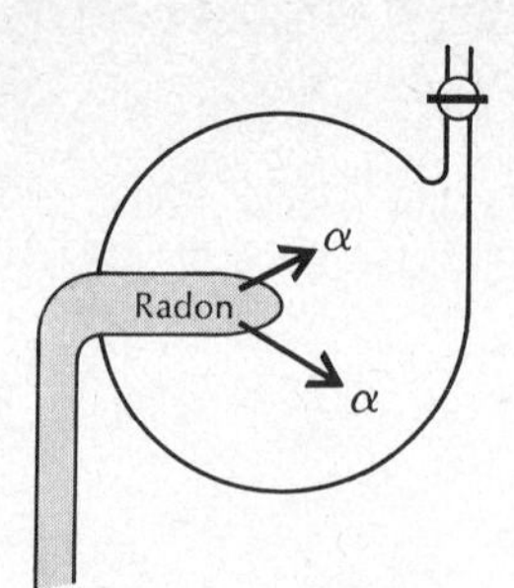

FIGURE 21.2 The experiment that demonstrated that α particles are helium nuclei. The α particles penetrate the thin tube containing the radon. After a week the gas in the large bulb is examined and found to be helium.

This experiment described above thus shows that the α particle carries away the mass of one helium atom. To conserve total mass, the atom left behind must be four mass units lighter than it was before, since the atomic mass of helium is about 4. That is, in the radioactive process, the product atom after the emission of an α particle must be an atom of an element four mass units lighter than the parent atom. Since the α particle has two positive charges, the atom left behind must lose two positive charges. When a β particle is emitted, one negative charge, but almost no mass, is lost from the atom.

We have shown in the discussion of the Rutherford model of the atom that the nucleus of an atom has a positive charge equal to the atomic number times the charge of the electron. Helium has the atomic number 2, and thus has two positive charges on its nucleus. Because the α particle has two positive charges and the mass of a helium atom, we can identify the α particle as the *nucleus* of a helium atom, missing both its electrons. (This identification was not made until after the development of the nuclear atom by Rutherford in 1911.)

In 1903 Rutherford and F. Soddy proposed a theory to account for the events occurring in radioactivity. The chief idea was that the atoms of a radioactive element change to atoms of a different element when an α or β particle is emitted. If emission of an α particle reduces the nuclear charge Z by 2, obviously, the atomic number is also reduced by 2. The emission of a negative electron (β particle) *increases* the net positive charge on the nucleus by 1. We shall postpone a further discussion of these radioactive transmutations until we have learned more about the structure of the nucleus.

21.3 THE STRUCTURE OF THE NUCLEUS

In Chapter 19 we stated that the α-particle-scattering experiment of Rutherford showed that the nucleus of an atom is very small compared with an atom, is positively charged, and contains almost all the mass of the atom. Rutherford's work, and also that of Mosely, showed that the positive charge of the nucleus is a whole number Z times the value of the charge of the electron e ($e = 1.6 \times 10^{-19}$ coulomb). The whole number Z is the atomic number of the atom. We thus have two quantities that can be used to characterize any particular nucleus: its atomic number, which gives the number of elementary charges, and its mass. To remind us of some of the values of these quantities, they are: for hydrogen, $Z = 1$, atomic mass = 1; for oxygen, $Z = 8$, atomic mass = 16; for sodium, $Z = 11$, atomic mass = 23; for mercury, $Z = 80$, atomic mass = 200.5; and for uranium, $Z = 92$, atomic mass = 238. Except for hydrogen, the atomic mass is always numerically greater than the atomic number.

As early as 1815 an English chemist, William Prout, proposed that all chemical elements were combinations of hydrogen, the lightest element. Since the atomic mass of hydrogen is 1, all atomic masses should then be whole numbers. Many, but not all, turned out to be very nearly whole numbers. Chlorine, for example, has an atomic mass of 35.5. Because of this result, Prout's idea was discarded.

After the discovery of the nuclear atom, this idea was revived in a slightly different form. It was proposed that the nucleus of the hydrogen atom was the basic building block of the nucleus, combined with enough electrons to make the charge come out right. The nucleus of the hydrogen atom is called the *proton*. There were supposed to be electrons *in* the nucleus, in addition to the planetary electrons of Bohr around the nucleus. For example, hydrogen has mass = 1 and $Z = 1$. Oxygen is found to have mass = 16 and $Z = 8$. Therefore the nucleus of oxygen was supposed to contain 16 protons and 8 electrons. This gives a mass of 16 and a positive charge of $16 - 8 = 8$. (The negative charge of the eight electrons would cancel the positive charge of eight of the protons.)

This idea still did not account for the existence of atomic masses which were not whole multiples of the atomic mass of hydrogen. The problem was resolved by the discovery of *isotopes* of elements. In 1912 Thomson was studying the atomic masses of the positive ions obtained in canal rays. His experiment gave the atomic masses of individual atoms. When he studied the gas neon, whose atomic mass, measured using large numbers of atoms, is 20.2, he found two masses: one 20 and the other 22. Because of problems which had arisen in the study of radioactivity, Soddy had earlier suggested that there might be atoms of the elements which had identical chemical properties but differed in other respects. These he called *isotopes*. Thomson's experiment showed the existence of isotopes of neon of atomic mass 20 and atomic mass 22. Later work showed that ordinary neon consists of 91 percent mass-20 isotope, and 9 percent mass-22 isotope, and a trace of mass-21 isotope.

The mass of 20.2 measured for large numbers of atoms is an average of the masses of the atoms of the individual isotopes.

The instrument now used to measure the masses and amounts of different isotopes is called a *mass spectrograph*. Almost all the elements have more than one stable (not radioactive) isotope. Figure 21.3 shows data for mercury. It shows significant amounts of isotopes with mass 198, 199, 200, 201, 202, and 204, with perhaps a trace of 196. The element tin has 10 isotopes.

Note that the masses of all of the individual isotopes are almost precisely whole numbers. With this observation it became reasonable to propose that each nucleus is a combination of a certain number of protons and electrons. Adding one proton and one electron increases the mass of a nucleus by 1, leaving the charge unchanged. Adding one proton increases the mass by 1 and the charge by 1. All the nuclear masses (with only a small correction for the mass of the electron) should then be multiples of the proton (hydrogen) mass. The nucleus of the isotope of neon with mass 22 was assumed to have two more protons and two more electrons than the nucleus of the mass-20 isotope. The difficulty with this whole scheme is that the idea that electrons are confined inside the nucleus was very difficult to understand. As time passed, it became increasingly clear that an electron cannot be confined to such a small region.

The problem was finally resolved in 1930 by an experiment performed by W. Bothe and H. Becker in Germany and interpreted by J. Chadwick in 1932. The experiment consisted of bombarding the metal beryllium (atomic number 4) with α particles from radioactive polonium. Radiation was given off from the beryllium. This radiation was far more

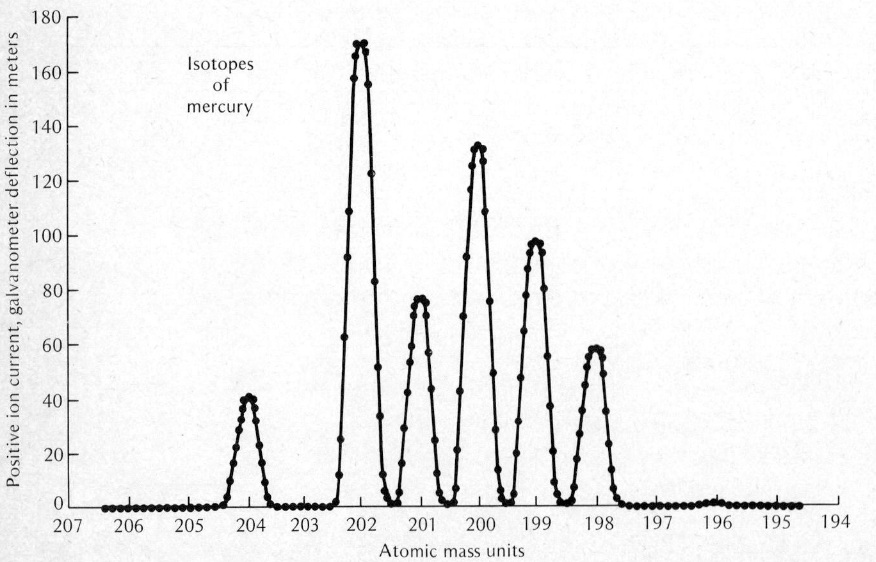

FIGURE 21.3 A mass spectrograph shows that mercury has at least six isotopes. The height of the peak is proportional to the amount of each isotope present. (*Courtesy A. O. C. Nier.*)

penetrating than either α or β particles. A first hypothesis was that it was a form of γ rays. A further observation served to disprove this: The emissions from the beryllium knocked protons out of a target containing hydrogen. Gamma rays would not have the kinetic energy required to dislodge a particle so massive as a proton. Chadwick resolved the problem by suggesting that the great penetrating power occurred because the object had no electric charge. He estimated the mass of this object by comparing its ability to knock protons out of a target with its ability to knock out heavier nuclei such as nitrogen. His conclusion was that the radiation consisted of uncharged particles somewhat heavier than the proton. These particles are called *neutrons*. The mass of the neutron is now known to be only about 0.1 percent more than that of the proton.

FIGURE 21.4 James Chadwick (1891–). (*Courtesy Burndy Library.*)

With the discovery of the neutron, it was relatively easy to understand the structure of the nucleus. Each nucleus consists of a certain number of protons and neutrons. The number of protons is given by the atomic number Z, because each proton has one unit of charge. The mass is the sum of the masses of the protons and neutrons. Since the mass of each of these is nearly 1 on the atomic-mass scale, the mass of each nucleus will be nearly a whole number. A useful concept applying this idea is the *mass number A*. The mass number is the total number of particles, protons plus neutrons. The mass number is also the nearest whole number to the exact atomic mass of the particular nucleus. With these two ideas we can show a symbolic way of writing any nucleus. The form is

$${}^{A}_{Z}\text{Chemical symbol}$$

For instance, ${}^{4}_{2}He$ is the symbol for the isotope of helium (He) with nuclear charge $Z = 2$ and mass number $A = 4$. The isotopes of mass 20 and 22 of neon, previously discussed, are written ${}^{20}_{10}Ne$ and ${}^{22}_{10}Ne$. The number of neutrons in a specific nucleus is called the *neutron number N* and is given by $N = A - Z$. ${}^{20}_{10}Ne$ has 10 neutrons, and ${}^{22}_{10}Ne$ has 12.

Example 21.1

How many protons and how many neutrons are there in the nuclei ${}^{226}_{88}Ra$ and ${}^{141}_{56}Ba$? The atomic number of ${}^{226}_{88}Ra$ is 88, and therefore there are 88 protons. The number of neutrons is given by the mass number, 226, minus the number of protons, 88.

$$226 - 88 = 138 \text{ neutrons}$$

Similarly, ${}^{141}_{56}Ba$ has 56 protons and $141 - 56 = 85$ neutrons.

Unsolved in this entire discussion is the question of what holds the protons and neutrons together in the nucleus. The forces holding them there are very strong—much stronger than electric or gravitational forces. This force is called the *nuclear force*, and will be discussed later. For now we shall merely assume that there exists such a force which holds the protons and neutrons together.

21.4 OTHER RADIOACTIVE SPECIES

William Crookes isolated, in 1900, a radioactive species from "pure" uranium, which was chemically unlike uranium. The radioactivity of this substance, which Crookes called uranium X, was of great intensity. From "pure" thorium he also isolated thorium X, a new substance whose radioactivity declined to one-half its original intensity in about 4 days. When thorium X was completely removed from the parent material, in a few days more thorium X was found in the original sample. The process could apparently be repeated indefinitely. This showed that a material is continuously being produced in thorium at a rate independent of the physical or chemical state of the original thorium. Rutherford and Soddy were led to propose in 1903 the startling (at that time) hypothesis that in the process of radioactivity, one element, thorium, in this example, is being transformed into another. This was the first proposal since the days of alchemists that allowed the possibility of one element being transformed into another.

21.5 RADIOACTIVE TRANSMUTATIONS

With the information now at hand, it is relatively easy to understand what happens to the nucleus of an atom in the process of radioactivity. In radioactivity, α, β, or γ rays are given off. The α particle was shown to be the nucleus of a helium atom, ${}^{4}_{2}He$. The β particle was shown to be an electron. For the moment we shall discuss just these two.

An α particle consists of two protons and two neutrons, since it is the nucleus ${}^{4}_{2}He$. If it leaves the nucleus of an atom, the nucleus remaining contains two fewer protons and two fewer neutrons than the original one. Loss of an α particle thus reduces the nuclear charge Z by 2, and the mass number A by 4. Let us consider the emission of an α particle by the isotope of uranium with mass 238. Uranium has $Z = 92$.

$${}^{238}_{92}U \rightarrow {}^{4}_{2}He + {}^{234}_{90}(\quad) \tag{21.1}$$

The chemical symbol left out is that for the element with atomic number 90, which is thorium (see Table 12.3). The loss of 4 mass units means that this is the isotope of thorium whose mass number is 234, ${}^{234}_{90}Th$. Therefore the emission of an α particle by ${}^{238}_{92}U$ leads to the formation of a nucleus of the element thorium. Thus this isotope of thorium will be continuously formed in a sample of uranium in the process of radioactive α decay. The isotope ${}^{234}_{90}Th$ has been identified as uranium X.

We can understand β emission or β decay in a similar fashion. The emission of a negatively charged electron by the nucleus increases the net positive charge of the nucleus by 1 unit. The way to understand this is to realize that the total electric charge of a system is the *algebraic* sum of the individual electric charges. Two positive and two negative charges with the same magnitude yield zero net charge. The removal of one negative charge then increases the positive charge of the nucleus.

The mass of the electron is so small that the mass number of the nucleus left behind in β decay is unchanged. $^{234}_{90}\mathrm{Th}$ is known to emit β particles. We have

$$^{234}_{90}\mathrm{Th} \rightarrow \beta + {}^{234}_{91}(\quad) \tag{21.2}$$

The element with $Z = 91$ is protactinium. The particular nucleus created is $^{234}_{91}\mathrm{Pa}$. β Decay increases the atomic number by one, and leaves the mass number unchanged.

Example 21.2

Find the nucleus which completes the reaction for the gas radon emitting an α particle.

$$^{220}_{86}\mathrm{Rn} \rightarrow {}^{4}_{2}\mathrm{He} + ?$$

The sum of atomic numbers, and thus charges, must be the same on the left as on the right. Therefore the missing element must have atomic number $86 - 2 = 84$. The total number of protons and neutrons as given by the mass number must remain constant also. Therefore the mass number of the unknown species is $220 - 4 = 216$. Since atomic number 84 is polonium, the complete reaction is written

$$^{220}_{86}\mathrm{Rn} \rightarrow {}^{4}_{2}\mathrm{He} + {}^{216}_{84}\mathrm{Po}$$

It is found that the nucleus $^{238}_{92}\mathrm{U}$ is the beginning of a long sequence of radioactive decays, each of which changes the nucleus involved to that of another element. Some of the steps in this chain are

$$^{238}_{92}\mathrm{U} \rightarrow {}^{4}_{2}\mathrm{He} + {}^{234}_{90}\mathrm{Th} \qquad \alpha \text{ decay} \tag{21.3}$$

$$^{234}_{90}\mathrm{Th} \rightarrow \beta + {}^{234}_{91}\mathrm{Pa} \qquad \beta \text{ decay} \tag{21.4}$$

$$^{234}_{91}\mathrm{Pa} \rightarrow \beta + {}^{234}_{92}\mathrm{U} \qquad \beta \text{ decay} \tag{21.5}$$

$$^{234}_{92}\mathrm{U} \rightarrow {}^{4}_{2}\mathrm{He} + {}^{230}_{90}\mathrm{Th} \qquad \alpha \text{ decay} \tag{21.6}$$

$$^{230}_{90}\mathrm{Th} \rightarrow {}^{4}_{2}\mathrm{He} + {}^{226}_{88}\mathrm{Ra} \qquad \alpha \text{ decay} \tag{21.7}$$

The final product of this chain is the isotope $^{206}_{82}\mathrm{Pb}$, lead 206. Between $^{238}_{92}\mathrm{U}$ and $^{206}_{82}\mathrm{Pb}$ eight α particles and six β particles are emitted. The total mass change of 32 units indicates that eight α's of mass 4 have been emitted. The six β's are necessary to bring the nuclear charge up to 82, since eight α's would reduce the charge from 92 to 76. Remember, each α emitted reduces the nuclear charge by 2, and each β increases it by 1.

Discussed above is what is called a radioactive-decay series. Since the masses of the nuclei change only by 4 when α particles are emitted (the β's cause no appreciable mass change), there can be four of these series. The one written above goes through numbers 238, 234, 230, 226, 222, 218, 214, 210, and 206. Other series begin at $^{232}_{90}Th$ and $^{235}_{92}U$. The fourth series, which was not found in nature but has been found since the creation of artificial isotopes, begins at $^{237}_{93}Np$.

In all of this discussion the emission of γ rays has been ignored. Gamma emission changes neither the nuclear charge nor the mass number. Gamma emission occurs after α or β emission, because the product nucleus has been left in an excited state, and must shed some energy. Gamma rays are electromagnetic waves of wavelengths as short or shorter than x-rays. Gamma rays and x-rays are identical. The different names are used more to indicate the source than to imply any intrinsic difference.

21.6 HALF-LIFE OF RADIOACTIVE ATOMS

One of the characteristics of radioactive materials is that their radioactivity seems to disappear in a very characteristic time. We discussed the material thorium X, which occurs in the decay series beginning with $^{232}_{90}Th$. Thorium X has been shown to be an isotope of radium, $^{224}_{88}Ra$. It decays with the emission of an α particle to radon, Rn.

$$^{224}_{88}Ra \rightarrow {}^{4}_{2}He + {}^{220}_{86}Rn \tag{21.8}$$

Radium 224 decays so that one-half of the radioactivity disappears in a few days (3.6 days to be exact). In the next period of 3.6 days another half disappears. The time in which the radioactivity of a sample decreases by one-half (3.6 days in this case) is called the *half-life* of the particular radioactive species. The half-lives of several naturally occurring radioactive isotopes are given in Table 21.1. We see that the isotopes $^{238}_{92}U$ and $^{235}_{92}U$ have different half-lives. The half-life is then not characteristic of the particular chemical element, but of each individual isotope.

The existence of a particular period of time in which half of any given sample of radioactive nuclei undergoes decay is a result of the nature of the decay process. For each radioactive nucleus there is a specific probability that it will decay in a given interval of time. Suppose this probability is $\frac{1}{100}$ per second. This means that in any given second a particular nucleus will decay once in 100 observations. If we observe 100 nuclei with this decay constant, on the average one will

TABLE 21.1 Half-lives of some radioactive isotopes

Element	Nuclide	Half-life	β or α decay
Uranium	$^{234}_{92}U$	2.44×10^5 years	α
	$^{235}_{92}U$	7.0×10^8 years	α
	$^{238}_{92}U$	4.5×10^9 years	α
Thorium	$^{232}_{90}Th$	1.39×10^{10} years	α
	$^{234}_{90}Th$	24.1 days	β
	$^{238}_{90}Th$	1.90 years	α
Radium	$^{224}_{88}Ra$	3.64 days	α
	$^{226}_{88}Ra$	1,622 years	α
	$^{228}_{88}Ra$	5.75 years	β
Polonium	$^{210}_{84}Po$	138 days	α
	$^{214}_{84}Po$	1.6×10^{-4} second	α
	$^{216}_{84}Po$	0.15 second	α, β
	$^{218}_{84}Po$	3.05 minutes	α
Lead	$^{206}_{82}Pb$	Stable	
	$^{207}_{82}Pb$	Stable	
	$^{208}_{82}Pb$	Stable	
	$^{210}_{82}Pb$	22 years	β
	$^{214}_{82}Pb$	26.8 minutes	β

decay every second. When the number of nuclei is reduced by this process to 50, one will decay on the average of every 2 seconds, because there are only one-half as many as there were. The rate of decay is cut in half, and the number of nuclei is also cut in half, so that it will take just as long to reduce the number from 50 to 25 as it did from 100 to 50.

The age of the earth can be estimated from radioactive half-lives. The half-life of $^{238}_{92}U$ is 4.5×10^9 years. (No one has waited this long. This value is found by measuring the rate of decay in a sample with a known number of atoms.) If the age of the earth were several times as large as this number, all uranium existing at the time of the formation of the earth would have decayed. There is also some additional information. The end product of the chain of radioactive decay is an isotope of lead, $^{206}_{92}Pb$. If one assumes that all lead 206 found in a sample of uranium mineral was formed by radioactive decay from uranium, then the original amount of uranium can be found. In this manner maximum and minimum ages for the earth can be estimated. These are 2.7 billion and 5.6 billion years. That is, it can be reliably stated, using the evidence from radioactivity alone, that the earth is not much older than 5.6 billion years or much younger than 2.7 billion years. Increasingly sophisticated use of these ideas has narrowed down this range. The present best estimate of the age of the earth is about 4.5 billion years.

Many other radioactive isotopes are used for dating. Carbon 14, with a half-life of a little over 5,000 years, is used to date archeological artifacts. Carbon dioxide in the atmosphere contains a certain amount of carbon 14 formed by cosmic rays. This carbon 14 is incorporated into living plants. When a plant dies, it adds no new carbon 14. Therefore the amount of carbon-14 radioactivity in a sample of plant material, like wood, decays with a half-life of 5,000 years. Measurements of carbon-14 activity have been used to date accurately charcoal and other wood objects as old as 25,000 years.

21.7 ARTIFICIAL TRANSMUTATIONS

In 1919 Ernest Rutherford, whose name permeates the story of nuclear physics from the very beginning, published the results of another study which had far-reaching importance. In studying the passage of α particles through various gases, such as oxygen and nitrogen, he found that α particles from a given radioactive material had a very characteristic range which they traveled in any gas. Beyond this range few α particles were found. But when α particles passed through nitrogen gas, a particle was found which traveled much farther than the α particles. He was able to convince himself that this particle was the nucleus of a hydrogen atom, the proton. He therefore concluded that the α particle had struck a nitrogen nucleus and ejected a different particle, the proton. In the process the nucleus of the nitrogen atom must have been changed to a different nucleus. We can schematically write the process as

$$^{14}_{7}\text{N} + {}^{4}_{2}\text{He} \rightarrow {}^{1}_{1}\text{H} + {}^{?}_{?}\text{X}$$

It is always found in nuclear processes that the total number of nuclear particles (protons plus neutrons) is left unchanged. Also left unchanged is the total electric charge. Therefore the sum of the atomic numbers on the left $(7 + 2)$ must equal the sum of the atomic numbers on the right $(1 + ?)$. The atomic number (number of protons) of X must be 8, which is the atomic number of oxygen. The sum of the mass numbers on the left $(14 + 4)$ must be the same as on the right $(1 + ?)$. Therefore the mass number of the nucleus which has been created is 17.

$$^{14}_{7}\text{N} + {}^{4}_{2}\text{He} \rightarrow {}^{1}_{1}\text{H} + {}^{17}_{8}\text{O} \qquad (21.9)$$

The impact of the α particle on the nitrogen nucleus has apparently created a different nuclear species. This was the first example of the controlled artificial transmutation of any nucleus by bombardment with a nuclear particle. For all nuclear processes ever observed the two basic

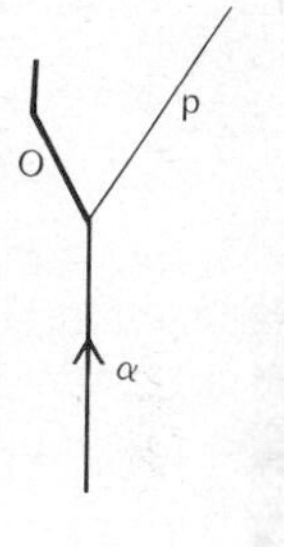

FIGURE 21.5 Cloud-chamber photograph of a collision of a particle with a nitrogen nucleus. α Particles enter from below. At the left a collision occurs. The thin track moving upward from the collision point is the proton. The thick track is the product oxygen nucleus. (*Courtesy Professor P. M. S. Blackett.*)

rules are found to hold: (1) The total number of particles (protons plus neutrons) is the same before and after, and (2) the total charge is the same before and after.

Another particle which has proved even more fruitful than the α particle in producing artificial transmutations is the neutron. The major reason for this is the fact that the neutron is uncharged and can penetrate to the positively charged nucleus more easily than the charged α particle or protons. Enrico Fermi was the first to realize the potentialities of the neutron. He and a group of co-workers in Rome between 1934 and 1938 bombarded as many of the known elements as possible with neutrons. In most cases new isotopes were formed. The predominant process is the capture of a neutron by the nucleus in question. Sometimes this new nucleus is radioactive and emits a β particle. It thus is possible to create radioactivity artificially by bombardment with neutrons or other particles.

For instance, ordinary hydrogen captures a neutron to form the mass-2 isotope of hydrogen called *deuterium*.

$$^1_1\text{H} + ^1_0\text{n} \rightarrow ^2_1\text{H} \qquad (21.10)$$

Another similar reaction which occurs is

$$^{65}_{29}\text{Cu} + ^1_0\text{n} \rightarrow ^{66}_{29}\text{Cu} \qquad (21.11)$$

In this case the product copper 66 is radioactive and emits a β particle to form zinc 66:

$$^{66}_{29}\text{Cu} \rightarrow \beta^- + ^{66}_{30}\text{Zn} \qquad (21.12)$$

FIGURE 21.6 Enrico Fermi (1901–1945). (*Courtesy University of Pennsylvania Library.*)

When Fermi and his group reached the end of the periodic table at uranium, their results were confusing. Apparently they had created an isotope, $^{239}_{92}\text{U}$, which they believed decayed with the emission of an electron to an isotope of an element with atomic number 93, according to the reactions

$$^{238}_{92}\text{U} + ^1_0\text{n} \rightarrow ^{239}_{92}\text{U} \qquad (21.13)$$

$$^{239}_{92}\text{U} \rightarrow \beta^- + ^{239}_{93}\text{Np} \qquad (21.14)$$

In fact, the processes which occur when uranium is bombarded with neutrons are much more complex. Fermi just missed discovering the process of nuclear fission, which will be discussed in Section 21.9.

21.8 THE ENERGY IN NUCLEAR PROCESSES

Very early in the study of radioactive materials Pierre Curie and A. Taborde observed that a sample of radium kept isolated from its surroundings was warmer than the surroundings. Radium apparently generates heat, and thus energy, continuously. Radium has a half-life of 1,622 years, and in 1,622 years 1.84 million calories of heat is evolved from each gram of radium. This is roughly the amount of heat obtained from burning 515 pounds, or about 240,000 grams of coal. Thus the amount of energy available in radioactive processes is enormous. In the decay of radium this energy shows itself as kinetic energy of the α particles which are emitted.

As early as 1905 Rutherford said, in an article in *Harper's Magazine*, "It is not impossible that under the influence of the very high solar temperature, the atoms of the non-radioactive elements may break up into the simpler forms with the evolution of large quantities of energy." Even before the discovery of the nucleus, he was well aware of the extraordinary amounts of energy available in radioactivity. In a lecture to the British Association for the Advancement of Science, in 1916, he pointed out prophetically that no one had yet succeeded in releasing at will the enormous energy of radium and, personally, he hoped this event would not occur until man could live peacefully with his neighbors.

The source of the tremendous energies available in radioactivity is what has come to be called *atomic energy* (nuclear energy is a better term). To understand this we must remember the result of Einstein's from Chapter 17. There it was pointed out that energy and mass are convertible one into the other. The conversion equation is Einstein's mass-energy relationship

$$E = mc^2 \tag{21.15}$$

Perhaps a more appropriate way to write this result is

$$E + mc^2 = \text{a constant} \tag{21.16}$$

This states that the sum of the energy associated with the mass and all other forms of energy, kinetic, potential, etc., is a constant. If some mass disappears, kinetic energy or some other form of energy must appear.

In Chapter 17 we calculated the energy equivalent of 1 kilogram of matter and found it to be 9×10^{16} joules—an enormous amount of energy. In the use of Equations 21.15 and 21.16 the units of energy are joules if we express m in kilograms and c in meters per second.

We can now find the source of energy in the decay of radium by investigating the masses of the various nuclei involved in the decay. Radium decays into two gases, radon and helium:

$$^{226}_{88}\text{Ra} \rightarrow {}^{4}_{2}\text{He} + {}^{222}_{86}\text{Rn} + \text{energy} \tag{21.17}$$

The exact atomic masses are: radium 226, 226.025 atomic mass units; helium 4, 4.003 atomic mass units; and radon 222, 222.018 atomic mass units. (These can be measured using a mass spectrograph.) The total mass of the products is $4.003 + 222.018 = 226.021$. This is less than the mass of the radium by 0.004 atomic mass unit. But suppose we deal with Avogadro's number of atoms. (See Section 12.7.) Then 226.025 grams of radium will decay to 226.021 grams of products, and 0.004 gram, or 4×10^{-6} kilogram, of mass will disappear. From Equations 21.15 and 21.16, we find that this yields an energy of

$$(4 \times 10^{-6} \text{ kilogram})(3 \times 10^{8} \text{ meters/second})^2 = 3.6 \times 10^{11} \text{ joules}$$

In all spontaneous radioactive decays the total mass of the products is less than the total mass of the starting nucleus. This loss in mass appears as kinetic energy of the products. If a reaction in which the products are more massive than the starting materials is contemplated, additional energy must be added in the form of kinetic energy of one of the starting materials.

21.9 NUCLEAR FISSION

In 1939, just before the outbreak of war, Otto Hahn and F. Strassman in Germany discovered and published a result of enormous significance. They bombarded uranium with neutrons, as Fermi had done. The heavy and most abundant isotope of uranium, $^{238}_{92}U$, captured a neutron to form an unstable isotope, $^{239}_{92}U$. But the isotope $^{235}_{92}U$ behaved in a more spectacular fashion. It captured a neutron to form $^{236}_{92}U$ and then also decayed. Among the decay products of uranium 236 was found the nucleus $^{141}_{56}Ba$, an isotope of barium. This indicated that the uranium had broken up into two (or possibly more) fragments, each much lighter than the original nucleus. When the complete reaction is studied, it works out as follows:

$$^{235}_{92}U + ^{1}_{0}n \rightarrow ^{236}_{92}U \rightarrow ^{141}_{56}Ba + ^{92}_{36}Kr + 3^{1}_{0}n + \text{energy} \qquad (21.18)$$

In addition to the two large fragments, three neutrons are among the products. Since only one neutron was needed to start the process, this gives the possibility of a sustained chain reaction in which the neutrons from one event trigger the next. The process is called *nuclear fission* because the nucleus breaks into parts. The products are not always barium and krypton. Sometimes they are other nuclei whose masses are near 141 and 92, but neutrons are always released.

If the masses of the starting and end products are compared, about 0.2 gram of mass disappears for 235 grams of uranium 235 which undergoes fission. This can only appear, as discussed above, as kinetic energy of the final products. It is this enormous release of energy which makes the process attractive for both useful and destructive purposes. The bombs used on Hiroshima and Nagasaki in World War II released an amount of energy roughly equivalent to 1 gram of matter in a period of time less than 0.001 second. The same reactions used in a controlled way in a nuclear pile, or reactor, can deliver heat energy for the generation of electric power or any other useful purpose.

At present less than 5 percent of United States electric energy is generated using nuclear reactors. There are predictions that by the end of the century perhaps one-half of our electric energy will come from reactors. There are also some very serious concerns and drawbacks involving the use of nuclear power.

The primary environmental concern arises from the fact that the fission products are intensely radioactive. The fission of uranium produces a variety of isotopes of differing half-lives. After a reactor has been operating for a period of time, the fission products interfere with the reaction, and the fuel must be purified. This results in large quantities of radioactive waste material. This material must be stored for thousands of years before its radioactivity will have died down to a safe level. The handling and storage of this extremely dangerous material poses enormous practical problems which have not yet been satisfactorily solved.

FIGURE 21.7 The Point Branch nuclear power plant in Wisconsin. An increasing share of our electric power will be produced by such plants in the future. The debate on risks and benefits will certainly continue. (*Courtesy Wisconsin Electric Power Co.*)

The dangers in nuclear radiation lie in the fact that the energies of the α, β, and γ rays are roughly a million times the energy of a chemical bond. So as one of these particles traverses a biological system, it breaks thousands or more chemical bonds. These bonds may be in enzymes the body needs, or in the DNA molecules which contain the genetic code, or in eggs or sperm necessary for the propagation of a species. If a person is subjected to intense doses of radioactivity, he dies immediately from a breakdown of the central nervous system. Much lighter doses can lead to cancers, including leukemia. Or if the irradiated individual is unharmed, genetic effects may be passed on to future generations. We should carefully total up the risks and hazards before deciding that nuclear energy is the only way to solve our energy needs.

A nuclear reactor cannot explode like an atomic bomb. If a reactor got completely out of control the result would be more like a "fizzle" than a bomb. Nevertheless, such a weak explosion might break the containment system of the reactor, and spread radioactive materials in the vicinity. It is also possible that a runaway reactor might melt its way through the floor of the building and deep into the ground before cooling. This could lead to a radioactive contamination of water supplies. Extensive safety systems are included in all reactor designs, but there are still doubts in the scientific community that these are adequate. The debate is continuing.

The uranium-235 isotope which fuels a reactor is only 0.7 percent

FIGURE 21.8 The world's first atomic bomb explosion in July 1945, at Alamagordo, New Mexico. The explosive force was approximately equivalent to 20,000 tons of TNT, and resulted from the conversion of roughly one gram of matter to energy. (*Courtesy Los Alamos Scientific Laboratory.*)

of uranium as it comes from the ground. The remainder is uranium 238, which cannot be directly used in a reactor. The amount of uranium 235 available is quite limited, and will be rapidly exhausted if we do not find means to obtain additional nuclear fuel. The proposed solution to this problem is a *breeder reactor.*

A breeder reactor creates as much or more fuel than it uses. This is achieved by the capture of neutrons in uranium 238. One of the neutrons from each fission is used to create the next fission in the chain reaction. The remainder of the neutrons (slightly more than one per fission), are captured by uranium 238. After two β decays this becomes ${}^{239}_{94}Pu$ (plutonium 239). Plutonium, which is not found in nature, can be used as a reactor fuel, or to make bombs. Since over 99 percent of the uranium in nature is uranium 238, the success of the breeder will increase the available nuclear fuel many fold.

The present status of breeder technology is open to debate. A number of breeders are nearing completion, but there is significant concern for their safety. A uranium-235 reactor is easier to control than a breeder reactor using plutonium as fuel. (The reasons for this are quite technical.) If breeder technology does not work, then nuclear fission is only a stopgap solution to the world's energy problems. There simply does not exist sufficient uranium 235 to supply all our energy needs. If the breeder does work, we will have available very large quantities of energy, but with very significant risks. It should be mentioned, in passing, that uranium 235 is the only naturally occurring reactor fuel. If the supply is used up in reactors that are not breeders, then there will be no way to start the breeding cycle. This might easily occur if present trends continue.

There is an additional, very significant danger. Uranium, as used in reactors, is enriched to 3 to 5 percent uranium 235 from the original 0.7 percent. To make a bomb requires enrichment to about 90 percent, which is very difficult. Therefore uranium reactor fuel, if stolen, could not easily be used to make a clandestine bomb. Plutonium 239, however, comes pure. If a significant fraction of our energy industry becomes based on plutonium breeder reactors, the amounts of plutonium that will be shipped about will be very large. The possibility that enough could be stolen to fabricate a crude bomb is not insignificant (a few kilograms is enough). This danger will increase as more and more plutonium comes into use. In a world where terrorism seems to be steadily increasing, this is not a comforting prospect.

21.10 NUCLEAR FUSION

At the other end of the periodic chart there exists another source of nuclear energy. Measurements show that the α particle is, roughly, 0.5 percent lighter than the sum of its parts (two protons and two neutrons). The fusion of two protons and two neutrons to form a helium atom should then also yield energy. This is the process of *nuclear fusion*. Its only known examples, at present, are the hydrogen bomb and the processes which generate energy in the stars. Since free neutrons do not exist, the simplest version of this process is the fusion of two deuterium atoms (the heavy isotope of hydrogen) to form one helium atom:

$$2\,{}^{2}_{1}\mathrm{H} \rightarrow {}^{4}_{2}\mathrm{He} + \text{energy} \qquad (21.19)$$

In fact, the process does not go this simply, but this is the net result of a more complex series of reactions.

In the sun the process is the fusion of four protons to form helium nuclei. The excess electric charge is carried off by positive electrons, called *positrons*

$$4\,{}^{1}_{1}\mathrm{H} \rightarrow {}^{4}_{2}\mathrm{He} + 2\,\beta^{+} + \text{energy} \qquad (21.20)$$

Again, Equation 21.20 is the sum of a complex series of reactions.

To achieve a fusion reaction, temperatures in excess of 10 million kelvins are required. This is necessary in order that the nuclei overcome the repulsion of their electric charges and get close enough together to react. In a hydrogen bomb this is achieved by using a fission bomb as the trigger. The fuel for a hydrogen bomb is the solid compound lithium hydride, LiH, composed of the particular isotopes ${}^{6}_{3}\mathrm{Li}$ and ${}^{2}_{1}\mathrm{H}$.

These react to form helium

$$^{6}_{3}\text{Li} + ^{2}_{1}\text{H} \rightarrow 2^{4}_{2}\text{He} + \text{energy} \qquad (21.21)$$

Again this is the net result of a more complex process.

Two major schemes have been proposed to produce a controlled fusion reaction. The first, which has been studied for more than 20 years, involves heating gas plasmas to very high temperatures and using magnetic fields to keep the gas from the walls of its container. This has turned out to be a much more difficult problem than originally imagined, and a working fusion reactor that could generate electric power is not yet in sight.

The second proposal is called *laser-induced fusion*. The idea is to use powerful lasers to heat a pellet of fuel (probably deuterium and tritium, the mass-3 isotope of hydrogen) so that it reaches the necessary temperatures very quickly. Energy generation in such a device would be a series of mini-explosions. To achieve these results lasers that deliver large amounts of energy in a time of about 10^{-9} second must be developed. This involves significant extensions of existing laser technology.

Since 0.015 percent of all hydrogen atoms are deuterium, the oceans of the world contain a nearly inexhaustible reservoir of fuel for fusion reactors. There has been a tendency to assume that obviously fusion will be harnessed, and, therefore, we need not really worry about an energy shortage. It should be noted, however, that enormous effort and money has already been expended on controlled fusion, and as yet there is no working reactor. Many surprising problems have turned up in this work. It would be folly to assume that all energy problems will be solved by controlled fusion until it is demonstrated that fusion can work in a practical way.

SUMMARY

Radioactivity was discovered by observing its effect on a photographic plate. The rays emitted in radioactivity are called α, β, and γ rays. α rays are the nuclei of helium atoms; β rays are negative electrons; γ rays are identical with x-rays.

The nucleus of an atom consists of Z protons (hydrogen nuclei) and N neutrons. Z is the atomic number, and N is the neutron number. The mass number A is the sum of these two. $A = N + Z$. The symbol for an individual nucleus is

$$^{A}_{Z}\text{Chemical symbol}$$

Isotopes are nuclei which have the same number of protons but different numbers of neutrons. Their chemical properties are identical.

α Decay leaves behind a nucleus whose atomic number is reduced by 2 and whose mass number is decreased by 4.

β Decay leaves behind a nucleus whose atomic number is increased by 1 and whose mass number is unchanged.

γ Decay removes energy only from the nucleus and leaves the mass number and charge unchanged.

The radioactive half-life of a nucleus is the time in which half the nuclei in a sample decay.

Rutherford and Fermi showed that bombarding the elements with α particles and neutrons produces different elements. This process is called artificial transmutation.

Einstein's mass-energy relation,

$$E + mc^2 = \text{a constant}$$

states that mass can be converted to energy and energy to mass, but that the sum of the two is constant.

In nuclear fission and nuclear fusion, the loss of mass in a nuclear reaction turns up as useful or destructive energy.

SELECTED READING

Fermi, Laura: "Atoms in the Family," University of Chicago Press, Chicago, 1954. A biography of Enrico Fermi by his wife, following him through the early work in Rome to his later work on nuclear reactors and the atomic bomb. The first nuclear reactor or pile was constructed by Fermi and his group and operated in 1941 in Chicago.

QUESTIONS

1. Do you believe that radioactivity would have gone undiscovered except for the "accident" of Becquerel?
2. We can obtain energy by fission of heavy nuclei, or fusion of light nuclei. Is there any other possible source of nuclear energy?
3. Nuclear energy has been proposed as the "cure" for the energy crisis. How many drawbacks of this "cure" can you list?
4. Since we can now change one element into another by bombardment with neutrons and α particles, why can we not create many new elements with superior properties?

SELF-TEST

__________ Becquerel

__________ Marie and Pierre Curie

__________ Rutherford

_____________ Prout
_____________ Einstein
_____________ Fermi
_____________ fluorescence
_____________ phosphorescence
_____________ radioactivity
_____________ polonium
_____________ radium
_____________ α ray
_____________ β ray
_____________ γ ray
_____________ radon
_____________ radioactive transmutation
_____________ proton
_____________ neutron
_____________ isotope
_____________ radioactive-decay series
_____________ half-life
_____________ artificial transmutation
_____________ nuclear fission
_____________ nuclear fusion

1 Doubly charged particle of mass = 4 atomic mass units
2 Gaseous by-product of radium decay
3 Combination of hydrogen nuclei to form helium and release energy
4 Originally proposed that all elements are created from hydrogen
5 Discovered radium and polonium
6 Conversion of a nucleus to that of a different element by bombardment with α particles or neutrons
7 Positively charged particle in the nucleus
8 Time required for half of a group of atoms to undergo radioactive decay
9 Emission of light caused by excitation with ultraviolet light
10 Discoverer of radioactivity
11 First new element discovered in radioactive samples
12 A chain of radioactive atoms, each decaying into the next
13 The emission of α, β, or γ rays.
14 Nuclei with identical chemical properties but different masses
15 Emission of light by a substance after the excitation source has been removed
16 Discovered artificial transmutation
17 When an α or β particle is emitted, the nucleus of a different element is formed

18 Particle in the nucleus that has no electric charge
19 Splitting of nuclei by the addition of neutrons
20 Developed the relationship between mass and energy
21 Negative electrons emitted in radioactivity
22 Emission in radioactivity identical to x-rays
23 Second new element discovered by the Curies
24 Induced radioactivity by using neutrons as the bombarding particle

PROBLEMS

Complete the following reactions (all are found in nature):

1 ${}^{212}_{84}\text{Po} \rightarrow {}^{4}_{2}\text{He}$ (α particle) + ?
2 ${}^{215}_{84}\text{Po} \rightarrow \beta^- + ?$
3 ${}^{234}_{91}\text{Pa} \rightarrow \beta^- + ?$
4 ${}^{227}_{89}\text{Ac} \rightarrow {}^{4}_{2}\text{He} + ?$
5 ${}^{230}_{90}\text{Th} \rightarrow {}^{4}_{2}\text{He} + ?$
6 ${}^{214}_{83}\text{Bi} \rightarrow \beta^- + ?$
7 Calculate the kinetic energy released, in joules and megaelectron-volts, for reaction 1 above. (Remember, 1 megaelectron-volt is 1 million electron-volts.)
8 Calculate the kinetic energy released, in joules and megaelectron-volts, for reaction 2 above. (The mass of the β^- is not negligible and cannot be neglected.)
9 Calculate the kinetic energy released, in joules and megaelectron-volts, for reaction 3 above. (The mass of the β^- is not negligible and cannot be neglected.)
10 Calculate the kinetic energy released, in joules and megaelectron-volts, for reaction 4 above.
11 Calculate the kinetic energy released, in joules and megaelectron-volts, for reaction 5 above.
12 How many α and β particles are emitted in the radioactive decay series which begins with uranium (${}^{235}_{92}\text{U}$) and ends with lead (${}^{207}_{82}\text{Pb}$)?
13 How many α and β particles are emitted in the radioactive decay series from thorium (${}^{232}_{90}\text{Th}$) to lead (${}^{208}_{82}\text{Pb}$)?
14 (*a*) If the mass of the β particles is considered negligible, how much energy is released when a single ${}^{235}_{92}\text{U}$ nucleus undergoes decay from uranium to lead?
(*b*) How much energy is released if 1 kg of uranium 235 undergoes this decay?
15 If the mass of β particles is considered negligible, how much energy is released in the decay of one thorium-232 nucleus all the way to lead?
16 The sun is estimated to convert about 6×10^{11} kg of hydrogen to helium each second. Assume that the effective reaction is

$$4\,{}^{1}_{1}\text{H} \rightarrow {}^{4}_{2}\text{He}$$

Calculate the energy in joules released each second in the sun. *Note:* The true sequence of reactions is more complex than this, but this is the effective result. The charge comes out right through the emission of positrons, or positive electrons, to be discussed in Chapter 22. The positron mass can be neglected here.

Isotopic masses for use in problems†

Isotope	Mass (atomic mass units) (1 atomic mass unit = 1.66×10^{-27} kg)
β^- (electron)	0.00055
$^{1}_{1}H$	1.007852
$^{4}_{2}He$	4.002603
$^{207}_{82}Pb$	206.97590
$^{208}_{82}Pb$	207.97666
$^{212}_{84}Po$	211.9889
$^{214}_{84}Po$	213.9952
$^{215}_{84}Po$	214.9994
$^{216}_{84}Po$	216.0019
$^{215}_{85}At$	214.9987
$^{216}_{85}At$	216.0024
$^{223}_{87}Fr$	223.0198
$^{226}_{88}Ra$	226.0254
$^{227}_{89}Ac$	227.0278
$^{230}_{90}Th$	230.0332
$^{232}_{90}Th$	232.0381
$^{234}_{91}Pa$	234.0433
$^{234}_{92}U$	234.0410
$^{235}_{92}U$	235.0439

†Taken from the "Chart of the Nuclides," 11th ed., Knolls Atomic Power Laboratory, operated by the General Electric Company, 1972.

22

PHYSICS TODAY: PARTICLES AND MORE PARTICLES

It has been possible up to this point to describe all nature in terms of four objects, or building blocks. These are the electron, the proton, the neutron, and the photon. Things would have been simpler if this had remained the state of affairs. But almost simultaneously with the finding of the neutron in 1932, another particle was discovered. This was only the first of a proliferation of new particles of various kinds. As a result, the array of fundamental building blocks, or elementary particles, has become complex, and the relationship among them unclear. In this chapter we shall describe the discovery and properties of some of these particles. The next chapter will describe some of the present-day theories about them.

In previous chapters we presented in detail the source of most of the experimental facts. In this chapter and the next it will be necessary to quote facts without such detailed discussion. The facts in these two chapters have been obtained through a wide variety of painstaking experiments carried out over a period of more than 30 years. The interested student can find additional details in the book listed in the Selected Reading.

22.1 COSMIC RAYS

Many of the discoveries to be discussed in this chapter have resulted from the study of very high energy particles entering the earth's atmosphere from above. In addition to being a source of information about subatomic physics, these *cosmic rays* are a fascinating phenomenon in their own right.

About 1900 it was observed that dry air has a small, but observable, electric conductivity. Electroscopes, for example, spontaneously lose their charge over a period of time. (Remember that the discharge of an electroscope was also used to detect radioactivity, by measuring the conductivity of the air.) At first this effect was thought to be ionization of the air due to radioactivity in the laboratory or nearby earth. But when electroscopes were sent above the earth's surface in balloons, the intensity of ionization increased, rather than decreased. To explain this observation, the existence of ionizing radiation of some unknown type incident on the earth from the outside was proposed; hence the name cosmic rays.

In the years since this discovery experimental observation has shown much about the properties of cosmic rays. It is possible to distinguish between two kinds, primary and secondary. Primary cosmic radiation is incident on the earth's atmosphere at enormous energies. It consists of the nuclei of atoms, stripped of all their electrons. About 77 percent of cosmic-ray primaries are protons, 20 percent are α particles, and the rest are very small amounts of the nuclei of elements of higher atomic number, including elements throughout the periodic table.

When a cosmic-ray particle enters the upper atmosphere, it may strike the nucleus of an atom. The result of such a collision is the creation of a number of other particles. The creation of particles will be discussed in more detail in several later sections. These particles, called secondary cosmic rays, then strike other nuclei and create more particles, leading to a shower of particles. All these particles are created at the expense of the kinetic energy of the incoming particle. The vast majority of cosmic rays observed at sea level are secondaries, not primaries. Among the most common secondaries are photons, electrons, positrons, muons, and pions. All these will be discussed in this chapter.

The number of secondary particles created in this cascade type of process is a rough measure of the energy of the original primary particle. A few showers consisting of 1 to 10 *billion* electrons have indicated that some cosmic-ray primaries have an energy as high as 10^{19} or 10^{20} electron-volts, or about 1 joule. This is 1 joule of energy in a single incoming particle.

22.2 THE DETECTION OF PARTICLES

Before beginning a detailed discussion of the various new particles, it is useful to describe in general how they are found and studied. The work on cosmic rays just discussed gives the basic idea. All energetic charged particles moving through matter leave behind them a trail of ions, atoms which have had electrons knocked off by the impact of the fast-moving particle. Most of the detection schemes for charged particles take advantage of this ionization. The detection methods of primary interest in this chapter are those that show the complete tracks of particles.

Many of the results of interest can be found either in cosmic rays or by the use of high-energy-particle accelerators. These were discussed briefly in Chapter 11. Energies presently attainable in particle accelerators are about 200 to 300 billion electron-volts, or 2×10^{11} to 3×10^{11} electron-volts. This is to be contrasted with energies as large as 10^{20} electron-volts in cosmic-ray primaries. The detection schemes discussed here can be used for either cosmic rays or particles from accelerators, although certain detectors are more convenient for certain types of experiments.

Except for photographic film, the oldest detection method is the Wilson *cloud chamber*. A cloud chamber is arranged so that the ions left in the wake of a particle form the nuclei of droplets of fog. The line of fog then shows the path of the charged particle. Figure 21.5 shows cloud-chamber tracks of α particles.

The *bubble chamber*, whose invention won for D. F. Glaser the Nobel Prize in physics in 1960, works in a somewhat similar way. A large chamber is usually associated with particle accelerators. A bubble chamber containing a liquid (frequently, but not always, liquid hydrogen) at its boiling point is placed in the beam of particles to be studied. When a charged particle passes through the chamber, the ions left behind serve as nuclei about which bubbles of vapor form. The track of the particle is shown by a line of bubbles, which can be photographed for future study.

The spark and the streamer detector are a pair of detectors of a somewhat different nature. In each of these the trail of ions left behind is detected because a spark jumps between two electrodes more easily where the gas is ionized than where it is not. The spark chamber is arranged so that a high voltage is applied between two or more metal plates. If a particle has passed through, a spark will leap along the path of the particle.

Experience in the use of these detectors indicates that it is possible

FIGURE 22.1 The 80-inch liquid-hydrogen bubble chamber at the Brookhaven National Laboratory. The stainless-steel chamber (80 inches long, 26 inches high, and 27 inches deep), which contains 900 liters of liquid hydrogen at a temperature of 20 K (−253°C), is surrounded by a vacuum chamber, large magnet coils, and a massive steel magnet yoke. The magnet, which requires 4 million watts of power, provides a uniform magnetic field throughout the chamber of 2 newtons/ampere-meter. One side of the chamber consists of a glass window. As a pulse of highly energetic particles is magnetically guided into the chamber, the liquid hydrogen is superheated by a sudden reduction in pressure. Charged particles passing through the chamber cause the superheated hydrogen to boil, leaving a track of tiny bubbles to mark their path. The magnetic field in the chamber deflects the charged particles and causes them to move in curved paths.

By measuring the curvature, length, and density of the tracks, the electric charge, momentum, mass, and other properties of the particles can be determined. (*Courtesy Brookhaven National Laboratory.*)

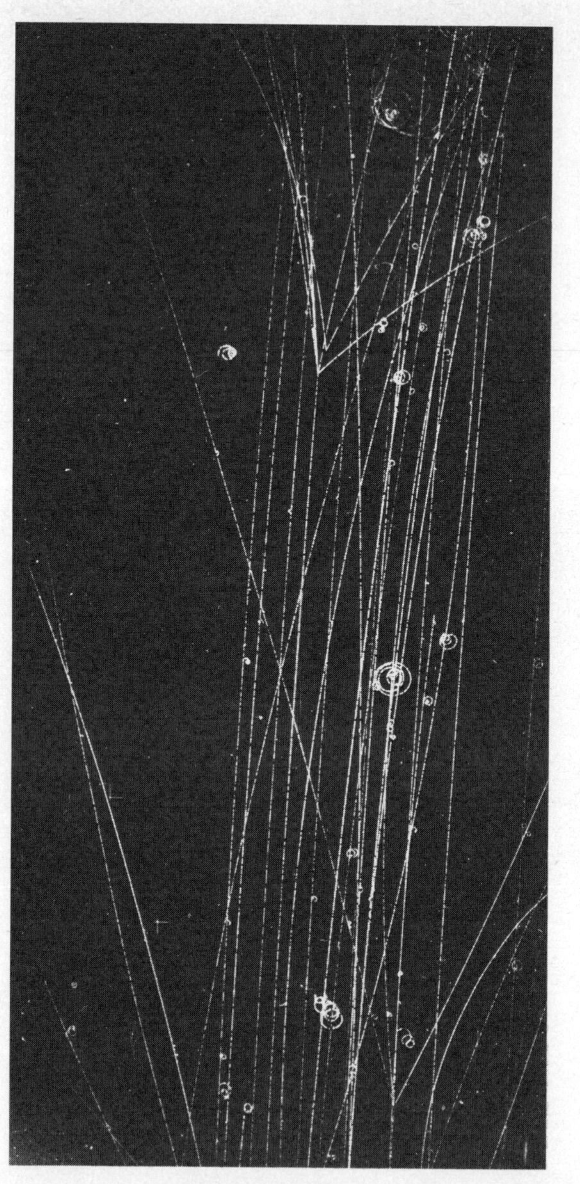

FIGURE 22.2 Particle tracks in a bubble chamber. About 15 high-energy particles enter from below. Near the top of the photograph one of the incident particles strikes a proton in the liquid hydrogen, creating several secondary particles. (*Courtesy Brookhaven National Laboratory.*)

to discriminate between one particle and another. Slower and more massive particles leave denser tracks of ions, and therefore denser tracks in bubble or cloud chambers. Another characteristic of most of the particles we shall discuss is their lifetime. In a finite and characteristic time the particle decays to other particles, similarly to the decay of a neutron to a proton and an electron (and a neutrino, as will be seen). All the particles we shall discuss travel at nearly the speed of light, so

FIGURE 22.3 Columbia University spark chambers located at the Brookhaven National Laboratory. These chambers were used to record interactions between neutrinos and nuclei (Section 22.9). The first chamber on the right is a 6-foot cube consisting of 200 $\frac{1}{4}$-inch aluminum plates spaced $\frac{3}{8}$ inch apart. The second chamber (on the left) is an 8-foot cube of 100 1-inch-thick aluminum plates $\frac{3}{8}$ inch apart.

In both chambers the space between the plates is filled with a mixture of helium and neon gas. A charged particle passing through the chamber ionizes the gas between the plates along its path; when a potential of 10,000 volts is applied between the plates, sparks jump between the plates along these ionized paths. The spark trails are then photographed, and their characteristics analyzed to identify the particles and their energies. (*Courtesy Brookhaven National Laboratory.*)

that the distance they travel before decaying is a direct measure of their lifetime.

Of course, none of the detection schemes discussed here will work for neutral particles, because they leave no trail of ions behind them. Neutral particles are detected when they decay into other charged particles or collide with a charged particle so that the charged particle moves as a result of the collision. The neutron was discovered by Chadwick in this way. The decay of the Λ^0 particle in Figs. 22.12 and 22.13 is an example of the detection of a particle by its decay products.

22.3 THE POSITRON

In 1932 C. D. Anderson investigated the characteristics of secondary electrons produced by cosmic rays, using a large cloud chamber in a horizontal magnetic field. The field caused electrons coming from above to travel in curved paths. Knowing the strength of the magnetic field and measuring the curvature of the paths along which the electrons moved, Anderson expected to determine the velocity, and thus the energy, of these electrons. He observed a number of tracks which were bent the wrong way. To explain what is meant by the wrong way, we must recall the results of Chapters 11 and 14. The magnetic force on a moving charged particle will cause it to move in a circular path. The direction of the force, and thus the direction of the curve, reverses if the sign of the charge is reversed or if the direction of motion of the charge is reversed. This is shown in Fig. 22.4 for positively and negatively charged particles moving in a magnetic field directed out of the paper.

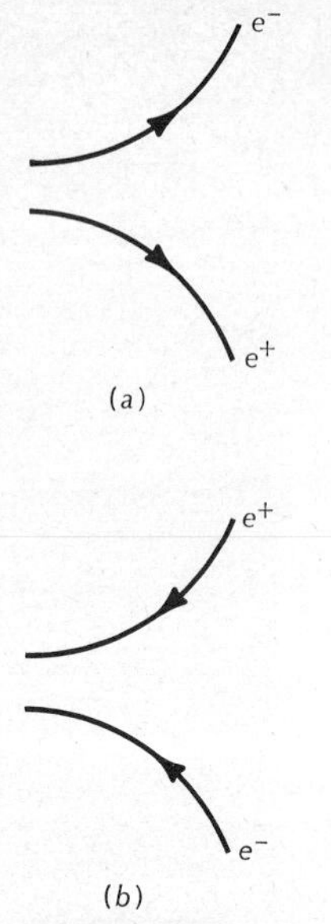

FIGURE 22.4 (*a*) Curved paths of positive and negative electrons moving to the right in a magnetic field directed out of the paper. (*b*) Curved paths of positive and negative electrons moving to the left in a magnetic field directed out of the paper.

The particle whose path curved in the wrong direction might have been a negative electron moving upward, or a positive particle moving downward. In order to determine the direction of motion, a lead plate was inserted in the cloud chamber. The particle was expected to lose energy in traversing the lead plate, and thus its path would be more tightly curved after going through the lead. This would give a clear indication of the direction of travel (Fig. 22.5). Anderson's result is shown in Fig. 22.6. The direction of motion of the unusual tracks showed them to be caused by a positive particle, but their other characteristics were those of the electron. This particle is the positive electron, or *positron*. Its mass is the same as the electron, but its electrical properties are reversed. It is probably true that positrons had been seen before in the work of others, but the significance of the observation was not realized. Anderson is therefore rightly given credit for the discovery.

The discovery of the positron had been preceded by about 2 years by theoretical work of P. A. M. Dirac, which seemed to indicate that such a particle should exist. With perfect hindsight it is easy to say that Dirac's theory had predicted the positron. As the matter developed, however, Dirac's theory had not spurred the experimental discovery of the positron. Only after the fact was the correlation clearly made. This occurred because, at the time, no one quite believed or understood all the implications of Dirac's theory, even though it was highly successful in accounting for the properties of the electron itself.

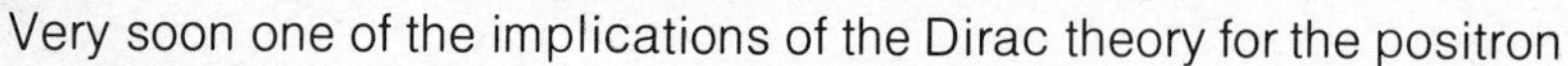

Very soon one of the implications of the Dirac theory for the positron

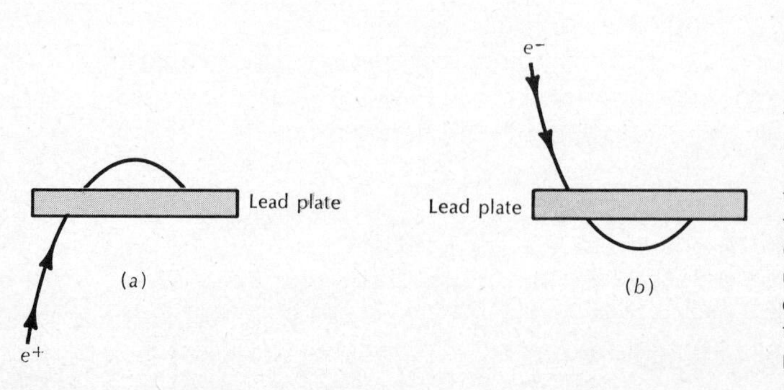

FIGURE 22.5 The technique for distinguishing a positive electron moving upward and a negative electron moving downward. The path is more tightly curved *after* the particle has passed through the lead plate. In (*a*) the particle is moving upward. In (*b*) it is moving downward. The magnetic field is directed out of the paper.

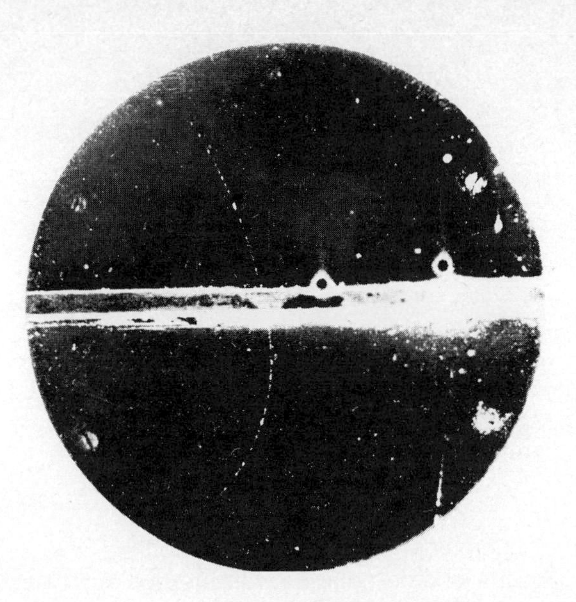

FIGURE 22.6 Anderson's original photograph of the positron in a cloud chamber. The track is more tightly curved below the lead plate, indicating that the particle entered from the top of the picture. Knowledge of the direction of the magnetic field then allows an assignment of the charge of the particle.

was verified. If a positron strikes an electron, the two particles are annihilated and *all* their mass disappears and appears as energy. This energy appears in the form of photons. Usually, the result is two γ-ray photons of equal energy:

$$e^+ + e^- \rightarrow 2\gamma \tag{22.1}$$

They divide the energy available from the total mass of one electron and one positron. Therefore each photon has the energy equivalent to the mass of one electron. The electron and the positron are called *particle* and *antiparticle*. The property of mutual annihilation is characteristic of the interaction between any particle and its antiparticle. The electron and positron differ in that their electrical properties are all reversed, and in some more subtle ways, which will be discussed in Chapter 23. The electron and positron have identical masses.

FIGURE 22.7 P. A. M. Dirac (1902–). *(Courtesy American Institute of Physics.)*

Example 22.1

Calculate the energy of the two γ rays from an electron-positron annihilation.

The two γ's share equally in the available energy. Therefore *each* has an energy equivalent to the mass of one electron. This energy is

$$E = m_e c^2$$

$$E = (9 \times 10^{-31} \text{ kilogram})(3 \times 10^{8} \text{ meters/second})^2$$

$$= 81 \times 10^{-15} \text{ kilogram-meter}^2\text{/second}^2$$

$E = 8.1 \times 10^{-14}$ joule

Since 1 electron-volt $= 1.6 \times 10^{-19}$ joule, this energy is

$$(8.1 \times 10^{-14} \text{ joule})\left(\frac{1 \text{ electron-volt}}{1.6 \times 10^{-19} \text{ joule}}\right) = 5 \times 10^{5} \text{ electron-volts}$$

or 0.5 megaelectron-volts (MeV). 1 megaelectron-volt is 10^6 electron-volts, or 1 million electron-volts of energy.

The positron and the electron can annihilate one another to produce energy. They can also be created if enough energy is available. A high-energy γ ray can produce an electron-positron pair:

$$\gamma \rightarrow e^+ + e^- \qquad (22.2)$$

This reaction occurs most commonly in matter. A simplified way of stating what occurs is that the photon collides with the nucleus of an atom, the electron-positron pair is created, and the photon disappears. The energy of the photon goes to create the mass of the two particles, and into the kinetic energy of the positron, electron, and struck nucleus. The minimum energy for a γ-ray photon to produce such a pair must be equal to the energy associated with the mass of two electrons, or about 1 megaelectron-volt. (The mass of either electron or positron is equivalent to 0.5 megaelectron-volt.) The creation of electron-positron pairs is shown in Fig. 22.8.

22.4 ANTIMATTER

Dirac's theory is not limited to electrons and positrons. It predicts that other particles also have antiparticles, in particular, protons and neutrons. The *antiproton* was predicted to be identical in mass with the proton, but its electrical properties, including its charge, to be opposite those of the proton. The neutron is electrically neutral, but it does possess magnetic properties. In the *antineutron* these are predicted to be reversed.

The discovery of the antiproton and antineutron did not occur until 1955. A particle accelerator called the Bevatron was specifically designed and built at the University of California at Berkeley to make it possible to perform this experiment, as well as many other experiments. Its name derives from the fact that the Bevatron accelerates particles to an energy of slightly more than 6 billion electron-volts, or BeV. The term used today is gigaelectron-volts (GeV, about 10^{-9} joule). This is somewhat more than the energy predicted as the minimum necessary to produce a proton-antiproton pair in the collision of two protons.

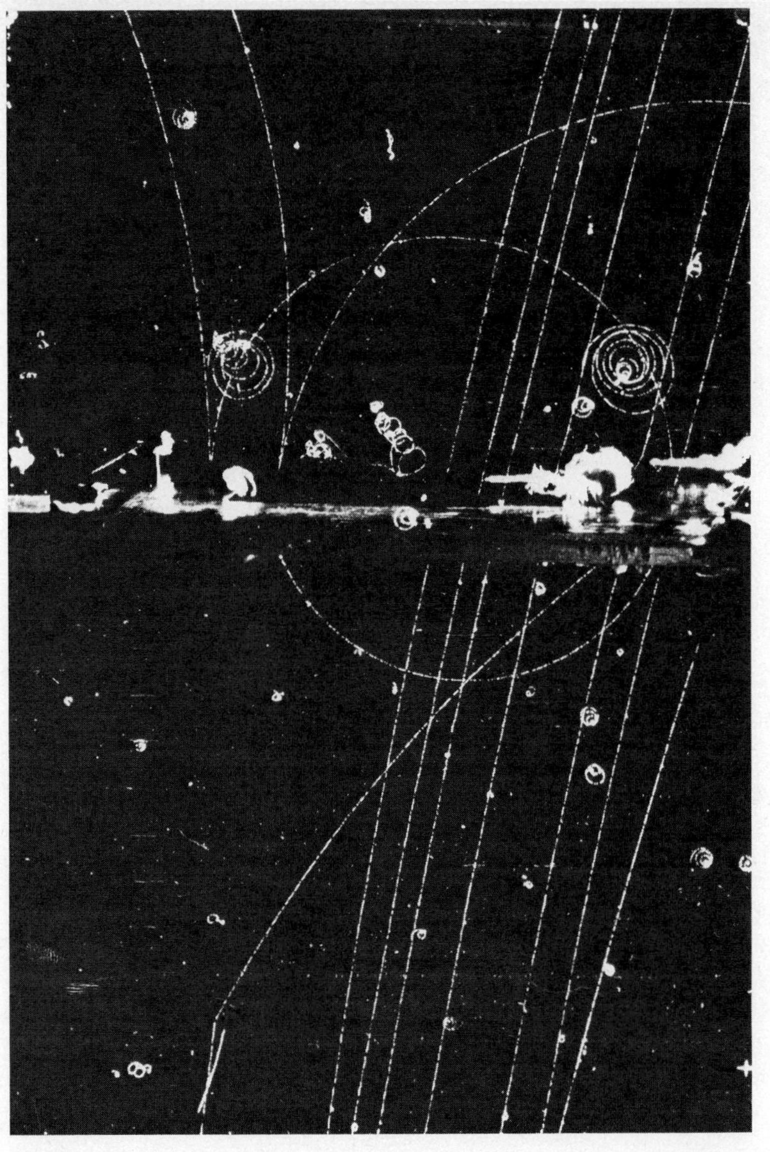

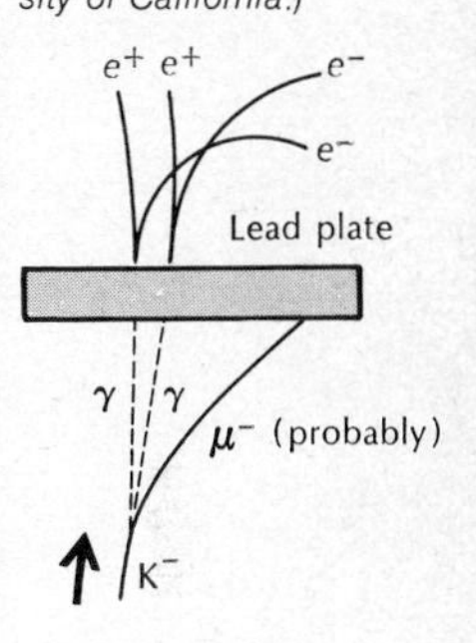

FIGURE 22.8 The production of electron-positron pairs. Two γ rays from the decay of the K meson (see Section 22.11) create two pairs in the lead plate inserted for that purpose. The γ's, being neutral, leave no tracks. (*Courtesy Lawrence Radiation Laboratory, University of California.*)

This reaction is

$$p + p + \text{kinetic energy} \rightarrow p + p + (p + \overline{p}) \tag{22.3}$$

The symbol $\overline{p}$ stands for the antiproton. A bar above the symbol for a particle always indicates the corresponding antiparticle. The two new particles are created from the kinetic energy of the collision between a proton which has been accelerated to 6 gigaelectron-volts and a stationary proton in the target. The two original particles are not destroyed. When producing antiparticles in this manner, a particle-antiparticle *pair* is always produced, never a single antiparticle. Since the antiparticle has a sign opposite that of the particle, this means that the *total* electric charge is never observed to change.

The experimental difficulties in proving the existence of the antiproton lay not in creating it, once sufficient energy was available, but in finding and identifying the antiprotons among all the other particles produced by the high-energy collisions of protons with protons.

A similar process led shortly afterward to the discovery of the antineutron. These two discoveries gave experimental verification of the idea that every particle has an antiparticle. When either an antiproton or antineutron comes into contact with an ordinary proton or neutron, the two annihilate each other. The products are *pions* (a new particle to be discussed shortly) and kinetic energy. A proton-antiproton annihilation is shown in Fig. 23.2.

As matters now stand, *every* particle in nature has an antiparticle. The antiparticle is identical in mass with the particle, but the signs of its electrical properties are reversed. Other properties of these pairs will be discussed in Chapter 23. A few particles, such as the photon and neutral pion, are their own antiparticles. We shall point out where this is the case, but we will not explore its implications further. Whenever a particle and an antiparticle collide, they annihilate, with the release of energy and other, less massive particles.

The existence of particles identical with the proton, neutron, and electron, except for the sign of their electric charge, leads to the idea that there may exist in the universe atoms composed wholly of antiparticles. A helium atom of antimatter would have a nuclear charge which is negative and equal to twice the charge of the electron. Rotating about this nucleus would be two positrons, in orbits identical with those of the electrons in an ordinary helium atom. If antiatoms exist, then perhaps some stars, or even whole galaxies, are constructed of antimatter. Without bringing them into contact with matter from our own system, it would be difficult to distinguish such stars from stars formed from ordinary matter. (Sophisticated proposals exist for determining this, but no such determination has been made.) One point of view is that our galaxy must be all matter, so that if there are regions of antimatter they are other whole galaxies. Hannes Alfvén believes, however, that there can be separate regions of matter and antimatter within our own galaxy.

It is possible to speculate that the universe is equally divided between matter and antimatter. This might mean that some galaxies are matter and others antimatter. There is something intrinsically satisfying in the symmetry of this idea. But in recent years physicists have learned to distrust ideas without experimental proof, despite their appeal based on symmetry or other grounds. We shall see why in the next chapter.

22.5 NUCLEAR FORCE

The forces that exist between protons and neutrons in the nucleus were a puzzle for many years. They do not behave like the coulomb force between an electron and a proton, for example. Experiments of a wide variety on nuclear structure have shown that nuclear forces have the following characteristics:

1. They have a very short range. Beyond a distance of about 10^{-15} meter, they have little or no effect. This can be shown in experiments like Rutherford's scattering experiment. If the two particles in the collision do not come closer than 10^{-15} meter, the coulomb force between them accounts for everything that happens. This is shown in Fig. 22.9.
2. They are very strong. Compared with the coulomb force between two charged particles at a distance of 10^{-15} meter, nuclear forces are about 100 times stronger. This can be shown from the energies involved.
3. They do not depend on the charge of the particles involved. The force between proton and proton, proton and neutron, neutron and neutron are apparently identical, or nearly so.

A Japanese physicist, Hideki Yukawa, proposed in 1935 that the source of the nuclear force is the exchange of a particle between the two nucleons involved. This idea is by no means as strange as it may seem. A few years earlier it had been shown that the major part of the force in a chemical bond arises from the exchange of electrons between

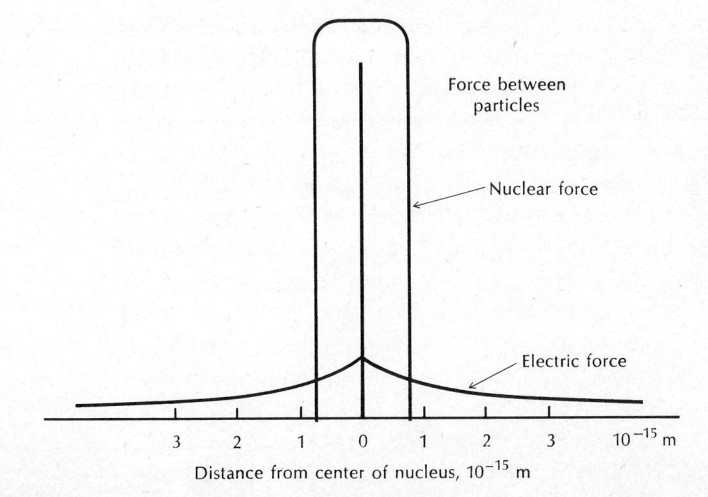

FIGURE 22.9 A comparison of the relative strength of electric and nuclear forces. Nuclear force is much stronger than electric force at small distances, but falls rapidly to zero at distances of about 10^{-15} meter.

the atoms bonded together. The basic notion is that this particle is passed back and forth between the protons and neutrons, and that the act of passing it back and forth leads to an attractive force. From the characteristics of the nuclear force Yukawa said that either positively or negatively charged types or a neutral version should be found, and that each should have a mass roughly 200 times that of an electron. When the accuracy of this proposal was demonstrated by the discovery of a particle with the proper characteristics, Yukawa received the Nobel Prize in 1949 for his work.

FIGURE 22.10 Hideki Yukawa (1907–). (*Courtesy American Institute of Physics.*)

22.6 THE MUON

In 1936 a particle was discovered in cosmic rays which seemed to have the properties of Yukawa's particle. It was originally called a *mu meson*, which has now been abbreviated to *muon* and designated by the Greek letter mu (μ). Its mass is 207 times the mass of an electron. But the Yukawa particle should interact strongly with nuclei, since it is the carrier of the nuclear force. The muon was found to be relatively indifferent to nuclear matter. As a result, muons from cosmic rays are not captured by nuclei, but can be found in mines as deep as 10,000 feet underground. At this depth the muon has passed directly through about 20,000 nuclei with relatively little effect. Therefore it does not have the strong interaction with nuclear matter the Yukawa proposal requires. Both positively and negatively charged muons exist. The particle is the negative muon, and the antiparticle is the positive muon. In Chapter 23 the basis of this assignment will be seen. Note that the magnitude of the electric charge is identical with that of the electron. The magnitude of the electric charge of all the particles discussed in this chapter is the same as the charge of the electron.

The muon has a finite lifetime. After an average time of 2×10^{-6} second (which is a long time as such matters go), a muon undergoes spontaneous decay. Like all cosmic-ray particles, muons travel at speeds near the speed of light, so that a muon travels (3×10^8 meters/second) $\times$ (2×10^{-6} second) = 600 meters in this time. (In fact, muons travel about 6,000 meters, because of the time dilation effect, discussed in Chapter 17.)

The decay of the muon is much like the decay of a free neutron to a proton and an electron. The only *visible* product of the decay of a muon is an electron or positron whose charge is the same as that of the original muon. Muon decay is illustrated in Fig. 23.1. We shall come back to this decay process in Section 22.9. Note that the μ^+, which is the antiparticle, decays to the e^+, which is also an antiparticle.

$$\mu^- \rightarrow e^-$$

or

$$\mu^+ \rightarrow e^+$$

One of the intriguing and unexplained facts about the muon is its remarkable similarity to the electron. Except for its mass, the muon behaves so much like an electron that it is sometimes called a "heavy" electron.

22.7 THE PION

The discovery of the muon in 1936, which resembles Yukawa's particle in some ways but not in others, led to a period of great confusion. The muon's mass is about what Yukawa had predicted, but it does not interact with matter in the way that the Yukawa particle should. The difficulty was resolved in 1947 with the discovery of another particle, the *pi meson* or *pion* (π). The pion mass is, roughly, 270 times the mass of the electron, and it has all the properties the Yukawa particle should have. It was first discovered in cosmic rays, and has since been created by using accelerators. Pions are produced as secondaries in cosmic rays. The pion interacts sufficiently strongly with matter, so that most of the pions in cosmic rays do not reach the surface of the earth. They collide with the nuclei of atoms in the atmosphere and are stopped. The lifetime of a charged pion is short—about 2.6×10^{-8} second. It decays after this time (if it has not already collided with a nucleus). The only *visible* product of a pion decay is a muon:

$$\pi^+ \rightarrow \mu^+$$
$$\pi^- \rightarrow \mu^-$$

There exists also a neutral pion, the π^0, whose lifetime is much less (about 10^{-16} second). The π^0 decays into two γ rays:

$$\pi^0 \rightarrow \gamma + \gamma$$

The pion has all the properties Yukawa predicted. The π^+ is the antiparticle of the π^-, and the π^0 is its own antiparticle.

22.8 THE NEUTRINO

In 1930 Wolfgang Pauli proposed a particle whose properties were so bizarre that only extremely strong arguments would lead anyone to conjecture its existence. Pauli proposed that there must exist a neutral particle which has no mass, no charge, no magnetic properties, and so little interaction with matter that it can traverse a solid wall 10^{14} miles thick before interacting with an atom. It is thus just about as close to nothing as is possible while still being something. It is considered something because it has energy, momentum, and angular momentum. Pauli originally called this particle the neutron, because of its electric neutrality. There was some confusion when another neutral particle, the neutron (with mass and magnetic properties) was discovered by Chadwick in 1932. Enrico Fermi, when asked by a group of students whether Chadwick's particle was Pauli's neutron, said that it was not, and that Pauli's neutral particle was a *neutrino* (which means little neutral one, in Fermi's native Italian).

Let us look at the arguments that led Pauli to propose the existence of something that seems so close to nothing. His reasoning was essentially that, without such a particle, the laws of conservation of energy, conservation of momentum, and conservation of angular momentum would all have to be abandoned. Even so, some physicists were more willing to allow violations of these laws than to concede the existence of such a peculiar object as the neutrino.

The experimental evidence that led Pauli to make this proposal came from the study of β decay, the emission of an electron in radioactive decay. The simplest example of β decay is the decay of a free neutron. A neutron not bound in a nucleus decays with a half-life of about 15 minutes into a proton and an electron. (See the discussion of half-life in Chapter 21.)

$$n \rightarrow p + e^- \tag{22.4}$$

When the decay of a neutron occurs in a nucleus, the effect is to increase the nuclear charge by one unit without changing the mass number, for example,

$$^{234}_{90}\mathrm{Th} \rightarrow e^- \ (\beta \text{ ray}) + {}^{234}_{91}\mathrm{Pa} \tag{22.5}$$

When very careful measurements of the mass of the original nucleus, the mass of the product nucleus, and the mass and energy of the emitted electron are made, an anomaly is found. A few of the emitted electrons have just the right amount of energy so that the energy of the starting nucleus (mass included) and the final energy of all products (mass included) are the same. But many more of the electrons have less than this amount, and no energy is found anywhere else. Therefore the evidence is that, either energy is not conserved, or some unseen particle is carrying off the missing energy. This particle must be electrically neutral to avoid detection. A photon has many of the properties required, but cannot penetrate matter. If a photon were carrying away the missing energy, it would be absorbed in the experimental apparatus, and its energy detected as a rise in temperature. Pauli proposed that the neutrino carries away the missing energy. The precise properties he gave to the neutrino were those necessary to account for the observations—electric neutrality, masslessness, and ability to penetrate matter.

A similar anomaly is found for the momentum. The momentum before the decay and afterward is found not to be the same. Again Pauli proposed that the neutrino carries away the missing momentum. If one takes into account the fact that electrons, protons, and neutrons have angular momentum, or spin—which we have not yet discussed—the neutrino must also carry away angular momentum, or angular momentum conservation will also fail.

The properties Pauli proposed for the neutrino are that it carries away energy, momentum, and angular momentum. It need have no other properties.

The experimental discovery of the neutrino occurred in 1956. We have discussed the β decay of a neutron. Let us write it again, showing that one of the products is also a neutrino. (Actually it is an antineutrino. We shall discuss in Chapter 23 how this is known.) The neutrino is indicated by the Greek letter nu (ν).

$$\mathrm{n} \rightarrow \mathrm{p} + e^- + \bar{\nu} \tag{22.6}$$

The process can also go, although very rarely, in the reverse direction.

$$\bar{\nu} + \mathrm{p} \rightarrow \mathrm{n} + e^+ \tag{22.7}$$

The products include a positron rather than an electron, because the total electric charge before and after must be the same.

To observe the reaction shown in Equation 22.7 requires a copious source of antineutrinos, since the reaction occurs so rarely. (Remember, a neutrino can penetrate 10^{14} miles of solid matter, on the average, before such an event will occur.) A nuclear reactor provides the best man-made source of neutrinos, from the β decay of the products of nuclear fission in the reactor. In 1956 Clyde Cowan and Fred Reines, working near a nuclear reactor at Savannah River, South Carolina, found evidence that reaction in Equation 22.7 occurred. About once every 20 minutes they observed a neutron and a positron appear *simultaneously* in their apparatus. These are exactly the products predicted if an antineutrino reacts with a proton as in Equation 22.7. Reines and Cowan found such products occurring only when the reactor was running, confirming that the reactor was the source of whatever they were seeing. They thus confirmed that neutrinos really did exist and were not a figment of Pauli's imagination.

22.9 THE MUON NEUTRINO

The decay of a pion or muon also requires a neutrino to be given off, in order that energy and momentum be conserved. A pion decays to a muon and a neutrino, and a muon to an electron and two neutrinos. (We show in Chapter 23 how this is known.)

$$\pi^{\pm} \rightarrow \mu^{\pm} + \nu \tag{22.8}$$

and

$$\begin{aligned} \mu^+ &\rightarrow e^+ + \nu + \bar{\nu} \\ \mu^- &\rightarrow e^- + \nu + \bar{\nu} \end{aligned} \tag{22.9}$$

The neutrino associated with these processes was first observed in 1962 at Brookhaven National Laboratory. The surprising result was that the neutrino associated with muon decay is *not* the same as the electron's neutrino. So there are two kinds of neutrinos, each with its own antineutrino. These will be designated as ν_e and ν_μ, and the reactions in Equation 22.9 written as

$$\mu^+ \rightarrow e^+ + \nu_e + \bar{\nu}_\mu \qquad (22.10)$$
$$\mu^- \rightarrow e^- + \bar{\nu}_e + \nu_\mu$$

In an experiment similar to that of Reines and Cowan, it is possible to try to detect the reverse reactions to those in Equation 22.10:

$$\nu_\mu + e^- \rightarrow \mu^- + \nu_e \qquad (22.11)$$

Since there are no free positrons in matter, the collision between a neutrino and a positron can be neglected. The spark detector of Fig. 22.3 was used to detect this reaction.

Figure 22.11 is a photograph of a detector set up in a mine 3,000 feet underground at Park City, Utah, to detect muons from reactions like (22.11). The size of this detector is about 40 × 30 × 20 feet. The large size is necessary because of the very small probability of neutrino reactions. The system is in a mine in order to filter out the background of other cosmic-ray events occurring above ground. This instrument detects muons moving toward the earth's surface, indicating that a neutrino has traveled through the earth and reacted as in Equation 22.11 in the last few thousand feet of rock. About two cosmic-ray neutrinos per year have been detected in this apparatus.

22.10 STRANGE PARTICLES

A number of additional particles have been found to exist. One convenient classification of them is according to their relative masses. (A better classification scheme is discussed in Chapter 23.) Those which are more massive than the proton and neutron are called *hyperons*. Those less massive are called *mesons*. We shall first describe the more massive group.

The first of the unusual, or *strange,* particles turned up in 1947. It was called the Λ^0 (lambda zero) particle because of the characteristic shape of the tracks in a cloud or bubble chamber. Figure 22.12 shows an example of Λ^0 decay. The neutral Λ^0 leaves no track. The Λ^0 is a neutral particle which can decay to a proton and a negative pion or a neutron and a neutral pion.

$$\Lambda^0 \rightarrow p + \pi^- \qquad (22.12)$$

$$\Lambda^0 \rightarrow n + \pi^0 \qquad (22.13)$$

Similar decays occur for the antiparticle, except that the products are an antiproton and an antineutron. Note that the total electric charge

FIGURE 22.11 Underground detector for muons. The large horizontal tubes are a type of spark detector which allow the position of the charged particle to be determined. Muons coming upward can only be caused by neutrinos, since only a neutrino can penetrate the entire earth. (*University of Utah photograph.*)

is the same before and after the decay in all of these equations. The tracks of the proton and pion produce the characteristic V-shaped cloud-chamber pattern. The mass of the Λ^0 is approximately 2,200 times the mass of the electron. (Remember, the masses of the proton and neutron are about 1,800 times the electron mass.) The Λ^0 has a lifetime of about 10^{-10} second.

FIGURE 22.12 The characteristic V-shaped track which gave the Λ^0 its name. The neutral Λ^0 is invisible; the tracks of its decay products, a π^- and a proton, are visible. The Λ^0 was created by the collision of a K^- with a proton in the liquid hydrogen of the bubble chamber. (*Courtesy Lawrence Radiation Laboratory, University of California.*)

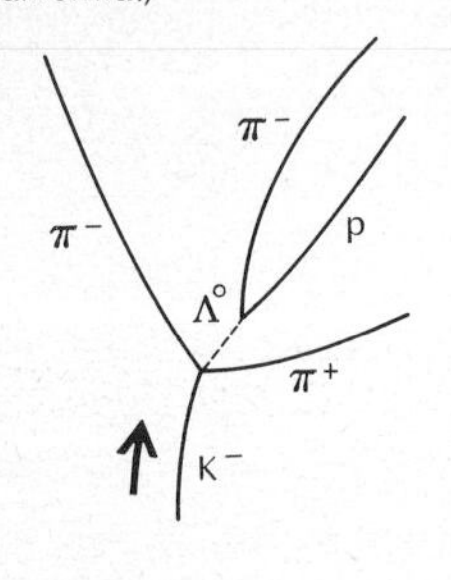

A second type of strange particle is the Σ (sigma) particle. There are three of them—positive, negative, and neutral—and an antiparticle of each. The mass of each member of this group is about 2,300 electron masses. Typical decays (all Σ particles decay in more than one way) are

$$\Sigma^+ \rightarrow p + \pi^0 \tag{22.14}$$

or

$$\Sigma^- \rightarrow n + \pi^- \tag{22.15}$$

and

$$\Sigma^0 \rightarrow \Lambda^0 + \gamma \tag{22.16}$$

The γ indicates a γ-ray photon. The charged Σ's decay in about 10^{-10} second. The Σ^0 decays in a much shorter period of time.

Another particle, somewhat more massive, is the Ξ (Greek xi) particle. Only Ξ^- and Ξ^0 and their antiparticles exist. Their mass is about 2,600 electron masses. Typical decays, which occur in about 10^{-10} second, are

$$\Xi^- \rightarrow \Lambda^0 + \pi^- \qquad (22.17)$$

$$\Xi^0 \rightarrow \Lambda^0 + \pi^0 \qquad (22.18)$$

Last in this group of strange particles is one which was predicted theoretically before it was found, the Ω^- (omega minus). Its mass is 3,300 electron masses, and its decay time is again about 10^{-10} second. Two possible decay schemes are

$$\Omega^- \rightarrow \Xi^- + \pi^0 \qquad (22.19)$$

$$\Omega^- \rightarrow \Xi^0 + \pi^- \qquad (22.20)$$

Note that there is thus a chain of decays from Ω^- through Ξ to Σ or Λ to either the proton or neutron. Each of these particles decays in several ways. Shown are some of the common modes. In each case the products are less massive than the initial particle. For example,

$$\Lambda^0 \rightarrow \mathrm{p} + \pi^- \qquad (22.12)$$

$$2{,}180m_e \rightarrow 1{,}836m_e + 273m_e$$

That is, the mass of the Λ^0 is 2,180 electron masses (m_e); the mass of the proton is $1{,}836m_e$; and that of the π is $273m_e$. The mass of the products is then $2{,}109m_e$, and an amount of mass equal to $2{,}180 - 2{,}109 = 71$ electron masses goes into kinetic energy of the products. (The exact masses come from Table 22.1.)

We can see from energy considerations alone that the decay

$$\Sigma^+ \rightarrow \Lambda^0 + \pi^+ \qquad (22.21)$$

cannot occur. The mass of the products is $2{,}180m_e$ for Λ^0 plus $273m_e$ for π^+, giving $2{,}453m_e$, while the mass of the Σ is only $2{,}315m_e$. Therefore the decay shown in Equation 22.21 cannot happen. For it to occur, about $138m_e$ of extra mass must be created. This can occur in an energetic collision of two particles, but never in the decay of a single particle. In Chapter 23 we shall discuss other rules for determining whether or not a specific decay process will occur.

22.11 KAONS AND THE ETA PARTICLE

There are two kinds of particles with mass intermediate between that of a pion and that of a proton. The first are K particles, or kaons. These exist in positive and neutral versions, K^+ and K^0, and their antiparticles. Kaon masses are about 970 electron masses. The mean lifetime of all kaons is in the vicinity of 10^{-8} second.

Family name	Particle	Symbol	Mass in electron masses	Charge	Lifetime
	Photon	γ	0	Neutral	Infinite
	Graviton		0	Neutral	Infinite
Electron family	Electron	e^-	1	Negative	Infinite
	Neutrino	ν_e	0	Neutral	Infinite
Muon family	Muon	μ^-	206.77	Negative	2.2×10^{-6} second
	Neutrino	ν_μ	0	Neutral	Infinite
Mesons	Pion	π^+	273.1	+	2.55×10^{-8} second
		π^-	273.1	−	2.55×10^{-8} second
		π^0	264.1	0	1.8×10^{-16} second
	Kaon	K^+	966.2	+	1.23×10^{-8} second
		K^0	974.0	0	10^{-10} to 10^{-8} second
	Eta	η^0	1,074	0	More than 10^{-22} second
Baryons	Nucleon	p (proton)	1,836.15	+	Infinite
		n (neutron)	1,838.68	0	1,013 seconds
	Lambda	Λ^0	2,183	0	2.6×10^{-10} second
	Sigma	Σ^+	2,237.5	+	8×10^{-11} second
		Σ^-	2,343.1	−	1.6×10^{-10} second
		Σ^0	2,333.6	0	About 10^{-20} second
	Xi	Ξ^-	2,585.7	−	1.7×10^{-10} second
		Ξ^0	2,573.2	0	3×10^{-10} second
	Omega	Ω^-	3,272	−	10^{-10} second

TABLE 22.1 Properties of the stable particles according to the family classification

Some typical kaon decays are (the K^- is the antiparticle of the K^+)

$$K^\pm \rightarrow \mu^\pm + \nu_\mu \qquad (22.22)$$

$$K^\pm \rightarrow \pi^\pm + \pi^0 \qquad (22.23)$$

$$K^0 \rightarrow \pi^+ + \pi^- \qquad (22.24)$$

and

$$K^0 \rightarrow \pi^0 + \pi^0 \qquad (22.25)$$

Kaons are created in association with other strange particles. Figure 22.13 is a bubble-chamber photograph of the production of a Λ^0 and a K^0. Both decay before leaving the bubble chamber.

The η, (eta) particle is a neutral particle that has such a short lifetime that one may wonder why it is considered a particle at all. It decays in about 10^{-19} second, most frequently as

$$\eta^0 \rightarrow \gamma + \gamma \qquad (22.26)$$

The η^0 lives about 1,000 times longer than a large group of objects called *resonances*. In comparison with these resonances, then, the η is a stable particle.

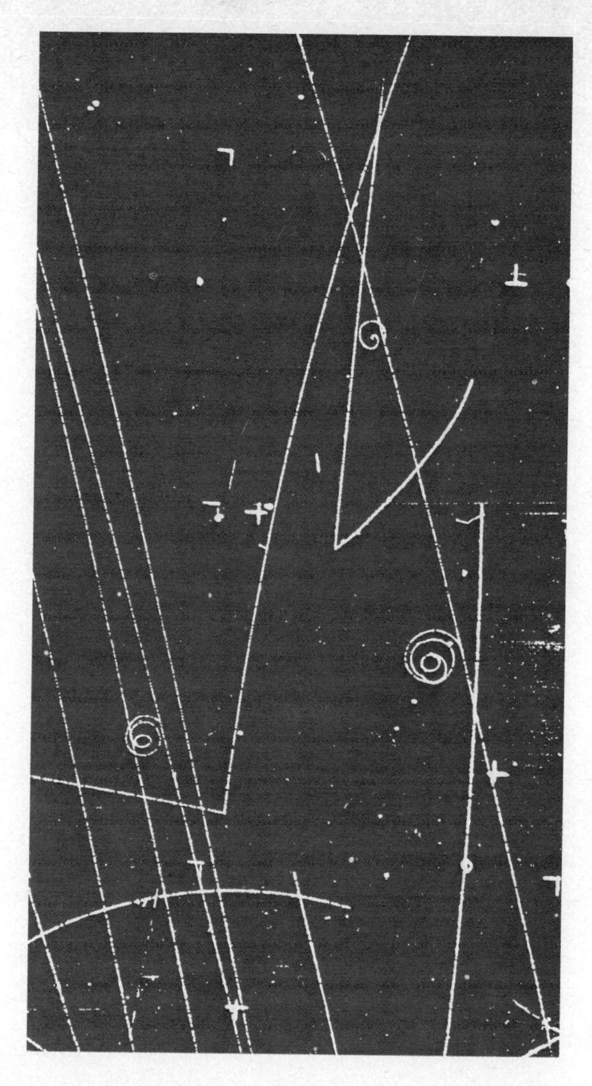

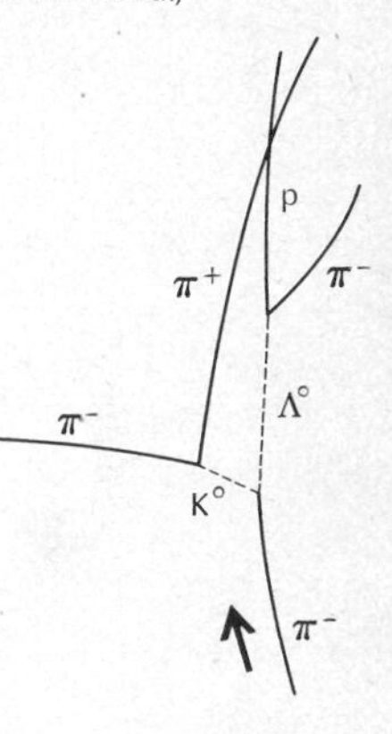

FIGURE 22.13 Associated production of a Λ^0 and K^0. Both are neutral and can be detected only by the tracks of their products. The reason that these particles are produced in pairs is discussed in Section 23.8. (*Courtesy Lawrence Radiation Laboratory, University of California.*)

22.12 RESONANCES

In addition to the particles discussed above, there are a large number (now numbered in the hundreds) of entities called resonances, whose lifetimes are of the order of 10^{-22} or 10^{-23} second. This is so short that it is tempting to take the point of view that resonances are not particles at all. It appears that nature has not been so kind as to keep things simple. The current view is that resonances have as much claim to the title of elementary particles as protons and neutrons. This probably means that the word elementary, when used to describe any particle, is misleading, in the sense of implying that that particle is a building block of matter.

Any complete theory of particles should account not only for stable particles, but also for resonances. Some significant progress, as seen in Chapter 23, has been made along these lines.

22.13 FAMILY CLASSIFICATION

One of the first steps in understanding the embarrassing wealth of information in this chapter is to find some scheme of classification. The following breakdown is a scheme which makes sense in terms of the physical interactions of the various particles.

The electron, muon, their antiparticles, and their neutrinos are called *leptons*, which means light particles. The pions, kaons, and the eta are called *mesons*, particles of intermediate mass. (Notice that the muon is *not* a meson under this classification.) The proton, neutron, Λ, Σ, Ξ, and Ω are classified together as *baryons*. For the present this classification scheme may seem arbitrary and mysterious, but in Chapter 23 we shall show some of the physical meaning and utility of this arrangement. Table 22.1 tabulates the properties of the stable particles according to this classification. Antiparticles have been omitted from the table.

SUMMARY

Primary cosmic rays are nuclei (mostly protons) arriving from outer space with enormous energies. Secondary cosmic rays are particles created by the collision of primaries with the nuclei of atoms in the upper atmosphere.

The positron was discovered in 1932. It has the same mass as the electron, but opposite charge. It is the electron's antiparticle. The antiproton and the antineutron are the antiparticles of the proton and neutron.

All known particles have antiparticles. (The photon and the π^0 are their own antiparticles.) The possibility exists that galaxies other than our own may consist of antimatter.

When a particle and an antiparticle collide, they annihilate into a burst of energy and other less massive particles.

The energy relationships when particles are created or annihilated are governed by Einstein's mass-energy relationship:

$$E + mc^2 = \text{a constant}$$

The force between protons and neutrons in the nucleus occurs because of the exchange of particles between them. The particles exchanged are mostly pions. The pion is the particle predicted by Yukawa.

The muon behaves almost exactly like an electron, except that its mass is 207 electron masses (m_e).

The existence of the neutrino was predicted by Pauli to preserve the laws of energy, momentum, and angular momentum conservation. There are two kinds of neutrinos, the electron neutrino and the muon neutrino.

The *strange* particles were first discovered in cosmic rays, and are now produced using several of the larger accelerators. The stable hyperons, and their masses, are: Λ^0, 2,200m_e; Σ^+, Σ^-, Σ^0, 2,300m_e; Ξ^-, Ξ^0, 2,600m_e; and Ω^-, 3,300m_e. The mesons are the kaons, 970m_e, and the eta, 1,074m_e.

Resonances are particles with very short lifetimes. Nevertheless they are considered to be as fundamental as any of the more long-lived particles and must be accounted for by any theory of particles.

SELECTED READING

Frisch, D. H., and Alan M. Thorndike: "Elementary Particles," Momentum Book 1, D. Van Nostrand Company, Inc., Princeton, N.J., 1964.

SELF-TEST

__________ primary cosmic ray
__________ secondary cosmic ray
__________ cloud chamber
__________ bubble chamber
__________ annihilation
__________ antimatter
__________ nuclear force
__________ baryon
__________ hyperon
__________ meson
__________ lepton
__________ neutrino
__________ Dirac
__________ Anderson
__________ Yukawa
__________ Pauli

1 Strange particles more massive than the proton and the neutron
2 Liquid at its boiling temperature shows the tracks of charged particles that have passed through
3 Particles created in the upper atmosphere by the action of cosmic rays

4 Discovered the positron
5 The group of light particles
6 Atoms composed of antiprotons, antineutrons, and positrons
7 The result of a particle and an antiparticle colliding
8 Arrangement in which particle tracks are shown as streaks of fog
9 The group of heavy particles, including protons and neutrons
10 Proposed that the nuclear force is carried by a particle, later shown to be the pion
11 The group of particles of intermediate mass, including pions, kaons, and the eta
12 The "little neutral one"
13 Developed a theory that "predicted" the positron and other antiparticles
14 Force between proton and neutron, proton and proton, and neutron and neutron
15 Particles incident on our atmosphere with enormous energies
16 Proposed the neutrino

23 PHYSICS TODAY: SYMMETRIES

In this chapter we shall attempt to show how some order is being wrenched from the chaos of particles and properties discussed in Chapter 22. The story is necessarily incomplete, because many of the basic problems are yet to be solved. But there are tantalizing clues that indicate that progress is being made. A complete solution to this problem would show why only those particles found in nature do exist, and why they have the particular properties they do.

We have already discussed three conservation laws and implied a fourth. These laws state that quantities such as energy, momentum, angular momentum, and electric charge are conserved. That is, they do not change unless there is outside interference. In this chapter we shall discuss several other conservation laws found to be obeyed by particles and their interactions. Some of the things which are conserved are rather abstract quantities, and so we must fall back to a previously stated position: *All physical laws are based on experiment.* Each of the principles discussed in this chapter is found to be true experimentally. Intrinsic beauty or reasonableness is insufficient reason for believing in the truth of a physical idea, although it may provide grounds for proposing theories to be subjected to experimental test. We shall find one example (parity) in which intrinsic reasonableness was no guarantee of truth. Physicists have learned to be wary of things that seem obvious. Too many of them have turned out to be untrue.

One basic principle seems always to be true in elementary-particle interactions. In simple language it is: *If something can happen, it will. All* the principles we shall discuss operate to tell us what cannot happen. They are rules prohibiting, more or less strictly, various kinds of results. For instance, energy conservation forbids any result which does not leave the total energy (mass included) of a system constant. Any process not specifically forbidden is almost always found to occur. If it does not, it may mean that a previously unknown conservation law is operating to prohibit it. Most of the conservation laws have been found by studying what did not occur and seeking reasons why it did not.

23.1 ENERGY AND MOMENTUM

We discuss energy and momentum conservation first because they are already familiar, and also because they show the connection between conservation laws and symmetry. This type of relationship is believed to hold for all conservation laws, although in many cases the symmetry principle is not yet known.

The conservation-of-energy principle can be demonstrated in many ways. Its primary basis is experiment: It works. But it is also possible to obtain it from the following idea. The laws of physics are not observed to change with time. A physical experiment done yesterday gives the same result as the same experiment performed today. If we believe that this is a general principle, a direct consequence is that energy should be conserved. We shall not attempt a proof of this statement. The point that is important is the relationship between the conservation law (energy is conserved) and the symmetry principle (physics is the same at all times). We shall encounter other conservation laws in which the relevant symmetry is not known but, on the basis of the close connection between symmetry and conservation laws where both are known, it is believed that a conservation law implies that there is a symmetry which may not yet be known.

Similarly, momentum conservation can be shown to be a direct consequence of the idea that the laws of physics do not change from point to point in space. (Again, we shall not attempt a detailed proof of this statement.) Take note that this does not mean that the same thing is occurring at every point in space. It means that the physical laws governing what is happening are the same everywhere. The law of conservation of momentum is a direct consequence of this spatial symmetry. If the laws of physics are the same at every point in space, the conservation of momentum is a direct consequence. The connection between symmetry and conservation laws is apparently very deeply embedded in the fabric of the universe.

Conservation of energy, when applied to particle physics, forbids any particle to decay to particles the sum of whose masses is greater than the parent particle. (The student might at this point wish to review the mass-energy equivalence discussed in Chapters 17 and 21.) Of course, in a collision of one particle with another, some of the kinetic energy of the collision may go to create the mass of a particle. But in this case the energy balance is all right so long as the starting energy (mass plus KE) is the same as the energy of the products.

As an example, a neutral kaon is forbidden to decay to four pions, because the mass of the kaon is not great enough.

$$\begin{array}{lrl} K^0 \rightarrow \pi^+ + \pi^- + \pi^0 + \pi^0 & \pi^+ = & 273.2m_e \\ & \pi^- = & 273.2m_e \\ & 2\pi^0 = & \underline{528.4m_e} \\ K^0 = 974.6m_e & & 1{,}074.8m_e \end{array} \tag{23.1}$$

Creation of the mass of 1,074.8 electrons from a starting mass of 974.6 without the addition of kinetic energy is not possible. A neutral kaon can decay to three pions (and is observed to do to), because the total mass on the right is less than that on the left.

$$K^0 \rightarrow \pi^+ + \pi^- + \pi^0 \tag{23.2}$$

Conservation of momentum forbids the decay of one particle into just one lighter particle. For instance, in the decay of a muon, the only visible product is an electron (Fig. 23.1). But for momentum to be conserved, there must be at least one other product of the decay. This is easily seen if one imagines the muon at rest. It decays to an electron

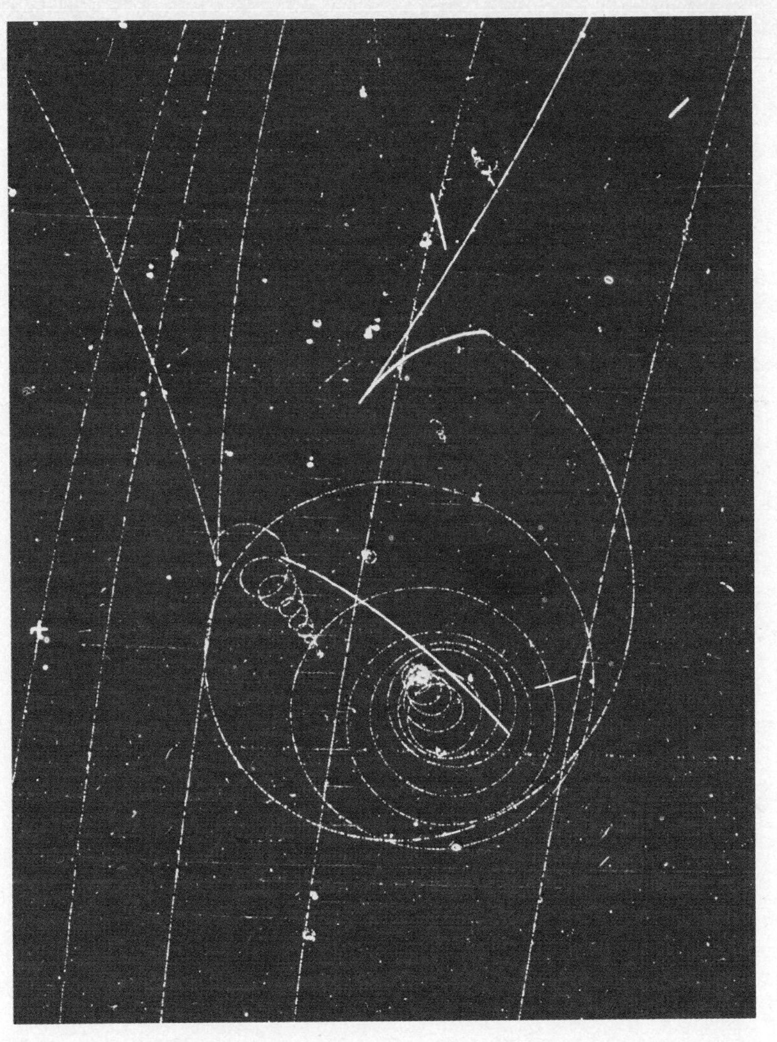

FIGURE 23.1 The decay of a muon to an electron and two neutrinos. The direction in which the invisible neutrinos must go is determined by momentum conservation. (*Courtesy Lawrence Radiation Laboratory, University of California.*)

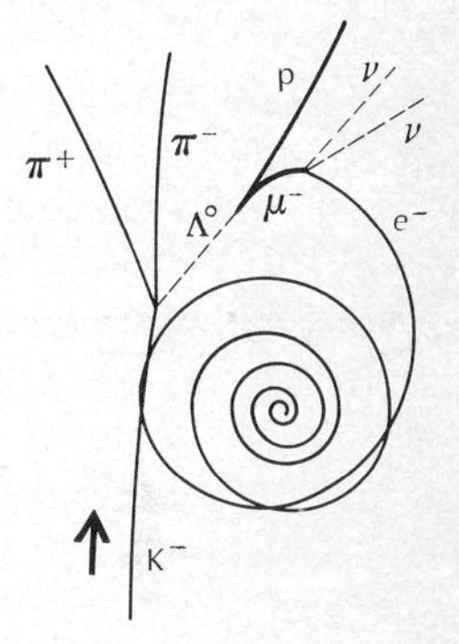

which is lighter, and therefore the electron will have some kinetic energy. Since the muon at rest had zero momentum, the system afterward must have none also. This can occur only if there is another particle, the neutrino, to balance the momentum of the moving electron. In fact, both a neutrino and an antineutrino are given off, for reasons we shall discover shortly.

$$\mu^- \rightarrow e^- + \nu_\mu + \bar{\nu}_e \tag{23.3}$$

23.2 ANGULAR MOMENTUM

Conservation of angular or rotational momentum can also be shown to follow from a symmetry principle. If one accepts the reasonable proposal that there is no preferred direction in space, angular-momentum conservation follows. The symmetry principle is, then, that space is the same in all directions, or that there is no unique direction. On the surface of the earth there is such a direction, the line which is up and down. But this exists because of the gravity of the earth. Away from any large and massive objects, there would be no preferred direction.

Angular momentum is important in elementary-particle interactions, because many particles are found to have an intrinsic angular momentum. This angular momentum of particles is called *spin*. The electron was first proposed to have spin by S. Goudsmit and G. Uhlenbeck in 1924, to account for some of the finer details of atomic spectra. The idea is, crudely, that the electron behaves as though it were spinning like a top. (This picture should not be taken too seriously.) It has since been shown that many other particles, including the proton, neutron, neutrino, muon, and all the baryons, have spin.

Bohr's proposals for hydrogen led to the notion that the angular momentum of electrons in an orbit comes in quantized units. The magnitude of the quantum of angular momentum he found to be $h/2\pi$. For spin angular momentum, the unit is found to be one-half as large, $\frac{1}{2}(h/2\pi)$. The spin angular momentum of the electron is $\frac{1}{2}(h/2\pi)$, and the electron is said to have a spin of $\frac{1}{2}$. A particle with spin 1 has a spin angular momentum of $1(h/2\pi)$, and one of spin $\frac{3}{2}$ has $\frac{3}{2}(h/2\pi)$. Most of the elementary particles we shall discuss have a spin of $\frac{1}{2}$ or 0.

If we look at the decay of a pion (which is known to have spin 0) into a muon (known to have spin $\frac{1}{2}$), we can see the need for another particle to make the angular momentum come out right.

$$\pi^- \rightarrow \mu^- + \bar{\nu}$$

$$\text{Spin:} \quad 0 \qquad \tfrac{1}{2} \qquad \tfrac{1}{2}$$

The extra particle is the neutrino, with spin $\frac{1}{2}$. The muon and the neutrino are "spinning in opposite directions," so their angular momenta add to zero.

23.3 CHARGE CONSERVATION

Charge conservation is a conservation law whose best justification is its experimental verification. This law means exactly what it says. Electric charge is never created or destroyed. But several instances have already been discussed in which charged particles are created out of kinetic energy (proton-antiproton) or a photon (electron-positron). These two reactions are

$$\gamma \rightarrow e^- + e^+ \tag{23.4}$$

and

$$\mathrm{p} + \mathrm{p} \rightarrow \mathrm{p} + \mathrm{p} + (\mathrm{p} + \bar{\mathrm{p}}) \tag{23.5}$$

In each instance a *pair* of charged particles is formed. In the reaction in

Equation 23.4 a positive and a negative electron or an electron-positron pair is formed. In the reaction shown in Equation 23.5 a proton-antiproton pair is formed. If we consider the total charge to be the sum of all individual charges, *including their sign*, no net charge has been created. In Equation 23.5 the total charge on the left is $+2e$ and on the right $+2e$. The creation of one negative and one positive object does not change the total charge, since $+1e + (-1)e = 0$. The experimental observation of particle creation shows that charged particles are always created in pairs, one positive and one negative. No experimental observation has ever shown a violation of this rule. Therefore we accept charge conservation as one of the inviolable conservation laws of nature.

The annihilation of particles and antiparticles obeys the same rule. When an electron and positron collide and annihilate, the products are γ rays, which are uncharged.

$$e^+ + e^- \rightarrow \gamma + \gamma \tag{23.6}$$

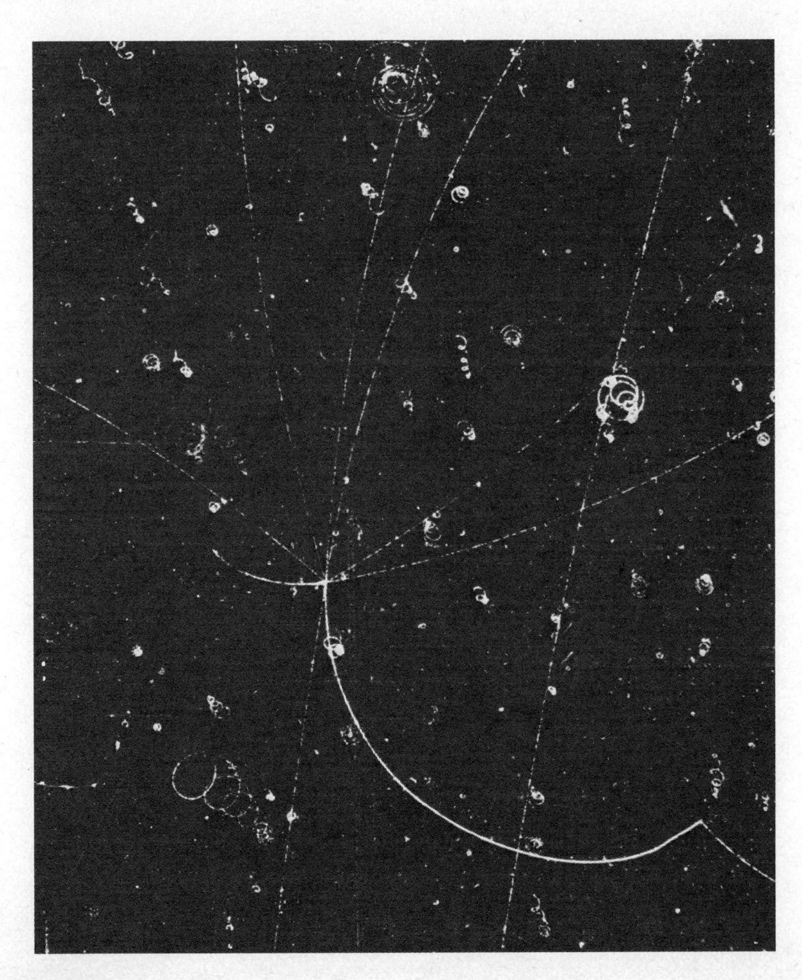

FIGURE 23.2 Conservation of electric charge in a proton-antiproton annihilation. The thick track curving up from the lower right is the antiproton. It annihilates with a proton in the bubble chamber and produces four π^+, four π^-, and perhaps several π^0. (*Courtesy Lawrence Radiation Laboratory, University of California.*)

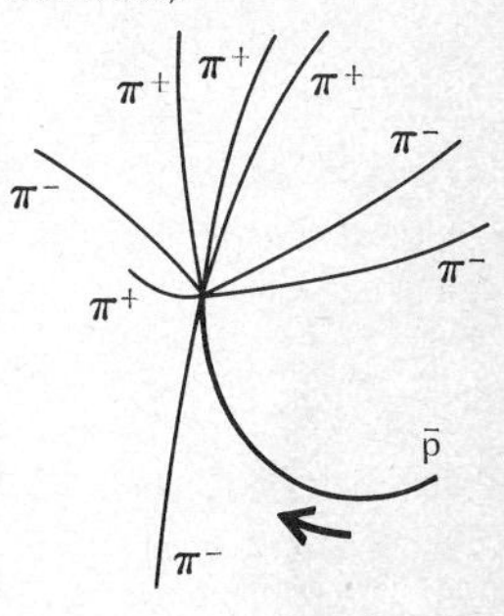

When a proton and antiproton annihilate, the result is a spray of π mesons, but the number of π^+ and π^- are exactly equal, so that again charge is conserved. This is shown in Fig. 23.2. Remember that it is the algebraic sum of all the charges which is conserved. Therefore, in the reaction in Equation 23.6, the total charge on the left is zero.

The symmetry principle associated with charge conservation cannot be stated in a fashion which can be easily understood at this level, so we shall leave charge conservation as an essentially experimental result.

23.4 CONSERVATION OF FAMILY MEMBERS

We have discussed in Chapter 22 the classification of particles into baryons, mesons, and leptons. Conservation laws are observed for some of these families, but not all. The important families in this discussion are the baryons, the muon family (the muons and their two neutrinos), and the electron family (electron, positron, neutrino, and antineutrino).

When properly defined, the total number of members of each of these families is not changed by any reaction. Consider first the decay of the neutron:

$$n \rightarrow p^+ + e^- + \bar{\nu}_e \qquad (23.7)$$

On the left there is one baryon, the neutron; and on the right, one also, the proton. On the left there are no members of the electron family; on the right there are the electron and an antineutrino. If we use the charge conservation to give us a clue, we see that the way to count members of a family is to count +1 for each particle and −1 for each antiparticle. In the reaction in Equation 23.7, then, the electron-family number is zero both before and after the reaction. In fact, this conservation law allows us to make statements as to whether neutrinos or antineutrinos are products in a given reaction—a question which seemed mysterious in Chapter 22.

Another example of this is the creation of a proton and antiproton pair:

$$p + p \rightarrow p + p + (p + \bar{p}) \qquad (23.8)$$

If we assign a baryon number of +1 to the proton and −1 to the antiproton, the baryon number is +2 on the left and +2 on the right, so that the baryon number is a constant. All the decays of the baryons discussed in the previous chapter show this result. The baryon number remains constant. For example,

$$\Lambda^0 \rightarrow p + \pi^- \qquad (23.9)$$

$$\Sigma^+ \rightarrow n + \pi^+ \qquad (23.10)$$

$$\Xi^- \rightarrow \Lambda^0 + \pi^- \qquad (23.11)$$

and

$$\Omega^- \rightarrow \Xi^0 + \pi^- \tag{23.12}$$

Decays of the antiparticles follow the same rule:

	$\overline{\Sigma}^+$	$\rightarrow$	$\overline{n}$	+	π^-	
Baryon number:	−1		−1		0	(23.13)
Charge:	−1		0		−1	

Remember that the antiparticle of Σ^+ has a negative charge. The baryon number is +1 for the proton, neutron, Λ^0, Σ, Ξ, and Ω, and −1 for their antiparticles. Baryon number is conserved in exactly the same way electric charge is conserved. The algebraic sum of the baryon numbers before and after a reaction is the same.

In the creation of the strange particles Λ, Σ, Ξ, and Ω, all of which are baryons, the baryon number is also conserved. For instance, in the collision of a π^- and a proton,

$$\pi^- + \mathrm{p} \rightarrow \Lambda^0 + \mathrm{K}^0 \tag{23.14}$$

There is one baryon on each side of the reaction.

A similar rule holds for the muon family—the two muons and their neutrino and antineutrino. In the decay of a π meson, a muon is the only visible product:

$$\pi^- \rightarrow \mu^- + \overline{\nu}_\mu \tag{23.15}$$

The neutrino associated with this proton *must* be the antineutrino in order that the number of muon-family members remain constant. Here again, an antiparticle counts as −1 and a particle +1 for the muon-family number.

If this rule is accepted, it becomes clear why two neutrinos are involved in the decay of a muon to an electron, as pointed out in Chapter 22.

	μ^-	$\rightarrow$ e^-	+ ν_μ	+ $\overline{\nu}_e$	
Muon-family number:	+1	0	+1	0	(23.16)
Electron-family number:	0	+1	0	−1	

The muon's neutrino keeps the muon-family number constant, and the electron's antineutrino keeps the electron-family number constant. For the decay of the μ^+,

	μ^+	$\rightarrow$ e^+	+ $\overline{\nu}_\mu$	+ ν_e	
Muon-family number:	−1	0	−1	0	(23.17)
Electron-family number:	0	−1	0	+1	

There is no conservation-of-family-members rule for the π mesons, the K mesons, or the η meson. Only the baryon group, the electron family, and the muon family seem to obey such a rule.

23.5 TYPES OF INTERACTIONS

Before it is possible to discuss the remaining conservation laws, it is necessary to digress slightly, for the next group of conservation principles are obeyed most of the time, but not always. To discuss the situations in which they are or are not valid, it is necessary to discuss the four types of forces, or interactions, found in nature.

All the forces, or interactions, presently known can be grouped into four categories. We have already discussed three of these. The first is nuclear force, which is the glue that holds protons and neutrons together in the nucleus. But nuclear force operates in a wider realm than this. All the baryons interact by nuclear force, which is also called *strong interaction*. In addition, all the mesons, π's, K's, and η, also show strong interaction. This group, baryons and mesons together, are called *hadrons*. All hadrons interact by means of strong interactions.

The second force, or interaction, is *electromagnetic interaction*. The simplest example of this is the coulomb force between charged particles. *All* charged particles show electromagnetic interactions. When compared with strong interactions, electromagnetic interactions are about 100 to 1,000 times weaker.

Third in line is the interaction responsible for β decay. This is called *weak interaction*. (Some of the ignorance which exists shows here in the inspired name.) Weak interactions are much, much weaker than strong interactions, by a factor of about 10^{-12}. All known particles (except the photon) show weak interactions.

The weakest force in nature, which has, so far as is known, no effect on nuclear or particle interactions, is gravity. It is, roughly, 10^{-40} times as strong as strong interactions.

Another way of discussing the strength of the various interactions is to discuss the time scale of processes involving them. Consider, for instance, a pion, traveling with nearly the speed of light, that passes through the nucleus of an atom. The speed of light is 3×10^8 meters/second, and the approximate size of the nucleus is 10^{-15} meter. Therefore the pion travels a distance roughly the size of a nucleus in less than 10^{-23} second. If measurements indicate that approximately one such collision is sufficient for the pion to interact and create other particles, we can say that the time scale for a pion to interact with a nucleus or any part of it is about 10^{-23} second. This is the time scale associated with strong interactions.

We can contrast the result above with the behavior of the muon, which can pass right through 10^4 to 10^5 nuclei before interacting. This experimental result shows that muons do not interact with nuclei as strongly as pions, and therefore do not interact with nuclei (or hadrons, in general) through strong interactions.

We can also talk about the relative time scales in terms of the lifetimes of the various particles. When the decay process is governed by strong interactions, the lifetimes found are about 10^{-23} second. These are observed in particles called resonances. When one of the conservation laws to be discussed forbids a decay governed by strong interactions, the lifetime of the particle will be longer. Decays governed by electromagnetic interactions yield lifetimes from 10^{-20} second up to about 10^{-17} second. Weak-interaction decays show lifetimes between 10^{-10} and 10^{-6} second. (Notice that the decay times in Table 22.1 are all in about this range.) It really does not make much sense to include gravity in this tabulation since, so far as is known, gravity does not participate in these processes.

23.6 OTHER SYMMETRIES

Now that we have stated the hierarchy of interactions, it is possible to discuss several other symmetries and their associated conservation laws. Some of these conservation laws can apparently sometimes be violated, but only when the interaction governing the particular process is weak, and not electromagnetic or strong.

CHARGE CONJUGATION

We have already discussed antiparticles and antimatter and proposed the speculative idea that perhaps there exist other galaxies composed entirely of antimatter. In such a world the role of every particle would be replaced by its antiparticle. For instance, the β decay of a neutron we write as

$$\mathrm{n} \rightarrow \mathrm{p} + e^- + \bar{\nu}_e \tag{23.18}$$

In an antimatter world each of these would be replaced by its antiparticle.

$$\bar{\mathrm{n}} \rightarrow \bar{\mathrm{p}} + e^+ + \nu_e \tag{23.19}$$

Remember that $\bar{\mathrm{p}}$ is the antiproton, which is negative, and $\bar{\mathrm{n}}$ is the antineutron.

Now imagine that the earth has somehow managed to communicate with a distant planet which is part of another solar system. Imagine also that the means to travel to this other planet exist. Since matter and antimatter annihilate with a violent release of energy if they come in contact, it is of great importance to the passengers and crew of the rocket about to take off to know whether their destination is a planet made of matter or antimatter. If it is antimatter, when their rocket made of matter comes into contact with antimatter from this planet, there will be a violent explosion.

The question can then be raised as to the possibility of finding out whether this other planet is matter or antimatter. More precisely, can we describe to the citizens of this new planet an experiment which they can perform and communicate to us, whose results will tell us whether their planet is matter or antimatter? *The charge-conjugation-symmetry principle states that no such experiment exists.* That is, any experiment or reaction involving ordinary matter will give results indistinguishable from those of a similar experiment in which each particle has been replaced by its antiparticle. This symmetry operation is called charge conjugation because, in a reaction involving antiparticles, all the signs for the electric charge of particles have been reversed. If charge-conjugation symmetry is valid, β decays (Equations 23.18 and 23.19) will be identical with respect to time constant, angular dependence, and every other property. Therefore it would not be possible to tell whether a given process has occurred in matter or antimatter, and therefore it is not possible to determine whether we live on an earth made of matter or antimatter from observations made on earth alone.

It should be pointed out that in β decay (Equation 23.18), an antineutrino is emitted and, in the reaction in Equation 23.19, a neutrino is emitted. If it is possible to distinguish neutrinos and antineutrinos, it will be possible, at least in principle, to distinguish a galaxy of matter from one of antimatter. This will depend on observation of the two kinds of neutrinos from distant galaxies. The neutrino is the only uncharged particle which will tell whether its source is matter or antimatter. To distinguish between neutrinos and antineutrinos involves a violation of parity symmetry, to be discussed shortly.

Charge-conjugation symmetry says, then, that processes involving matter and antimatter are identical, so that the only way of finding the difference is to bring them together. The student should note that if Ben Franklin had defined the two kinds of electric charge with reversed signs, our world would still be the same, and any physical experiments done on this earth would still be the same. There is no way someone on earth could define positive and negative charge to someone on another planet, unless he happened to know that the other planet was composed of matter. Then it could be done. If the composition, matter or antimatter, of the other planet were unknown, positive and negative could not be defined.

The importance of charge-conjugation symmetry does not lie in what it says about communication with other planets. Rather, the important question is whether charge-conjugation symmetry is a law which governs the processes in the real physical world or not. This question can be answered only experimentally, and the answer appears to be that it is not a perfectly obeyed law. This point will be taken up in the next section.

PARITY

Another type of symmetry that can also be described in terms of communication with another planet is called *parity*. Try to imagine the problem of explaining to an inhabitant of another planet which is your right

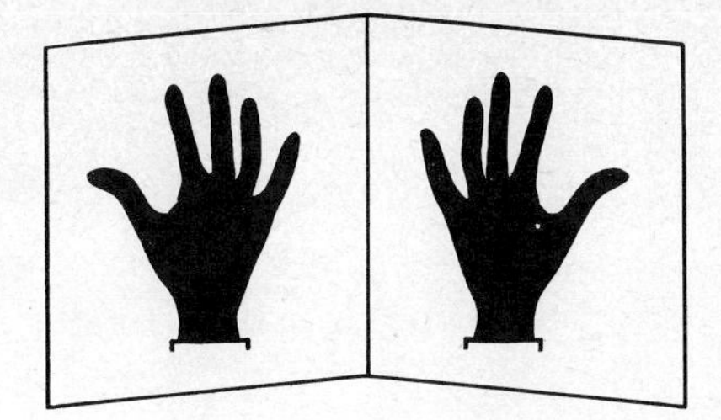

FIGURE 23.3 Right and left hands are mirror images of each other. If one is real and the other its mirror image, which is which?

hand without being able to go there. We all know that the right and left hands are different from one another; they are in fact mirror images, as in Fig. 23.3. But how is it possible to describe the difference? The only possible method would be to find some physical process which has a natural right- or left-handedness. Those who have studied biology know that such a handedness is found in biology as it exists on earth, but this might be exactly the opposite on some other planet. Therefore that idea will not work. The parity-symmetry principle states the idea that no physical process can exist that has a built-in, natural right- or left-handedness. The validity of this idea *must* be subjected to experimental test.

TIME REVERSAL

The third in this set of three symmetry principles is called *time reversal*, which means exactly what it says. Time-reversal symmetry says that, if any physical process can occur, the same process reversed in time can also occur. Like charge conjugation and parity, time reversal has an intrinsic reasonableness. But the only true test is a direct experimental one.

29.7 VIOLATIONS OF SYMMETRY

The three symmetries just discussed have an innate appeal. After all, why should a left-handed process be different from a right-handed one? Why should processes involving matter be different from those involving antimatter? Because of the obvious nature of these symmetries, they were accepted as valid for some time. Then, in 1956, the boom was lowered.

Two theoreticians—C. N. Yang and T. D. Lee—suggested that possibly, in some processes involving weak interactions, parity symmetry might be violated. The proof of this involved finding a physical process which was *not* the same as its mirror image. If such a process could be found, it would be possible to describe this experiment to the hypothetical inhabitant of another planet in such a way that he would know what is meant by right and left.

The experiment suggested by Lee and Yang could have been per-

FIGURE 23.4 T. D. Lee (1926–) and C. N. Yang (1922–).

formed by any of a number of people during the previous 5 or 10 years. In fact, one experimental result in 1928 was not believed, because it violated common sense in that it violated the parity-symmetry principle. Lee and Yang proposed that if parity were not a valid symmetry principle for weak interactions, the result might be other than expected. As a result, the experiment described below, which demonstrated the correctness of their idea, was performed by Chien-Shiung Wu and several others.

The experiment was a study of the β decay of the nucleus $^{60}_{27}Co$, cobalt 60. This nucleus is one of the more dangerous components of radioactive fallout from nuclear weapons testing. The cobalt-60 nucleus, like most others, possesses an intrinsic angular momentum, or spin. We can therefore represent this nucleus as a small spinning top. Like any moving charge, this spinning nucleus generates a magnetic field (Fig. 23.5). At ordinary temperatures, the cobalt nuclei in a sample are oriented randomly in space, as in Fig. 23.6*a*. But at extremely low temperatures, these nuclei can be lined up in a magnetic field, as in Fig. 23.6*b*. When this is done, more of the electrons given off in β decay are observed to go down than up. It is this preferential emission of electrons in one direction that violates the parity-symmetry principle. Because, if we examine the experiment, as it occurred, and its mirror image, we

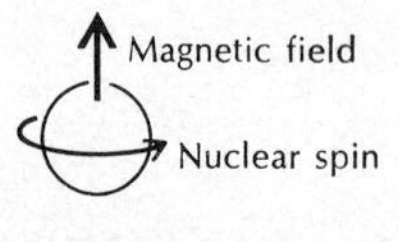

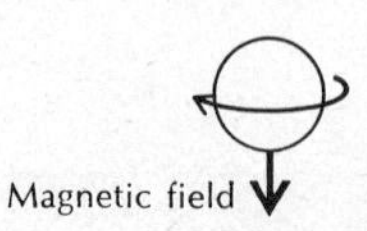

FIGURE 23.5 Because the nucleus is charged and spins, it creates a magnetic field. If the nucleus is turned over, the direction of the magnetic field is reversed, as shown.

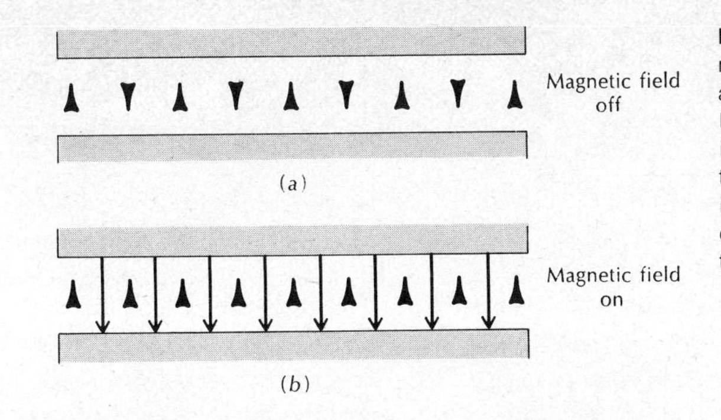

FIGURE 23.6 In a magnetic field at low temperatures, cobalt-60 nuclei all line up in one direction. Now, if more electrons are found to be emitted up than down, it means that electron emission from cobalt violates the parity-symmetry principle.

see they are different. This is shown in Fig. 23.7. The difference lies in the fact that, although the sense of rotation of the nucleus is reversed, the β-decay electrons continue to be emitted in the same direction. We can see clearly that the two situations are different if we apply a type of left-hand rule to relate the direction of rotation and the direction of preferential electron emission. The same rule applied to the mirror-image case clearly does not hold, since it predicts emission in the opposite direction. The second case obeys the mirror-image right-hand rule. Therefore a process is found whose mirror image is not the same as the process itself. With this experiment, parity symmetry was shown to be violated in weak (β-decay) interactions.

TCP THEOREM

There exists a general theorem, believed to be true on the basis of strong arguments, that says that the product of the three symmetries, time reversal T, charge conjugation C, and parity P, should be a valid symmetry.

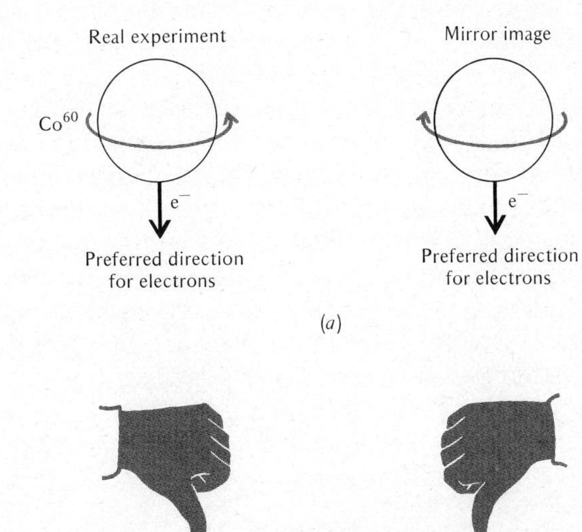

FIGURE 23.7 (*a*) If electrons are preferentially emitted in one direction, the mirror image of this experiment is not identical to the experiment itself, in the same sense that right and left hands are not identical. (*b*) Left and right hands used to show the different symmetry. The fingers curl with the direction of rotation, and the thumb points in the direction of electron emission.

To understand the meaning of this statement, consider again a process such as the β decay of the neutron.

$$n \rightarrow p + e^- + \bar{\nu}_e \qquad (23.18)$$

Now take the charge conjugate of this reaction,

$$\bar{n} \rightarrow \bar{p} + e^+ + \nu_e \qquad (23.19)$$

Now reflect in a mirror the observations on the process in Equation 23.19. This gives the product of the two operations *C* and *P*. This product is designated *CP*. Now, if time is allowed to run backward for the process which results from the *CP* product, we shall have performed all three symmetry operations. *If* the resulting process is a valid physical process whose results can be predicted precisely from a knowledge of the results of the original process (Equation 23.18), we say that this process is symmetric under the product of the three operations *TCP*.

No violation of *TCP* symmetry has yet been found, but there exists experimental evidence that charge conjugation *C*, parity *P*, and the product *CP* are not perfectly obeyed symmetries in weak interactions. There is a suspicion that time reversal may not be a good symmetry property as well. Evidence is not yet strong enough to prove the case.

23.8 STRANGENESS

The first of the strange particles, the Λ^0, was discovered in 1947. The properties of this particle and of the other strange particles led to the discovery of another conservation law. The Λ^0 is produced with sufficient ease so that it is clear that it is produced through strong interaction. That is, the probability of producing a Λ^0 in a single collision of two particles, such as two protons, is large enough so that the process must be occurring on the time scale of strong interactions. But the decay time of a free Λ^0 is about 10^{-10} second, which is much too long for a decay governed by strong interaction. It was proposed by Murray Gell-Mann, and independently by K. Nishijima, that there is another conservation law obeyed by strong interactions, but not by weak interactions, which prevents the rapid decay of the Λ^0. This law is called the *conservation of strangeness*. Each particle is assigned a number, called a *strangeness number*. This number is conserved in strong interactions, in the same sense that electric charge is conserved. The algebraic sum of strangeness numbers remains constant. Strangeness numbers are assigned by studying the experimental results.

The Λ^0 is assigned a strangeness number of -1, and the proton and neutron, 0. In the production of a Λ^0 by a collision between a pion and proton, strangeness will not be conserved unless another strange particle with strangeness $+1$ is created simultaneously. This was shown in Fig. 22.13.

$$\pi^- + p \rightarrow \Lambda^0 + K^0$$

	π^-	p	Λ^0	K^0
Strangeness number:	0	0	−1	+1
Charge:	−1	+1	0	0
Baryon number:	0	+1	+1	0

(23.20)

In the decay of the Λ^0, shown below, we find that strangeness is not conserved.

$$\Lambda^0 \rightarrow p + \pi^-$$

	Λ^0	p	π^-
Strangeness number:	−1	0	0
Charge:	0	+1	−1
Baryon number:	+1	+1	0

(23.21)

The strangeness on the left is −1, and on the right, 0. Therefore strangeness conservation is violated, and the Λ^0-decay process must go by the weak, or slow route. That it does go "slowly" is borne out by the fact that the decay time is 10^{-10} second, which is characteristic of weak decay processes.

A careful study of the production and decay of the strange particles has led to a table of strangeness numbers, Table 23.1. It should be borne in mind that this table is a result of experiments, not theory. A strangeness number of −2 does not mean that a particle is more strange than if it has a strangeness of −1. If a particle has a strangeness of +1, its antiparticle has a strangeness of −1. The sign is always reversed.

23.9 THE EIGHTFOLD WAY

The most important result of the accumulation of experimental data and conservation laws has been the emergence of some degree of order, or organization, in the zoo of particles. We shall be unable to explain why the particular organization given ought to work. We can only describe the results as they now exist.

Particle	Strangeness number
p	0
n	0
π^0, π^+, π^-	0
Λ^0	−1
Σ^0, $\Sigma^\pm$	−1
K^0, K^+	+1
Ξ^0, Ξ^-	−2
Ω^-	−3

TABLE 23.1 Strangeness numbers. The antiparticle in each case has the opposite sign. These numbers are found by studying the creation and decay of the strange particles

Particle	Multiplicity
Proton, neutron	2
Σ^+, Σ^0, Σ^-	3
Λ^0	1
Ξ^-, Ξ^0	2
π^+, π^0, π^-	3
Ω^-	1
N*	4
Y*	3
Ξ^*	2

TABLE 23.2 Some of the families of particles and their multiplicity. The N*, Y*, and Ξ^* are resonances discussed in the text

We have seen that there are groups of similar or related particles. This similarity has been recognized by giving them names such as Σ^+, Σ^0, and Σ^-, indicating that the main·difference is in the value of their electric charge. The proton and neutron are a pair of nearly identical particles (the mass difference is about 0.1 percent), except for their electric charge. As a result of this observation, it has been proposed that all particles occur in such families in which the number of particles in the family is called the multiplicity. Table 23.2 shows several of these families and their multiplicity.

An extension of the above ideas, called the *eightfold way*, was proposed by Gell-Mann in 1961. The basic idea is that there are larger family groups which are also nearly identical. The larger families are less identical than members of the family groupings discussed above. The number of particles in the possible families can be predicted from the precise mathematical structure of the theory. (An everyday analogy is the fact that, if we wish to cover a floor with tiles, each of which has the same shape, three and only three shapes are allowed, as seen in Fig. 23.8.) The family sizes allowed by the mathematics of the theory are 1, 8, 10, etc. The name eightfold way arises from the family with eight members. All the members of a given family have the same spin, and the values of their electric charge and strangeness numbers are related as shown in Fig. 23.9. In this figure are shown two families of eight: proton, neutron, Λ, Σ, Ξ, with spin $\frac{1}{2}$, and π, η, K, $\overline{\text{K}}$, with spin 0. Each particle is entered in this chart at the point corresponding to its charge and strangeness number. The similarity in the two groupings indicates that this scheme has some significance.

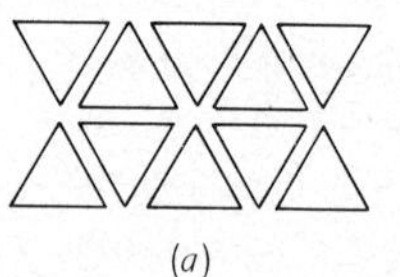

(*a*)

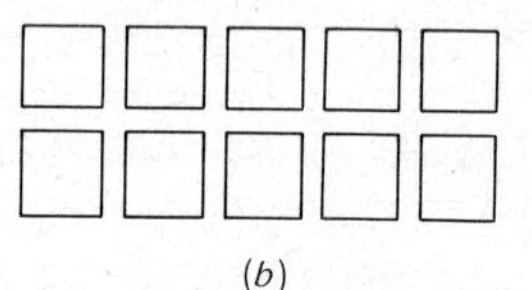

(*b*)

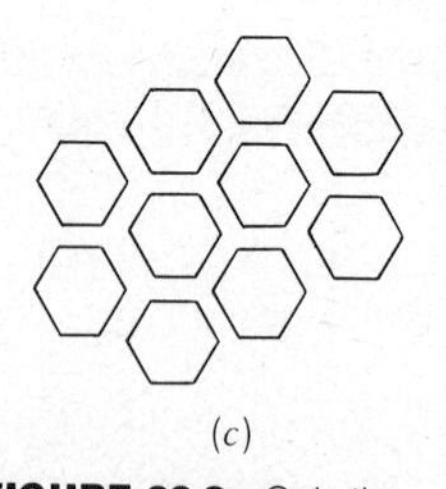

(*c*)

FIGURE 23.8 Only the three geometric figures, (*a*) triangles, (*b*) squares, and (*c*) hexagons, will completely cover a flat surface. This is a mathematical property of flat surfaces. In the same sense, the mathematical structure of Gell-Mann's theory limits the number of members of each of the family groups it predicts.

It might be asked, What good is all this? If we remember the construction of the periodic table by Mendeleev, we know that he was able to predict the existence of elements which had not been discovered. A similar process, using a chart similar to that of Fig. 23.9, led to the prediction and discovery of a new particle, the Ω^-. Figure 23.10 shows a plot of a group of three families of resonances, N* (mass = $2{,}420m_e$), of which there are four (with electric charge +2, +1, 0, −1), Y* (m =

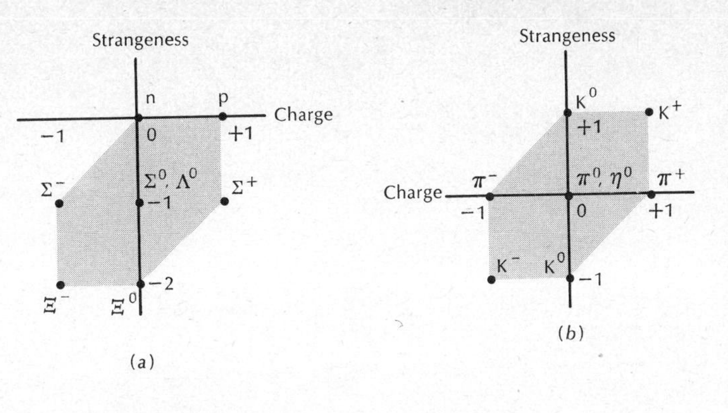

FIGURE 23.9 (*a*) The charge and strangeness of the particles n, p, Λ, Σ, and Ξ are plotted in this diagram. For example, Ξ is at charge −1 and strangeness −2. (*b*) The K, π, and η mesons are plotted in a fashion similar to (*a*). All particles in (*a*) have spin $\frac{1}{2}$; all particles in (*b*) have spin 0.

$2{,}705m_e$), which has three members (+1, 0, −1), and Ξ^* ($m = 2{,}992m_e$), with two members (0, −1). All these have a spin of $\frac{3}{2}$. Above, it was stated that the number of members in a family could be 1, 8, 10, etc. This group already has nine. A logical guess is that there is a tenth member at $Q = -1$, $S = -3$. If we note that the mass differences are $Y^* - N^* = 285m_e$ and $\Xi^* - Y^* = 282m_e$, we might guess that the mass difference between the new particle and Ξ^* is also about $285m_e$. This suggests a mass for the new particle of about $3{,}274m_e$. This particle is the Ω^-, which was found to have the mass, strangeness, and charge predicted. The measurement of its spin (which is a very difficult experiment) has not yet been accomplished, but in the words of one well-known theorist, "If it turns out not to be $\frac{3}{2}$, like the other members of the family, all theorists will leap off the Golden Gate Bridge." Notice that in this group of 10 we have both resonances, and one longer-lived particle. This is one piece of evidence that we cannot ignore resonances just because they have such short lifetimes.

The physics of the atom is in a satisfying state because we can show that all atoms are built up from only three elementary particles: the proton, neutron, and electron. Gell-Mann has attempted to account for all the known particles as combinations of three quarks.† Quarks are objects with charge $\frac{2}{3}e$ or $\frac{1}{3}e$. Two quarks of charge $+\frac{2}{3}e$ and one of

†The name comes from a line in James Joyce's "Finnegan's Wake." The line begins, "Three quarks for Muster Mark."

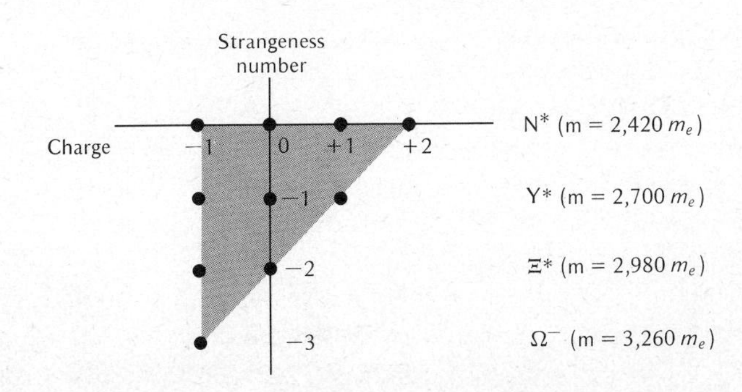

FIGURE 23.10 The resonances N*, Y*, Ξ*, and the Ω^- particles fall into a group of 10. The existence of the Ω^- was predicted from this type of plot. Because of the regularity in mass differences, the mass of the Ω^- was also predicted. All these particles have spin $\frac{3}{2}$.

charge $-\frac{1}{3}e$ will yield an object of charge $+1e$. As of this writing no isolated quark has been observed. The family relationships we have been discussing can be accounted for with a quark model, but there may be other ways of accounting for these relationships. Exact understanding of these things is not yet available. Much of the experimental work at the world's largest accelerators is directed to these problems.

23.10 THE NEW PARTICLES

Since the early 1960s discoveries in particle physics have come more slowly than before. There has been some discouragement, even though new and more detailed experimental information about the various particles has been accumulated.

In the early 1970s several new accelerators were completed, permitting experiments at higher energies than before. In November 1975, two groups announced the discovery of a new particle, of mass 6,200 electron masses. The lifetime of this new particle is about 10^{-20} second, which is fantastically long for a particle of such a heavy mass. Clearly the decay of this particle, called ψ (psi), is inhibited by some conservation law. Shortly after the original discovery, another similar particle was found whose mass is 7,400 electron masses with an equally long lifetime. These discoveries are considered to be the most exciting events in particle physics in 20 years.

There has been much theoretical activity attempting to understand these new objects. Since the answer is not yet in, we can only comment on some of the ideas being considered. One is a modification of the quark model. The quark scheme, as originally proposed, has three quarks (and their antiparticles). It has been suggested that more than this must exist, and that the new quarks must have some property, like strangeness, which inhibits the fast decay of particles containing them. Two of these proposed properties have been called *charm* (a long-lived particle has a charmed life) and *color*. The idea is similar to that associated with strangeness. A decay that changes the charm or the color of the quarks is prevented from going rapidly, and must go slowly via the weak interaction.

Because the ψ particles are so new, their place in the structure of physics is unclear, and the effect their discovery will have on existing theories is unknown. They form one more step in the struggle to show how all of the particles are formed from a few elementary building blocks. Perhaps they are simply the forerunners of a large number of new particles to be found, with new confusions to come. Perhaps they will lead to a final resolution of the puzzle. What is clear is that physics is still a live and developing science, and that new discoveries are still occurring. It is this lure of the unknown that keeps scientists enthusiastic in their work, and provides much of the impetus for the progress of science.

SUMMARY

All symmetry and conservation principles tell us what is forbidden in elementary-particle interactions.

Conservation of energy forbids particle decay where the total mass of products is greater than the mass of the original particle. Conservation of momentum forbids decay of a particle to a single-particle product.

Angular-momentum conservation forbids a decay that does not conserve the spin of particles.

Charge conservation requires that the total charge (algebraic sum of all charges) not change in any nuclear or particle process. Therefore, in any creation process, pairs of particles of opposite charge are created.

The number of particles in the baryon family, muon family, and electron family are conserved in particle processes. Particles count as +1 and antiparticles count as −1.

There appear to be four interactions in nature. Strongest is the strong interaction, or nuclear force. Next is electromagnetic interaction. Third, and much weaker, is the weak interaction, responsible for β decay. Last, and vastly weaker than the others, is gravity.

Charge-conjugation symmetry says that decay reactions with all particles turned to their antiparticles (and thus charge-reversed) will behave exactly the same as the original reaction. It appears that this law may be occasionally violated.

Parity states that reactions and their mirror images ought to be indistinguishable. It is violated in certain β-decay processes.

Time-reversal symmetry asserts that time-reversed processes should occur, and that their features should be directly calculable from the forward reaction. It, too, may not be a perfectly obeyed law.

Strange particles are created in high-energy collisions. The strangeness number is conserved in the creation process, but is not conserved in the decay of the strange particles.

The eightfold way is an organizational scheme for particles based on their charge and strangeness. Its existence implies some underlying order.

SELECTED READING

Ford, Kenneth W.: "The World of Elementary Particles," Blaisdell Publishing Company, New York, 1963. A very good survey, at an elementary level, of the whole elementary-particle field.

SELF-TEST

__________ energy conservation

__________ momentum conservation

__________ angular-momentum conservation
__________ spin
__________ charge conservation
__________ muon family
__________ electron family
__________ baryon number
__________ strong interaction
__________ quark
__________ weak interaction
__________ electromagnetic interaction
__________ charge-conjugation symmetry
__________ parity symmetry
__________ time-reversal symmetry
__________ *TCP* theorem
__________ strangeness number
__________ eightfold way

1 Group of particles consisting of muon, antimuon, and two muon neutrinos
2 Interaction involved in processes with lifetimes between 10^{-6} and 10^{-10} second
3 Number attached to strange particles to account for associated production. Conserved in production but not in decay reactions
4 Mirror-image symmetry
5 Intrinsic angular momentum of a particle
6 Number attached to baryons to account for their decays
7 The nuclear force
8 The electron, positron, and two neutrinos
9 Electric charge is neither created nor destroyed in particle processes
10 In particle decays the mass of products must be less than the mass of the original particle
11 Interaction somewhat weaker than the strong interaction
12 Symmetry law associated with replacing all particles in a process with their antiparticles
13 Law that governs the spins of particles in allowed reactions
14 Symmetry theorem involving charge conjugation, parity, and time reversal
15 Attempt to organize the various particles into groups based on charge and strangeness
16 Symmetry associated with running a process backward in time
17 Forbids decay of a particle to a single particle
18 Proposed as a building block for particles in Gell-Mann's theory

REVIEW
OF ALGEBRA

This appendix is a review of what the student is expected to know in order to use this book, not a complete program for learning algebra. Therefore, if there are portions of this material which are unfamiliar, they should be reviewed at the beginning of the course. Algebra is arithmetic in which letters are used to represent numbers; that is, instead of $3 \times 5 = 15$, we write $ab = c$, which means a times b equals c. This has the distinct advantage that we can manipulate the numbers a, b, and c *before* we know their numerical values. In a very real sense algebra is a shorthand for writing down relationships which would be much more cumbersome in any other form.

A.1 POSITIVE AND NEGATIVE NUMBERS

The use of positive and negative numbers occurs frequently in physics. Consider one example used early in the text. If we discuss a rock which is thrown or dropped from the top of a tall building, it might be of interest to know where the rock is after a period of time. If the answer is given as 50 feet from the starting point, this is not sufficient information to indicate where the stone actually is. It is also necessary to state whether this is 50 feet up or 50 feet down from the building roof. The problem is solved by using positive distances to represent positions *above* the starting point, and negative distances to represent positions below the starting point, as shown in Fig. A.1. Once it has been agreed which direction to call positive and which negative, the sign of the distance tells us the *direction* from the starting point.

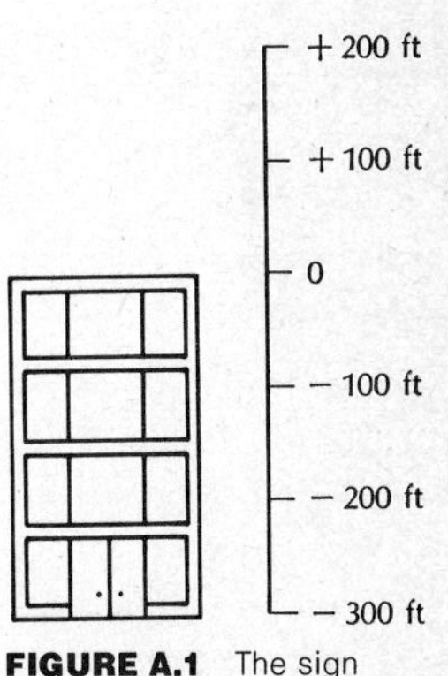

FIGURE A.1 The sign of the distance tells whether it is up (positive) or down (negative) from the point chosen as zero. In this example the roof of the building is chosen as zero.

Addition of signed numbers: The addition of a negative number is equivalent to subtraction. The addition of two positive or two negative numbers proceeds normally; the two numbers are added, and the sign of the answer is the sign of the two numbers.

$$5 + (-3) = +2$$
$$(-5) + (-3) = -8$$
$$(-5) + (+3) = -2 \qquad \text{note the sign of this answer}$$

Subtraction: Subtracting a negative number is equivalent to adding that number with its sign reversed.

$$7 - (-3) = 7 + (+3) = +10$$
$$-8 - (-4) = -8 + (+4) = -4$$
$$-9 - (+5) = -9 + (-5) = -14$$

Multiplication: The multiplication of numbers with like signs gives a positive result. The multiplication of numbers with unlike signs gives a negative result.

$$(+5) \times (+4) = +20$$
$$(-5) \times (-4) = +20$$

$(-5) \times (+4) = -20$

$(+5) \times (-4) = -20$

Division: Since division by any number a is equivalent to multiplication by $1/a$, the rules are similar to those for multiplication.

$+20 \div (+4) = +5$

$+20 \div (-4) = -5$

$-20 \div (+4) = -5$

$-20 \div (-4) = +5$

A.2 EXPONENTS

The use of exponents is a device of mathematical notation. The exponent is used to write things in a more condensed or convenient fashion. The student should note that *all* the properties of exponents are a direct consequence of their definition.

In the quantity a^n, n is the exponent, and this quantity is *defined* to mean a multiplied by itself n times. For example,

$a^3 = a \times a \times a$

$2^4 = 2 \times 2 \times 2 \times 2$

a^2 is called a squared; a^3 is called a cubed; and a^n is a to the nth power.

Multiplication: The multiplication of numbers with exponents follows directly from the definition above.

$a^2 \times a^3 = (a \times a) \times (a \times a \times a) = a^5$

Therefore, to multiply any number to some power n times the same number to some other power m, the exponents n and m are added.

$a^n \times a^m = a^{n+m}$

This does not work if we try to multiply a^3 times b^2. The result here is just

$a^3 \times b^2 = (a \times a \times a) \times (b \times b) = a^3 \times b^2$

Division: Again, using the definition and a specific example, the general law can be found.

$$a^4 \div a^2 = \frac{a \times a \times a \times a}{a \times a}$$

$$= a \times a = a^2$$

Therefore $a^4 \div a^2 = a^{4-2} = a^2$. When dividing, the exponents are subtracted: $a^n \div a^m = a^{n-m}$.

Negative exponents: If we compare $a^4 \div a^2$ and $a^4 \times a^{-2}$, we see that the result is the same, a^2. That is,

$$\frac{a^4}{a^2} = a^2$$

and

$$a^4 \times a^{-2} = a^2$$

Therefore $a^{-2} = 1/a^2$. This is a general result for a number with a negative exponent. Any time a number with an exponent is moved from the top to the bottom of a fraction, the sign of the exponent must be changed.

Using these results it is possible to find out what a^0 means.

$$\frac{a^n}{a^n} = a^{n-n} = a^0 = 1$$

Any number to the zero power is equal to 1.

Roots and fractional exponents: A root of a number is defined as the number which, multiplied by itself the stated number of times, yields the original number. For example:

The second root, or square root, of 16 is 4, since $4 \times 4 = 16$.

The third root, or cube root, of 27 is 3, since $3 \times 3 \times 3 = 27$.

The fifth root of 32 is 2, since $2 \times 2 \times 2 \times 2 \times 2 = 32$.

From the definition of a root and the multiplication of exponential numbers, the meaning of a fractional exponent can be shown.

$$(a^{1/3})^3 = a^{1/3} \times a^{1/3} \times a^{1/3}$$

$$= a^{1/3+1/3+1/3}$$

$$(a^{1/3})^3 = a^1 = a$$

Therefore $a^{1/3}$ is the number which, multiplied by itself three times, gives a. By definition, $a^{1/3}$ is the third, or cube, root of a. In general, $a^{1/n}$ is the nth root of a.

A.3 POWER-OF-TEN NOTATION

In physics it is frequently necessary to discuss very large numbers. In order to avoid awkwardness, we use a convenient notation. From the definitions above we know

$$10^6 = 1{,}000{,}000$$

$$10^9 = 1{,}000{,}000{,}000$$

The exponent tells the number of zeros. For negative exponents we have

$$10^{-6} = \frac{1}{1{,}000{,}000} = 0.000001$$

$$10^{-3} = 0.001$$

We use these results to write large numbers. The basic idea is to use 10 to some power to indicate where the decimal place is. For example,

TABLE A.1 The powers of ten

$10^6 =$	1,000,000
$10^5 =$	100,000
$10^4 =$	10,000
$10^3 =$	1,000
$10^2 =$	100
$10^1 =$	10
$10^0 =$	1
$10^{-1} =$	0.1
$10^{-2} =$	0.01
$10^{-3} =$	0.001
$10^{-4} =$	0.0001

500,000 is the same as $5 \times 100{,}000$, which can be written 5×10^5, since $10^5 = 100{,}000$.

Some other examples are

$$2{,}000{,}000 = 2 \times 10^6$$
$$310{,}000{,}000 = 3.1 \times 10^8$$
$$0.00027 = 2.7 \times 10^{-4}$$
$$0.0000038 = 3.8 \times 10^{-6}$$

Exercises

Express the following in power-of-ten notation. The answers to the exercises are given at the end of the appendix.

1 230
2 4,600
3 95,000
4 7,260,000
5 0.0062
6 0.25
7 0.0000000029
8 0.002

Multiplication of numbers in power-of-ten notation: The procedure is most easily seen by an example. Find the product of 2×10^5 and 3×10^6.

$$\begin{aligned}(2 \times 10^5) \times (3 \times 10^6) &= 2 \times 3 \times 10^5 \times 10^6 \\ &= 6 \times 10^5 \times 10^6 \\ &= 6 \times 10^{5+6} \\ &= 6 \times 10^{11}\end{aligned}$$

The 10's with their exponents multiply according to the rules for exponential numbers. The numbers in front multiply normally.

$$(6.2 \times 10^4) \times (2 \times 10^{-3}) = 12.4 \times 10^{4-3}$$
$$= 12.4 \times 10^1$$
$$= 1.24 \times 10^2$$

Answers are customarily written as shown, with one figure to the left of the decimal point.

Exercises

Multiply

9 $(4 \times 10^6) \times (5 \times 10^7)$
10 $(4 \times 10^6) \times (5 \times 10^{-7})$
11 $(2.2 \times 10^{11}) \times (7 \times 10^{-12})$
12 $(6 \times 10^{-10}) \times (7 \times 10^{-12})$

Division proceeds similarly:

$$\frac{4 \times 10^6}{2 \times 10^3} = \frac{4}{2} \times 10^{6-3} = 2 \times 10^3$$

Exercises

Divide

13 $(8 \times 10^6) \div (4 \times 10^3)$
14 $(9 \times 10^8) \div (3 \times 10^{12})$
15 $(6 \times 10^{-7}) \div (4 \times 10^{10})$
16 $(8 \times 10^{-11}) \div (4 \times 10^{-3})$

Addition and *subtraction* of numbers in power-of-ten notation: Add 2×10^4 to 3×10^5. To see how this works we must write these numbers out in normal notation.

$$2 \times 10^4 = 20{,}000$$
$$3 \times 10^5 = \underline{300{,}000}$$
$$320{,}000 = 3.2 \times 10^5$$

Therefore $(2 \times 10^4) + (3 \times 10^5) = 3.2 \times 10^5$. The proper way to do this is to express both numbers in terms of the *same* power of ten, and add.

$$\begin{aligned} 2 \times 10^4 &= 0.2 \times 10^5 \\ 3 \times 10^5 &= \underline{3.0 \times 10^5} \\ & 3.2 \times 10^5 \end{aligned}$$

Note that the power of ten does not change.

A.4 ALGEBRAIC PROCEDURES AND EQUATIONS

The use of parentheses: Parentheses ( ) means that the quantity within is to be treated as a single quantity in any operation. That is, $a(b + c)$ means a times the sum of b and c. For instance, $3(4 + 5)$ is the same as $3 \times (9)$, which equals 27.

Working with equations: An equation remains an equality if the same quantity is added to or subtracted from both sides.

Example 1

$$\begin{aligned} 2x - 5 &= +15 \\ +5 &= +\ 5 \\ \hline 2x \quad &= +20 \end{aligned}$$

Example 2

$$\begin{aligned} 2x + 5 &= +15 \\ -5 &= -\ 5 \\ \hline 2x \quad &= +10 \end{aligned}$$

This procedure is used to obtain a simpler equation, with quantities involving x on the left and numbers on the right.

A common method of handling the same result is the following. To move any quantity from the right- to left-hand side of an equation, or the reverse, change its sign. In Example 1 above,

$$2x \; \textcircled{-5} = +\ 15 + 5$$
$$2x = +20$$

An equation remains an equality if both sides are multiplied by the same quantity.

Example 3

$\frac{x}{2} = 10$. Multiply both sides by 2.

$$2\left(\frac{x}{2}\right) = 2 \times 10$$

$$x = 20$$

This procedure may be used to simplify equations involving fractions such as $\frac{x}{2}$.

Example 4

$2x = 10$. Multiply by $\frac{1}{2}$.

$$\tfrac{1}{2}(2x) = \tfrac{1}{2} \times 10$$

$$x = 5$$

This is equivalent to dividing both sides of the equation by 2. It gives a result in x rather than in $2x$ in the example above. If the equation to be solved were $5x = 25$, the proper procedure would be to divide both sides by 5, or multiply both by $\frac{1}{5}$, which is equivalent.

A process which frequently is useful when dealing with algebraic fractions is *cross multiplication*. If the equation to be solved is

$$\frac{x}{y} = \frac{2}{5}$$

one method of proceeding is to multiply both sides by $5y$. This gives

$$5y\left(\frac{x}{y}\right) = 5y\left(\frac{2}{5}\right)$$

$$5x = 2y$$

The result is the same as if we had cross-multiplied. Multiply the top of one fraction by the bottom of the other.

$$\frac{x}{y} = \frac{2}{5}$$

$$5x = 2y$$

An equation remains an equality if both sides are raised to the same power or the same root is taken of both sides.

Example 5

$x^2 = 4$. Take the square root of each side.

$x = +2$ or -2 note that $(-2) \times (-2)$ is also $+4$

Example 6

$x = 4$. Square both sides.

$$x^2 = 4 \times 4$$

$$x^2 = 16$$

Multiplication of quantities enclosed in parentheses proceeds in a straightforward fashion.

Example 7

$$(a + b) \times (a + b) = \begin{array}{r} a + b \\ \underline{a + b} \\ ab + b^2 \\ \underline{a^2 + ab } \\ a^2 + 2ab + b^2 \end{array}$$

Example 8

$$(a + b)(c + d) = \begin{array}{r} a + b \\ \underline{c + d} \\ ad + bd \\ \underline{ac + bc } \\ ac + ad + bc + bd \end{array}$$

It is impossible to collect terms further, since ac, ad, bc, and bd are different quantities.

Example 9

$$(a - b)(a - b) = (a - b)^2$$

$$= a^2 - ab - ab + b^2$$
$$= a^2 - 2ab + b^2$$

We shall frequently use the rules discussed here to solve equations involving only algebraic quantities.

Example 10

$xyz = a^2b$. Solve for y. Divide both sides by xz:

$$\frac{xyz}{xz} = \frac{a^2b}{xz}$$
$$y = \frac{a^2b}{xz}$$

Example 11

$\frac{xy}{zb} = \frac{A}{r^2}$. Solve for b. Multiply both sides by b, to bring it into the numerator of the fraction.

$$\frac{bxy}{zb} = \frac{bA}{r^2}$$
$$\frac{xy}{z} = \frac{bA}{r^2}$$

Multiply by $\frac{r^2}{A}$.

$$\frac{r^2}{A}\left(\frac{xy}{z}\right) = \frac{r^2}{A}\left(\frac{bA}{r^2}\right)$$
$$\frac{r^2xy}{Az} = b$$

Example 12

Given the two equations below, eliminate A.

$$\frac{xy}{z} = \frac{A}{r^2}$$
$$xy = Ar$$

$$\frac{xy}{z} = \frac{A}{r^2} \quad \text{yields} \quad \frac{xyr^2}{z} = A$$

$$xy = Ar \quad \text{yields} \quad \frac{xy}{r} = A$$

Two quantities each equal to a third quantity are equal to each other. Therefore

$$\frac{xyr^2}{z} = \frac{xy}{r}$$

Divide by xy: $\frac{r^2}{z} = \frac{1}{r}$

Cross-multiply: $r^3 = z$

Exercises

---

17 $d = \frac{1}{2}at^2$. Solve for a.
18 $15x + 25 = 100$. Solve for x.
19 $\frac{5x}{y} = \frac{3}{4}$. Solve for x.
20 $v^2 = 2ad$. Solve for d.
21 $G\frac{m_1m_2}{r^2} = \frac{m_1v^2}{r}$. Solve for v.
22 $\frac{xy}{z} = \frac{AB}{c}$. Solve for c.

A.5 GRAPHS AND PROPORTIONALITY

It will frequently be useful to deal with relationships of the following type:

$$x = y$$
$$x = 2y$$
$$x = ay$$

In each of these three equations x is proportional to y because, if y is doubled, x is doubled. The *proportionality constants*, the factors relating the two quantities, are 1, 2, and a, respectively. A way of writing this type of relationship is

$$x \propto y$$

The expression is read "x is proportional to y." What is implied is that, if y is doubled, x is doubled. We might also find a relationship such as

$$x \propto y^2$$

This says that x is proportional to y^2. Therefore y^2 must be doubled if x is to be doubled. It is not true in this case that doubling y doubles x. In fact, doubling y quadruples x.

From the equation $y = 2x$ it is possible to make a small table showing the values of x that correspond to various values of y.

y:	2	4	6	8	10
x:	1	2	3	4	5

It is also possible to present these data in graphical form, as shown in Fig. A.2. We let the distance of a point from the vertical axis represent the value of x, and the distance from the horizontal axis represent the corresponding value of y. The solid black circles represent the points tabulated above. Note that all the points lie in a straight line. This is a general result. Whenever two quantities are directly proportional, the graph of their relationship, such as Fig. A.2, is a straight line passing through the point corresponding to $x = 0$, $y = 0$, the origin. Conversely, the existence of such a straight-line graph implies the direct proportionality.

The slope-intercept method: Another type of equation frequently encountered is of the form

$$y = mx + b$$

Here m and b are numerical constants. For example,

$$y = \tfrac{1}{2}x + 3$$

means that, for every value of x, there corresponds a value of y that is equal to half the value of x plus 3. A graph of this equation is given in Fig. A.3. We see that the line intersects the vertical axis at $y = +3$. The value of b in the preceding equation is $+3$, and the value of b is always the intercept on the vertical axis.

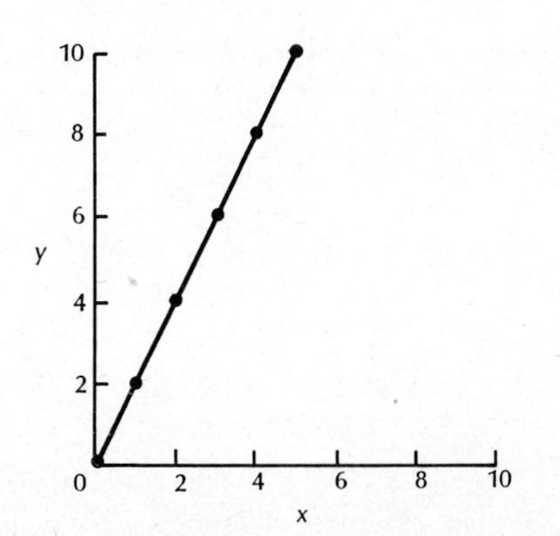

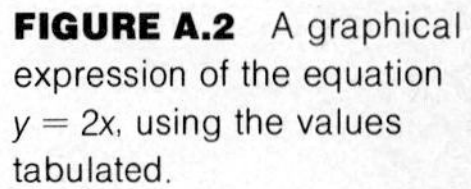

FIGURE A.2 A graphical expression of the equation $y = 2x$, using the values tabulated.

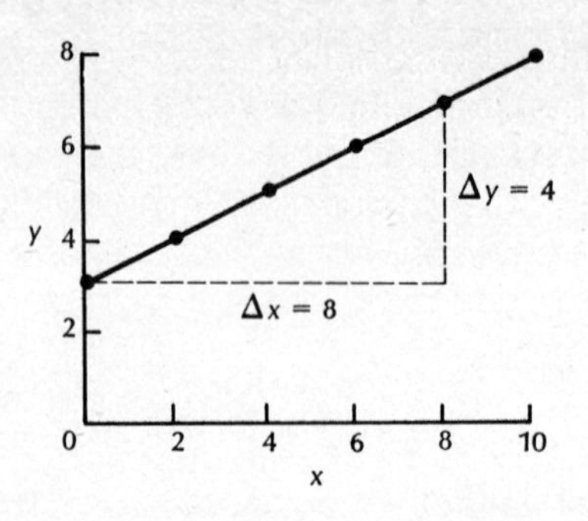

FIGURE A.3 A graphical representation of the equation $y = \frac{1}{2}x + 3$.

The slope of the line in Fig. A.3, which is *defined* as the change in y corresponding to 1 unit of change in x, is given by

$$\text{Slope} = \frac{\Delta y}{\Delta x} = \frac{4}{8} = \frac{1}{2}$$

Comparison with the original equation shows that $m = \frac{1}{2}$. Therefore, for any equation of the form

$$y = mx + b$$

m gives the slope of the line, and b its intercept on the y axis.

ANSWERS TO EXERCISES

1 2.3×10^2
2 4.6×10^3
3 9.5×10^4
4 7.26×10^6
5 6.2×10^{-3}
6 2.5×10^{-1}
7 2.9×10^{-9}
8 2×10^{-3}
9 2.0×10^{14}
10 2
11 1.54
12 4.2×10^{-21}
13 2×10^3
14 3×10^{-4}
15 1.5×10^{-17}
16 2×10^{-8}
17 $a = 2d/t^2$
18 $x = 5$
19 $x = 3y/20$
20 $d = v^2/2a$
21 $v^2 = Gm_2/r$
$v = (Gm_2/r)^{1/2}$
22 $c = ABz/xy$

ANSWERS TO PROBLEMS

These answers have been calculated using the rounded-off values for the numerical constants and conversion factors given on the inside front cover of the book. At most, three significant figures have been kept in the answer, and where appropriate, less. As a result, slight discrepancies between your answer and the one given here should not be cause for concern.

Chapter 2

1 1.28×10^7 (= 12,800,000) m
1.28×10^4 (= 12,800) km

2 131 mi/hour

3 0.94 mi

4 88 km/hour

5 75 mm

7 90 m

8 0.9 second

9 (*c*) 1 mi/hour
(*d*) $\frac{1}{2}$ mi/hour

10 3.83 hours $\simeq$ 3 hours, 50 minutes

12 (*a*) 22.9 ft/second
(*b*) 7.04 m/second
(*c*) 1,500 m

13 5.64 m/second

14 3.32 m/second

15 (*a*) 7.8×10^{15} seconds $\simeq$ 2.5×10^8 years
(*b*) About 20

16 19.8 ft/second

17 (*a*) 70 mi
(*b*) 46.7 mi/hour

18 11,000 ft $\simeq$ 2.1 mi

19 5.74 m/second

20 18 mi/second

21 3,570 mi/hour
5,240 ft/second

22 35.8 mi/hour

23 30 mi/hour

24 -2.0 ft/second^2

25 5 seconds

26 (*a*) 3,000 m/second
(*b*) 10,000 ft/second

27 12.5 m/second

28 (*a*) 200,000 ft/second^2
(*b*) 0.005 second

29 256 ft; 78.4 m
128 ft/second; 39.2 m/second

30 (*a*) 29 ft/second
(*b*) 13.8 seconds
(*c*) 2.1 ft/second^2

31 (*a*) 2 ft/second^2
(*b*) 40 ft/second

32 (*a*) 73 seconds
(*b*) -2.6 ft/second^2

33 202.5 ft

34 171 seconds

35 (*a*) 0.67 second
(*b*) Yes; 1.7 seconds

36 (*a*) 2.24 seconds
(*b*) 71.7 ft/second

Chapter 3

1 3,270 lb

3 20 kg

4 \$1.53/lb

5 22.5 newtons
2.3 kg

6 562 newtons

7 (*a*) 5 m/second^2
(*b*) 5,000 newtons

8 0.1 m/second^2

9 111 m/second^2

10 (*a*) 400,000 m/second^2
(*b*) 24,000 newtons
(5,300 lb)

11 128 ft/second^2 or 39.2 m/second^2

12 306 newtons

13 205 lb

14 22.5 newtons

15 2,880 lb

16 75 newtons

17 940 lb

18 40 m/second

19 (*a*) 0.6 m/second^2
(*b*) 30 m

20 4,500 newtons

21 (*a*) 10,000 newtons
(*b*) 5 m/second^2
(*c*) 19,600 newtons

22 (*a*) 11 seconds
(*b*) 302 m

23 (*a*) 1.07 ft/second^2
(*b*) 7.5 seconds
(*c*) 9.6 newtons $\simeq$ 2.1 lb

24 (*a*) 0.67 second
(*b*) 45 m/second^2
(*c*) 90 newtons

25 (*a*) 1.12 m/second^2
(*b*) 26.8 seconds
(*c*) 1,340 newtons

26 (*a*) 5 m/second^2
(*b*) 62.5 m
(*c*) 5,000 newtons

27 6.75 m or 22.5 ft

28 (*a*) 0.05 m/second^2
(*b*) 0.5 m/second

29 (*a*) 16 m/second^2
(*b*) 80 m/second
(*c*) 80 newtons
(*d*) 49 newtons

30 (*a*) 50 m/second^2
(*b*) 625 m
(*c*) 250 m/second
(*d*) 100 m/second

Chapter 4

1 3.5 ft/second^2

2 40 ft/second

3 (*a*) 32 ft/second
(*b*) 4 lb

4 0.04 m/second^2

5 2.33×10^4 ft/second

6 1.0×10^{22} m/second^2

7 24.5 m/second

8 22.4 m/second

9 5.33 m/second
1.07×10^4 newtons

10 (*a*) 0.52 m/second
(*b*) 1.8 m/second^2
(*c*) 1.8×10^{-3} newton, inwards

11 4.7 years

12 16 hours

13 0.125 years $\simeq$ 46 days

14 (*a*) 13.3 years
(*b*) 35.2 AU $\simeq$ 3.27×10^9 mi
(*c*) 2.81×10^4 mi/hour

15 2.7×10^{-12} newton

16 1.2×10^{-4} newton or 2.7×10^{-5} lb

17 4×10^{-12} newton

18 7.4×10^{-10} newton

19 1.8×10^{4} ft/second

20 $R = 0.0098$

21 (*a*) 2.8×10^{3} m/second
(*b*) 1.1×10^{5} seconds $\simeq$ 30.6 hours

22 (*a*) 6.0×10^{-3} m/second^2
(*b*) 3.6×10^{22} newtons
(*c*) 2.0×10^{30} kg

23 (*a*) 24 hours
(*b*) 2.66×10^{8} m
(*c*) 3.08×10^{3} m/second
(*d*) 0.22 m/second^2
(*e*) 5,100 km, which is below the earth's surface

24 (*a*) 4.0×10^{2} newtons
(*b*) 8.4×10^{2} m/second
(*c*) 4.4×10^{7} m/second
(*d*) 5.2×10^{4} seconds

25 (*a*) 1.0×10^{3} m/second
(*b*) 2.41×10^{9} m
(*c*) 2.4×10^{6} seconds = 27.8 days
(*d*) 1.9 m/second^2

26 (*a*) 4.6×10^{3} m/second
(*b*) 2.2×10^{9} m
(*c*) 4.8×10^{5} seconds = 5.6 days
(*d*) 2.3 m/second^2

27 (*a*) 8.9×10^{2} newtons
(*b*) 0.89 m/second^2
(*c*) 2.5×10^{3} m/second
(*d*) 1.7×10^{4} seconds $\simeq$ 4.7 hours
(*e*) 0.89 m/second^2

28 (*a*) 1.6×10^{21} newtons
(*b*) 0.11 m/second^2
(*c*) 1.1×10^{4} m/second
(*d*) 6.3×10^{5} seconds $\simeq$ 7.3 days

29 (*a*) 4.01×10^{7} m
(*b*) 4.64×10^{2} m/second
(*c*) 3.37×10^{-2} m/second^2

30 23.4 newtons

31 (*a*) 5.8×10^{22} kg
(*b*) 1.1×10^{3} m/second

32 (*a*) 3.3×10^{11} m/second^2
(*b*) 1.3×10^{5} m
(*c*) 8.2×10^{7} m/second
(*d*) 1.6×10^{-3} (= 0.0016) second

Chapter 5

1 15,000 (= 1.5×10^{4}) kg·m/second

2 50,000 (= 5.0×10^{4}) kg·m/second

3 1.8×10^{29} kg·m/second

4 2.0×10^{3} m/second

5 1.2×10^{9} kg·m/second

6 1.6×10^{28} kg·m/second

7 7×10^{25} kg·m/second

8 10 kg·m/second

9 17 kg·m/second

10 4.5×10^{4} kg·m/second

11 $1\frac{2}{3}$ kg

12 600 newtons

13 0.25 m/second

14 (*a*) 200 mi/hour
(*b*) 0
(*c*) 0

15 1.3 m/second, eastward
16 14.3 m/second
17 2.22 m/second, north
18 12.5 m/second
19 1.5 m/second
20 340 m/second
21 1.2×10^4 newtons
22 7.8 ft/second
23 10 mi/hour, north
24 3 mi/hour, north
25 100,000 ($= 10^5$) mi/hour

Chapter 6

1 37.5 joules
2 1.8×10^{-18} joule
3 7.4×10^{-17} joule
4 1×10^4 joules
5 7.1×10^5 joules
6 2.66×10^{33} joules
7 (*a*) 2.0×10^3 m/second
(*b*) 1.8×10^{-24} joule
8 3.9×10^{32} joules
9 3.5×10^{28} joules
10 200 joules
11 (*a*) 70 foot-pounds
(*b*) 11 foot-pounds
12 (*a*) 588 joules
(*b*) 90 joules
13 (*a*) 128 m
(*b*) 12,500 joules, if the window is taken as zero
14 81.6 m
15 1.3 m
16 (*a*) 11.8 joules
(*b*) 11.8 joules
(*c*) 11.8 joules
(*d*) 23.6 newtons
17 (*a*) 9.65×10^{-6} joule
(*b*) 9.65×10^{-6} joule
(*c*) 1.27×10^{-2} newton
18 (*a*) 10 m
(*b*) 1.47×10^3 joules
(*c*) 14 m/second
(*d*) At the lowest point
19 (*a*) 13.9 ft/second
(*b*) 5.2 ft/second
20 (*a*) 17.5 joules
(*b*) 0.64 meter
21 (*a*) 2 joules
(*b*) 157 joules
(*c*) 159 joules
(*d*) 8.9 m/second
22 (*a*) 54,250 joules
(*b*) 7.2 meters
23 (*a*) 6.67 m/second
(*b*) 1,500 kg · m/second
(*c*) 5,000 joules
(*d*) 6.67 m/second
24 (*a*) 73.5 joules
(*b*) 58.5 joules
(*c*) 4.8 m/second
(*d*) 29.5 newtons
25 (*a*) 44 joules
(*b*) 5.4 m/second
26 (*a*) 44 joules
(*b*) 14.7 newtons
27 (*a*) 6.7 m/second
(*b*) 7.5×10^6 joules
(*c*) 6.7×10^6 joules
(*d*) 0.8×10^6 ($= 8 \times 10^5$) joules
28 (*a*) 2.4 m/second
(*b*) 3.8×10^2 joules
(*c*) 3.7×10^2 joules
(*d*) 10 joules
(*e*) No
29 (*a*) 20 ft/second, north
(*b*) 31 foot-pounds
30 (*a*) 11.2 m/second
(*b*) 3.3×10^5 joules

(*c*) 62 foot-pounds
(*d*) 31 foot-pounds, from the chemical energy of the explosion

(*c*) 2.4×10^5 joules
(*d*) 0.9×10^5 ($= 9 \times 10^4$) joules

31 3.3×10^4 joules

32 2.75×10^5 joules

33 (*a*) 11.4 m/second
(*b*) 1.7×10^5 joules

34 (*a*) 19.6 joules
(*b*) 6.9 m/second

35 (*a*) 3,840 lb
(*b*) 7.1×10^5 foot-pounds
(*c*) 0
(*d*) 7.1×10^5 foot-pounds; it went to smash the cars

36 (*a*) 26.7 mi/hour, north
(*b*) 9.6×10^4 foot-pounds
(*c*) 4.3×10^4 foot-pounds
(*d*) 5.3×10^4 foot-pounds

37 (*a*) 32 foot-pounds
(*b*) 0
(*c*) 32 foot-pounds
(*d*) 1 ft

Chapter 7

1 440 calories

2 6000 calories

3 600 calories

4 0.30

5 0.50

6 2,500 joules; 595 calories

7 0.5 kg

8 8×10^{-4} joule

9 $0.07 = 7\%$

10 33%
1.33×10^8 joules of useful work
2.67×10^8 joules rejected

11 (*a*) 9.8 joules
(*b*) 4.2 m/second

12 500 K; 227°C

13 (*a*) 4 joules
(*b*) 0.41 kg

14 (*a*) 26 m/second
(*b*) 1.6×10^5 joules
(*c*) 3.8×10^4 calories

15 9×10^{-4}°C $= 0.0009$°C

16 3.5×10^{-3}°C $= 0.0035$°C

17 2.3×10^{-2}°C $= 0.023$°C

18 1.2×10^{-2}°C $= 0.012$°C

19 1.46°C

20 91.7°C

Chapter 8

1 11.7 years

2 0.6%; 1.75%

3 Approx. 1000 watts (for a 10×10 m roof)

Chapter 9

1 30 meters

2 $+2 \times 10^6$ newtons

3 6.7×10^{-11} coulomb, opposite signs

4 (*a*) 1×10^{-5} coulomb
(*b*) Same sign

5 +90 newtons
6 −0.27 newton
7 1.05×10^{-5} coulomb
8 4.5×10^{-3} m/second^2, apart
9 7.7×10^{22} m/second^2
10 $+2.3 \times 10^{-12}$ newton
11 (*a*) -2.3×10^{-8} newton
(*b*) 2.6×10^{22} m/second^2, toward the proton
12 (*a*) $+2.3 \times 10^{-28}$ newton
(*b*) Apart
(*c*) Together
(*d*) Apart
13 Electric force = 2.3×10^{-8} newton
Gravitational force = 1.7×10^{-44} newton
Gravity is much, much smaller

Chapter 10

1 -1.4×10^{-3} newton/coulomb
2 $+2.2 \times 10^{3}$ newtons/coulomb
3 -1.4×10^{11} newton/coulomb
4 +0.25 newton/coulomb
5 (*a*) 1×10^{-7} joule
(*b*) 0.1 volt
6 6×10^{-6} joule
7 3,000 watts
8 (*a*) 9×10^{-3} newton/coulomb
(*b*) 90 m/second^2
9 (*a*) 1.1×10^{-15} coulomb
(*b*) 0.7 m
(*c*) Away
10 (*a*) 3.6×10^{5} newtons/coulomb
(*b*) 3.6×10^{-7} newton
(*c*) Away
11 (*a*) 1.6×10^{-14} newton
(*b*) 8.9×10^{-30} newton
(*c*) No
12 (*a*) 1.6×10^{-15} joule
(*b*) 1.6×10^{-15} joule
(*c*) 5.9×10^{7} m/second
13 0.45 ampere
14 4.5 amperes
15 (*a*) 100 watts
(*b*) 6,000 joules
16 (*a*) 7,700 watts
(*b*) 6.3 ohms
(*c*) 1.4×10^{7} joules
3.85 kilowatthours
(*d*) 9.6¢
17 (*a*) 1,650 watts
(*b*) 5.9×10^{6} joules
1.65 kilowatthours
(*c*) 4.1¢
18 (*a*) 8.3 amperes
(*b*) 1.4 ohms
(*c*) 83 amperes
19 (*a*) 0.12 ampere
(*b*) 1.3×10^{6} joules
0.36 kilowatthours
(*c*) 0.9¢
20 (*a*) 1.82 amperes
(*b*) 60.4 ohms
(*c*) 2¢
(*d*) 2.88×10^{6} joules

21 1.6×10^{-19} joule

22 1.6×10^{-16} joule

23 8×10^{-16} joule

24 None

25 4×10^{-17} joule

26 (*a*) 1.6×10^{21} newtons/coulomb
(*b*) 2.5×10^{2} newtons
(*c*) Toward

27 (*a*) 4.0×10^{-15} joule
(*b*) 4.0×10^{-15} joule
(*c*) 9.4×10^{7} m/second

28 (*a*) 4.0×10^{5} newtons/coulomb
4.0×10^{5} volts/m
(*b*) 4.0×10^{-10} newton
(*c*) 8.0×10^{-12} joule
(*d*) 1.26×10^{5} m/second

29 (*a*) 6.4×10^{-17} joule
(*b*) 6.4×10^{-17} joule
(*c*) 1.2×10^{7} m/second
(*d*) 4×10^{5} newtons/coulomb
4×10^{5} volts/m

30 (*a*) 1 volt
(*b*) 1×10^{-15} joule
(*c*) 1×10^{-15} joule
(*d*) 4.5×10^{2} m/second

31 (*a*) 1.3×10^{-14} newton
(*b*) 1.3×10^{-13} joule
(*c*) 1,300 volts

32 (*a*) 1.4×10^{2} newtons
(*b*) 1.15×10^{12} newtons/coulomb
(*c*) 1.84×10^{-7} newton

33 (*a*) 3.6×10^{-2} newton/coulomb
(*b*) 3.6×10^{-2} newton/coulomb
(*c*) None
(*d*) Zero

Chapter 11

1 0.3 newton

2 1.2 newtons

3 0.02 newton/ampere·m

4 (*a*) 5×10^{-3} newton
(*b*) 1.1×10^{-3} pound

5 Out of the paper

Chapter 12

1 11% H; 89% O

2 0.40 kg C; 1.10 kg O

3 222 g Cu; 28 g O

4 39% Na; 61% Cl

5 22% S; 78% F

6 187 g O; 163 g N

7 123 g C; 27 g H

8 33% H; 67% C

9 56% O; 44% P

10 3.4×10^{-25} kg

11 3×10^{26} atoms

12 1.5×10^{27} atoms

13 1.8×10^{-25} kg

14 3.75×10^{23} atoms

15 2.81×10^{24} atoms

Chapter 13

1 3.8 m^3
2 3 atmospheres
3 0.45 atmosphere
4 0.084 m^3
5 0.33 m^3
6 8 atmospheres
7 819 K; 546°C
8 0.94 g/liter
9 4.5 g/liter
10 12 m^3
11 0.02 m^3
12 4.3 m^3
13 400 K; 127°C
14 0.36 g/liter
15 5.76 g/liter

Chapter 14

1 2
2 3.3×10^{-19} coulomb
3 60
4 4.9×10^{-17} kg
5 1×10^4 m/second
6 2×10^5 m/second
7 0.5×10^{-17} ($= 5 \times 10^{-18}$) coulomb
8 (*a*) 2×10^{-19} coulomb
(*b*) 10, 8, 7, 6
9 0.33×10^{-18} ($= 3.3 \times 10^{-19}$) coulomb

Chapter 15

1 1.24×10^8 m/second
2 1.97×10^8 m/second
3 2.3×10^{-4} second
4 2.56 seconds
5 Greater in medium 2

Chapter 16

1 (*a*) 7.5×10^{14} hertz
(*b*) 3.4 m
2 6×10^{14} hertz
3 (*a*) 0.02 m; 2 cm
(*b*) 0.015 m; 1.5 cm
4 5.3×10^2 m
5 3.3×10^{-7} m; 2.1×10^{-7} m

Chapter 17

1 2.29×10^{-6} second
2 120 m
3 183 kg
4 0.8*c*
5 (*a*) 1.50×10^{-30} kg
(*b*) 2.06×10^{-30} kg
(*c*) 6.38×10^{-30} kg
6 (*a*) 5.7×10^{-8} second
(*b*) 13.7 m
(*c*) 5.7×10^{-8} second
7 (*a*) 6.25×10^{-9} second
(*b*) 2.5 m
(*c*) 1.04×10^{-8} second
8 (*a*) 1.01×10^5 years
(*b*) 1.41×10^4 years
(*c*) 1.41×10^4 light years
9 (*a*) 2×10^7 m/second
(*b*) 9.98 m
(*c*) 4.99×10^{-7} second
10 (*a*) 1.6 m
(*b*) 6.7×10^{-9} second
(*c*) 2 m

11 (*a*) 4.5×10^{-12} joule
(*b*) 6.75×10^{14} joules
(*c*) $2.14 \times 10^{5} = 214,000$

Chapter 18

1 3.0×10^{-19} joule
2 4.0×10^{-19} joule
3 2.0×10^{-15} joule
4 6.6×10^{-28} joule
5 2.0×10^{-15} joule
6 3×10^{-32} joule·second
7 2×10^{-19} joule
8 3.4×10^{-19} joule

Chapter 19

1 (*a*) 4.5×10^{-10} m
(*b*) 3.2×10^{-34} joule·second
2 2×10^{-10} m
3 4.2×10^{-34} joule·second

Chapter 20

1 7.3×10^{-11} m
2 4.1×10^{3} m/second
3 1.0×10^{-14} m
4 7.2×10^{-10} m

Chapter 21

1 ${}^{208}_{82}Pb$
2 ${}^{215}_{85}At$
3 ${}^{234}_{92}U$
4 ${}^{223}_{87}Fr$
5 ${}^{226}_{88}Ra$
6 ${}^{214}_{84}Po$
7 1.44×10^{-12} joule
9.0 MeV
8 2.24×10^{-14} joule
0.14 MeV
9 2.61×10^{-13} joule
1.63 MeV
10 8.06×10^{-13} joule
5.04 MeV
11 7.76×10^{-13} joule
4.85 MeV
12 7α; 4β
13 6α; 4β
14 (*a*) 7.44×10^{-12} joule
(*b*) 1.9×10^{13} joule
15 6.84×10^{-12} joule
16 3.9×10^{26} joules

INDEX

INDEX